The book provides an extensive theoretical treatment of whistler-mode propagation, instabilities and damping in a collisionless plasma. This book fills a gap between oversimplified analytical studies of these waves, based on the cold plasma approximation, and studies based on numerical methods. Although the book is primarily addressed to space plasma physicists and radio physicists, it will also prove useful to laboratory plasma physicists. Mathematical methods described in the book can be applied in a straightforward way to the analysis of other types of plasma wave. Problems included in this book, along with their solutions, allow it to be used as a textbook for postgraduate students.

T0254165

Cambridge atmospheric and space science series

Whistler-mode waves in a hot plasma

Cambridge atmospheric and space science series

Editors

Alexander J. Dessler

John T. Houghton

Michael J. Rycroft

Titles in print in this series

M. H. Rees, *Physics and chemistry of the upper atmosphere*

Roger Daley, *Atmospheric data analysis*

Ya. L. Al'pert, *Space plasma, Volumes 1 and 2*

J. R. Garratt, *The atmospheric boundary layer*

J. K. Hargreaves, *The solar–terrestrial environment*

Sergei Sazhin, *Whistler-mode waves in a hot plasma*

Whistler-mode waves
in a hot plasma

Sergei Sazhin

Department of Physics, The University of Sheffield

CAMBRIDGE
UNIVERSITY PRESS

CAMBRIDGE UNIVERSITY PRESS
Cambridge, New York, Melbourne, Madrid, Cape Town, Singapore, São Paulo

Cambridge University Press
The Edinburgh Building, Cambridge CB2 2RU, UK

Published in the United States of America by Cambridge University Press, New York

www.cambridge.org
Information on this title: www.cambridge.org/9780521401654

First published 1993
This digitally printed first paperback version 2005

A catalogue record for this publication is available from the British Library

Library of Congress Cataloguing in Publication data
Sazhin, S.S. (Sergei Stepanovich)
Whistler-mode waves in a hot plasma / Sergei Sazhin.
p.   cm. – (Cambridge atmospheric and space science series)
Includes bibliographical references and index.
ISBN 0-521-40165-8
1. Plasma waves.   2. High temperature plasmas.   3. Space plasmas.
4. Magnetosphere.   I. Title.   II. Series.
QC718.5.W3S39   1993
530.4'4–dc20   92-3873 CIP

ISBN-13 978-0-521-40165-4 hardback
ISBN-10 0-521-40165-8 hardback

ISBN-13 978-0-521-01827-2 paperback
ISBN-10 0-521-01827-7 paperback

Contents

Contents

Acknowledgements

The idea of writing this monograph was first suggested by Prof. M. J. Rycroft, editor of the series. I deeply appreciate his encouragement, support, and useful suggestions about the draft version of the manuscript.

The work has been completed in the University of Sheffield, where I was invited by Prof. T. R. Kaiser and Dr K. Bullough and arrived in April 1988. I am grateful to both of them for this invitation and their continuous support. Dr Bullough was the first to read the whole manuscript and to comment on it both from the point of view of physics and as regards the quality of the English language.

Historically this book resulted from the extensive development of Chapter 1 of a projected more comprehensive book 'Whistler-mode waves in the magnetosphere', on which I began work in 1983 together with Dr H. J. Strangeways. I am indebted to him for his cooperation, which strongly stimulated many ideas developed in my book, although our major common project has never been completed.

I wish to thank Prof. R. J. Moffett, Dr N. M. Temme, Mr L. J. C. Woolliscroft, Prof. O. A. Pokhotelov, Dr D. R. Shklyar, Dr D. Nunn, Dr A. E. Sumner and Mr R. Body for their very useful comments on selected chapters of the manuscript. My special thanks are to Mr M. Behrend, the text editor, for his careful reading of the manuscript and important comments which helped to improve considerably the presentation of the book.

I would like to thank all my colleagues in the Department of Physics of the University of Sheffield for providing me with varied assistance throughout the whole period of the work on this monograph.

It is not possible to enumerate all my personal friends in Sheffield and other parts of the United Kingdom who helped me and my family to find new roots in England. Without this the whole idea of writing the book would never have materialized.

Acknowledgements

Finally I would like to acknowledge financial support from the SERC and NERC during the whole period of my work in the University of Sheffield.

Sergei Sazhin
August 1992

Introduction

Following Helliwell (1965) we can define whistlers as radio signals in the audio-frequency range that 'whistle'. Usually a whistler begins at a high frequency and in the course of about one second drops in frequency to a lower limit of about 1 kHz, although the duration of the event may vary from a fraction of a second to two or three seconds. Occasionally this 'lower' branch of a whistler's dynamic spectrum is observed simultaneously with the 'upper' branch where the frequency of the signal increases with time, so that the whole dynamic spectrum appears to be of the 'nose' type. Typical dynamic spectra of such whistlers observed at Halley station in Antarctica ($L = 4.3$) are shown in Fig. I.

The energy source for a whistler is a lightning discharge where the waves are generated over a wide frequency range in a very short time. These waves propagate from their source in all directions. Part of their energy propagates in the Earth–ionosphere waveguide with a velocity close to the velocity of light and almost without frequency dispersion. These signals are called

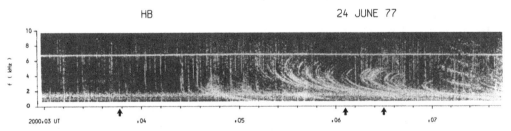

Fig. I The spectra of atmospherics (or sferics) and whistlers recorded at Halley on 24 June 1977 in the frequency range from 0 to 10 kHz (see the scale of the ordinate). The arrows indicate the causative sferic and the corresponding whistlers which will be used for diagnostics of magnetospheric parameters in Chapter 9.

1

atmospherics or 'sferics' and are recorded as almost vertical lines as shown in the dynamic spectrogam of Fig. I. The time delay of atmospherics with respect to the original lightning usually does not exceed 40 ms. Another part of the wave energy produced during a lightning discharge penetrates through the ionosphere into the magnetosphere. Part of the energy of these waves may propagate almost parallel to the magnetic field lines of the magnetosphere in a duct of enhanced plasma density and eventually reach the conjugate point on the Earth's surface in the opposite hemisphere. The travel time of the signal propagating through the magnetosphere is much greater than the travel time of the signal propagating in the Earth–ionosphere waveguide and is of the order of a few seconds. Moreover, the time delay of the signal is frequency-dependent, which results in the characteristic whistler frequency/time profile shown in Fig. I.

The dispersion of a whistler signal is that of electromagnetic waves propagating through the magnetosphere (Storey, 1953). These waves are known as whistler-mode waves and will be of primary interest in this book. These waves are responsible not only for the transfer of wave energy from lightning discharges from one hemisphere to another but also for the transfer of energy from artificial radio transmitters and many types of natural radio emissions (Sazhin, 1982a). Moreover, it is now generally believed that these waves play an important role in the loss of the electrons from the radiation belts in the Earth's magnetosphere (Kennel & Petschek, 1966) and presumably in the magnetospheres of other planets (see e.g. Coroniti *et al.*, 1987; Gurnett *et al.*, 1990; Menietti *et al.*, 1991). They can exist in the atmospheres of stars and in laboratory plasmas (see e.g. Wharton & Trivelpiece, 1966). They have been observed in metals (Kaner & Skobov, 1971) and semiconductors (Baynham & Boardman, 1971), where they are known as helicon waves. This accounts for the attention which researchers in different fields of plasma and space physics pay to this particular type of wave.

The first overview of whistler-mode waves was given by Helliwell (1965). Since then, many original papers have been published, some of which have dramatically changed our understanding of the phenomenon. Nobody has attempted to review all these papers, i.e. to produce a modern version of Helliwell's book (although reviews related to particular aspects of this problem have been presented, e.g. by Mikhailovskii, 1975; Hasegawa, 1975; Alexandrov, Bogdankevich & Rukhadze, 1978; Al'pert, 1980; Cuperman, 1981; Carpenter, 1988; Timofeev, 1989; Breizman, 1990; Rönnmark, 1990; Stix, 1990; Petviashvili & Pokhotelov, 1991; Rycroft, 1991, and others) and it is not my task to do this. Instead, an approximate theory of whistler-mode propagation (including damping or amplification) in a hot anisotropic plasma is considered and applied to the interpretation of whistler-mode phenomena in the magnetosphere. From this point of view this book is complementary to the monographs of Ginzburg (1970) and Budden (1985), who restricted their analyses mainly to the cold plasma approximation, and to the monographs of Akhiezer *et al.* (1975), Melrose (1986), Al'pert (1990) and Walker (1992), who considered the theory of whistler-mode waves within the more general frame of plasma wave theory. The theory under consideration has mainly been developed in my papers published during the last 20 years, and its aim is to fill a gap between oversimplified models of whistler-mode propagation, e.g. those based on a cold plasma approximation, and numerical analyses of whistler-mode propagation in more sophisticated plasma models. Although oversimplified models are very convenient for practical applications and have been widely used, the range of their applicability was often not justified, a fact which could lead to the erroneous interpretation of particular phenomena. On the other hand, the high accuracy of numerical models is often unnecessary for the interpretation of wave phenomena in the magnetosphere of the Earth; there the accuracy of determination of the input parameters, such as plasma density and temperature, is at least an order of magnitude worse. The accuracy of the results predicted by our theory is, as a rule, less than the accuracy of numerical methods, but is better than the accuracy of the determination of the input parameters for magnetospheric conditions. At the same time this theory is far more convenient for practical applications when compared with numerical methods. In many cases its results allow a straightforward physical interpretation and yet it is much more rigorous when compared with oversimplified models. This makes it particularly useful for the interpretation of wave phenomena in the magnetosphere of the Earth. However, it can easily be applied to the analyses of wave phenomena in astrophysical and laboratory plasmas as well, although description of details of applications to these plasmas are beyond the scope of the book. This theory is presented not in the form of a review of previous results but as a course (including problems and the solutions to them) which can be used by a wide range of specialists and postgraduate students in different fields of plasma physics, physics of the magnetosphere and ionosphere, and radio physics. It might be useful even to seismologists interested in radio emissions of seismic origin (Gokhberg, Morgunov & Pokhotelov, 1988).

In Chapter 1 we derive the main equations for wave propagation in a homogeneous plasma, including the expressions for the elements of the plasma dielectric tensor in different approximations. In Chapter 2 we give an overview of the main results relating to whistler-mode propagation in a cold plasma. In Chapters 3–6 we give a linear theory of whistler-mode propagation in a hot anisotropic plasma in different limiting cases. Growth and damping of the waves are analysed in Chapter 7. Some non-linear effects related to whistler-mode waves are briefly described in Chapter 8. Applications of our theory to the interpretation of particular wave phenomena in the magnetosphere are considered in Chapter 9.

The Gaussian system of units is used throughout the book, although in most cases the form of equations will not depend on the system of units.

1

Basic equations

1.1 Approximations

In view of the applications of our theory to the conditions of the Earth's magnetosphere the following assumptions are made:

(1) The plasma is assumed to be homogeneous in the sense that its actual inhomogeneity does not influence its dispersion characteristics, instability or damping at any particular point, although in general these characteristics can change from one point to another. For low-amplitude waves this assumption is valid when the wavelength is well below the characteristic scale length of plasma inhomogeneity, a condition which is satisfied for whistler-mode waves propagating in most areas of the magnetosphere (except in the lower ionosphere). For finite amplitude waves the condition for plasma homogeneity depends on wave amplitude, but the discussion of these effects is beyond the scope of the book (see e.g. Karpman, 1974).

(2) The plasma is assumed to be collisionless in the sense that we neglect the contribution of Coulomb collisions between charged particles as well as collisions between charged and neutral particles leading to charge exchange. More rigorously this assumption can be written as:

$$\frac{q_\alpha^2}{\langle r_{12} \rangle T_\alpha} \ll 1, \tag{1.1}$$

where q_α is the particle's charge (index α indicates the type of particle: $\alpha = e$ for electrons, $\alpha = p$ for protons; $\langle r_{12} \rangle$ is the average distance between particles; T_α is the particles' temperature in energy units.

The physical meaning of (1.1) is obvious: the average energy of interaction between charged particles is well below their average kinetic energy. Condition (1.1) also means that in an 'absolutely cold' plasma ($T_\alpha = 0$) the interaction between particles can never be neglected. This condition is almost always satisfied in the magnetosphere (except the lower ionosphere) and will allow us to describe plasma kinetics in terms of a single particle distribution function in phase space (i.e. we reduce the chain of Bogolubov–Born–Green–Kirkwood–Yvon (BBGKY) equations for multiparticle distribution functions to the Vlasov equation for the single particle distribution function: see Akhiezer *et al.* (1975) for details).

(3) We consider mainly the contribution of electrons to the process of wave propagation, instability or damping. The contribution of protons will make only small perturbations to the wave parameters in an electron plasma. This assumption is satisfied for whistler-mode waves at frequencies well above both the proton gyrofrequency and proton plasma frequency.

(4) The plasma is assumed to be weakly relativistic so that the electron thermal velocity w is assumed to be well below the velocity of light c, i.e.

$$w \ll c. \tag{1.2}$$

Condition (1.2) is satisfied in most areas of the magnetosphere. Note that this condition does not automatically guarantee the validity of the non-relativistic approximation. This point will be discussed later in this chapter.

(5) We shall take into account only the contribution of the electromagnetic field to particle motions. The contribution of other fields, e.g. the gravitational field, will be neglected; this is justified for both protons and electrons in the magnetosphere. Moreover, we shall not consider the influence of external electric fields on whistler-mode waves, although it is not at first evident whether they can be neglected in magnetospheric conditions (see e.g. Das & Singh (1982) for a discussion of this problem).

The theory considered in this book, as with all modern theories of wave propagation, instability or damping in a plasma with non-zero temperature (this temperature cannot be strictly equal to zero, as this would contradict the inequality (1.1)) is based on the pioneering paper by Landau (1946). There the problem of the development of the initial electrostatic perturbation in a collision-free unmagnetized plasma was considered. The main ideas of this paper will be discussed in the next section. This will enable us to clarify other approximations used in the theory under consideration and help us to approach a more sophisticated theory of electromagnetic wave (in particular, whistler-mode) propagation in a magnetized plasma.

1.2 Development of an initial perturbation

As was mentioned in Section 1.1, condition (1.1) allows us to describe plasma particle dynamics by the Vlasov equation. Restricting our analysis to electrostatic disturbances propagating in an unmagnetized non-relativistic electron plasma (protons are assumed to form a neutralizing background), we can consider the Vlasov equation for the electron distribution function f_e alone and write it in the form:

$$\frac{\partial f_e}{\partial t} + \mathbf{v}\frac{\partial f_e}{\partial \mathbf{r}} + \frac{e}{m_e}\nabla\Phi\frac{\partial f_e}{\partial \mathbf{v}} = 0, \qquad (1.3)$$

where e is the modulus of the electron charge, m_e is the electron mass, and Φ is the electrostatic potential (the electric field is determined by $\mathbf{E} = -\nabla\Phi$). Provided that there are no external electric fields the only source of non-zero $\nabla\Phi$ might be a self-consistent field due to the disturbances of f_e. In this case Φ is determined by the Poisson equation:

$$\nabla^2\Phi = 4\pi e n_e \int (f_e - f_{0e})\mathrm{d}\mathbf{v}, \qquad (1.4)$$

where f_{0e} is an unperturbed distribution function and the integration is performed over all velocities $\mathbf{v}$.

As follows from (1.3) the distribution function f_e can be perturbed in the direction $\nabla\Phi$ only. Hence, we can integrate this equation with respect to all velocities in the direction perpendicular to $\nabla\Phi$, thus obtaining the functions $\bar{f}_e$ and $\bar{f}_{0e}$, depending on scalar velocity v and scalar distance r in the direction parallel to $\nabla\Phi$. Introducing the notation $\bar{f}_{1e} = \bar{f}_e - \bar{f}_{0e}$ and neglecting the contribution of second and higher order terms, we can write equations (1.3) and (1.4) in a more compact form:

$$\frac{\partial \bar{f}_{1e}}{\partial t} + v\frac{\partial \bar{f}_{1e}}{\partial r} + \frac{e}{m_e}\frac{\partial \Phi}{\partial r}\frac{\partial \bar{f}_{0e}}{\partial v} = 0, \qquad (1.5)$$

$$\nabla^2\Phi = 4\pi e n_e \int_{-\infty}^{\infty} \bar{f}_{1e}\mathrm{d}v. \qquad (1.6)$$

The condition $|f_{1e}| \ll |f_{0e}|$, which allowed us to neglect the deformation of f_{0e} and to derive equations (1.5) and (1.6), is known as a linear approximation. This approximation will be used throughout most of the book (Chapters 1–7). Theories taking into account the effects of deformation of f_{0e} under the influence of the waves will be briefly discussed in Chapter 8.

Using equations (1.5) and (1.6) we solve the general problem of the development of an initial perturbation in a plasma: $\bar{f}_{1e}(r,v,t)|_{t=0} = g(r,v)$, where $g(r,v)$ is a known function of the coordinates and velocity. To do this we use the Fourier space and Laplace time transformations of the functions $\bar{f}_{1e}$ and Φ (following Landau, 1946):

$$a_{kp} = \frac{1}{2\pi}\int_0^{\infty} \mathrm{d}t\, \exp(-pt) \int_{-\infty}^{+\infty} \mathrm{d}r\, a(r,v,t)\exp(-ikr), \qquad (1.7)$$

where a_{kp} are the Fourier space and Laplace time components of the disturbances of the distribution function ($a = \bar{f}_{1e}$) or of the potential ($a = \Phi$).

Remembering (1.7) we obtain from (1.5) and (1.6):

$$(p + ikv)\bar{f}_{1kp} + i\frac{e}{m_e}\Phi_{kp}k\frac{\partial \bar{f}_{0e}}{\partial v} = g_k, \tag{1.8}$$

$$k^2\Phi_{kp} = -4\pi e n_e \int_{-\infty}^{\infty} \bar{f}_{1kp}dv, \tag{1.9}$$

where g_k is the Fourier component of the function $g(r, v)$:

$$g_k = \frac{1}{2\pi}\int_{-\infty}^{+\infty} g(r, v)\exp(-ikr)dr.$$

From (1.8) and (1.9) after taking the inverse Laplace transformation we obtain:

$$\Phi_k(t) = \frac{-1}{2\pi i}\int_{\sigma - i\infty}^{\sigma + i\infty} dp\, e^{pt}\frac{4\pi e}{k^2}\left\{\frac{\int_{-\infty}^{+\infty} g_k(v)/(ikv + p)dv}{1 - \frac{i\Pi^2}{k}\int_{-\infty}^{+\infty}\frac{d\bar{f}_{0e}(v)/dv}{ikv+p}dv}\right\}, \tag{1.10}$$

where $\Pi = \sqrt{4\pi n_e e^2/m_e}$ is the electron plasma frequency. The contour of integration in (1.10) lies to the right of all the singularities of the integrand.

Then, after taking the inverse Fourier transformation of (1.10), we obtain the function $\Phi(t)$. In a similar way we could obtain the function $\bar{f}_e(t)$ which would complete the solution of the problem under consideration. However, in most practically important cases we do not need this complete solution but are interested in the asymptotic behaviour of $\Phi(t)$ for sufficiently large t. This can be obtained directly from (1.10).

The integrand in (1.10) is determined for $\Re p > p_0$, where p_0 is determined by the condition of convergency of the integral (1.7). Hereafter we assume that $p_0 = 0$. However, in many practically important cases this integrand can be analytically continued to the whole complex p plane. For example, if $g_k(v)$ and $d\bar{f}_{0e}(v)/dv$ can be analytically continued to the whole complex v plane and their analytical continuations are entire functions of v (i.e. have no singularities in this plane; this will be assumed in the whole book), then the only singularities in the internal integrands in (1.10) are simple poles at $v = ip/k$. In order that these integrals at $\Re p \leq 0$ may be analytical continuations of the corresponding integrals at $\Re p > 0$, the initial contours of integration along the real v axis should be deformed in such a way that the poles at $v = ip/k$ remain above these contours. As a result the analytical continuation of the whole expression in curled brackets in (1.10) appears as a ratio of two entire functions in the whole complex p plane. The only singularities of this expression are poles corresponding to zeros of denominator. In this case we can move the contour of integration with respect to p in (1.10) to the left by $a + \sigma$ ($a > 0$) and rewrite the expression for $\Phi_k(t)$ as:

$$\Phi_k(t) = \Sigma c_\nu e^{p_\nu t} + \frac{1}{2\pi i} \int_{-a-i\infty}^{-a+i\infty} \hat{\Phi}_{kp} e^{pt} dp, \qquad (1.11)$$

where $\hat{\Phi}_{kp}$ is the analytical continuation of the function Φ_{kp}, and c_ν are the residues of the function Φ_{kp} at the poles p_ν inside the strip $-a < p_\nu < \sigma$. For sufficiently large values of a and t the integral on the right hand side of (1.11) vanishes and we can neglect the contribution of all the exponential terms except those corresponding to the maximal value of $\Re p_\nu$. Remembering our assumption that $g_k(v)$ is an entire function then, as follows from (1.10), the poles p_ν are determined by the following equation:

$$D(p, k) = 1 - \frac{i\Pi^2}{k} \int_L \frac{d\bar{f}_{0e}(v)/dv}{ikv + p} dv = 0, \qquad (1.12)$$

where the contour of integration L lies below all the singularities of the integrand. Now we change variables $p \equiv -i\omega = -i(\omega_0 + i\gamma) = -i\omega_0 + \gamma$, assuming that $|\gamma| \ll \omega_0$. This allows us to consider the contour of integration L along the real v axis except in the immediate vicinity of the point $v = \omega_0/k$ which should be encircled by this contour from below. The part of the integral along the real axis will give us the principal part of the integral, while the integral along the semicircle will give us half of the residue of the integrand. As a result equation (1.12) can be written as:

$$D(\omega, k) = 1 - \frac{\Pi^2}{k^2} \left[P \int_{-\infty}^{+\infty} \frac{(d\bar{f}_{0e}/dv)dv}{v - \omega/k} + i\pi \frac{d\bar{f}_{0e}}{dv} \bigg|_{v=\omega_0/k} \right] = 0. \quad (1.13)$$

Formally the same equation (1.13) could be obtained if we assumed from the very beginning that both $\bar{f}_{1e}$ and Φ were proportional to $\exp[i(kr - \omega t)]$ (i.e. if we used Fourier rather than Laplace transformations with respect to t, but with $\omega = \omega_0 + i\gamma$, $|\gamma| \ll \omega_0$), provided that we went around the singularity in the integrand along a semicircle from below. In other words, we assume the following condition:

$$\int_{-\infty}^{+\infty} \frac{A(v)dv}{v - \omega/k} = P \int_{-\infty}^{+\infty} \frac{A(v)dv}{v - \omega/k} + i\pi A(v) \bigg|_{v=\omega_0/k}. \qquad (1.14)$$

Condition (1.14) allows us to use Fourier rather than Laplace transformations not only for the analysis of equations (1.5) and (1.6) but also for the analysis of more complicated equations for electromagnetic wave (in particular, whistler-mode) propagation in a magnetized plasma, considered in the next chapters. The analysis of this section based on the Laplace transformation will always be assumed, but never repeated.

Now we return to equation (1.13) which can be considerably simplified if we take into account our assumption $|\gamma| \ll \omega_0$. In this case $D(\omega, k)$ can be expanded in a Taylor series along the section $[(\omega_0, \gamma), (\omega_0, 0)]$ in the complex plane (ω_0, γ) to obtain:

$$
\begin{aligned}
\Re D(\omega, k) &= \Re D(\omega_0, k) + i \Im D(\omega_0, k) + \gamma \frac{\partial \Re D(\omega, k)}{\partial \gamma} \bigg|_{\gamma=0} \\
&+ i\gamma \frac{\partial \Im D(\omega, k)}{\partial \gamma} \bigg|_{\gamma=0} = 0.
\end{aligned}
\tag{1.15}
$$

Then, assuming that wave damping or amplification does not influence wave propagation, i.e. $\partial \Re D / \partial \gamma = 0$ (influence of growth and damping on wave (whistler-mode) propagation has been considered in the recent paper by Sazhin (1992)), and using the Cauchy–Riemann theorem ($\partial \Im D / \partial \gamma = \partial \Re D / \partial \omega_0$), we obtain from (1.15):

$$
\Re D(\omega_0, k) = 0,
\tag{1.16}
$$

$$
\gamma = -\frac{\Im D(\omega_0, k)}{(\partial / \partial \omega_0) \Re D(\omega_0, k)}.
\tag{1.17}
$$

$\gamma < 0$ corresponds to wave damping, while $\gamma > 0$ describes wave growth. Were we to assume that both $\bar{f}_{1e}$ and Φ were proportional to $\exp[i(\omega t - kr)]$, i.e. to the function conjugate to $\exp[i(kr - \omega t)]$, then we should put $\omega = \omega_0 - i\gamma$. In this case $\gamma < 0$ would again correspond to wave damping, while $\gamma > 0$ would describe wave growth.

The system of equations (1.16) and (1.17) is a very general one and, in the rest of the book, it will be applied to much more complicated functions $D(\omega_0, k)$ provided that $|\gamma| \ll \omega_0$. For an isotropic Maxwellian plasma, when

$$
\bar{f}_{0e} = \frac{1}{\sqrt{\pi} w} \exp(-v^2 / w^2)
\tag{1.18}
$$

equation (1.13) can be simplified to:

$$
D(\omega, k) = k^2 \lambda_D^2 - \frac{dZ(\xi)}{d\xi} = 0,
\tag{1.19}
$$

where $\lambda_D = w / \Pi$ is the Debye length, $\xi = \omega / kw$, and

$$
Z(\xi) = \frac{1}{\sqrt{\pi}} \int_{-\infty}^{+\infty} \frac{\exp(-t^2)}{t - \xi} dt
\tag{1.20}
$$

is the plasma dispersion function (Fried & Conte, 1961).

In general $Z(\xi)$ is a complicated complex function of a complex argument ξ and it needs to be analysed numerically (e.g. Poppe & Wijers, 1990) and/or be approximated by simpler (rational) functions (Fried, Hedrick & McCune, 1968; Brinca, 1973; Martín & González, 1979; Martín, Donoso & Zamudio-Cristi, 1980; Németh, Ág & Paris, 1981; Sato, 1984; McCabe, 1984; De Jagher & Sluijter, 1987; Robinson & Newman, 1988; Sazhin, 1989d, 1990b). In many practically important cases of wave propagation we can assume that $|\Im\xi| \ll |\Re\xi| \equiv |\xi_0|$. Then remembering (1.14) and with some rearrangement we can rewrite the expression for $Z(\xi)$ in a more convenient form:

$$Z(\xi_0) = -2 \int_0^{\xi_0} \exp(-\xi_0^2 + t^2)dt + i\sqrt{\pi} \exp(-\xi_0^2). \qquad (1.21)$$

Plots of $\Re Z = -2 \int_0^{\xi_0} \exp(-\xi_0^2 + t^2)dt$ versus ξ_0 and $\Im Z = \sqrt{\pi} \exp(-\xi_0^2)$ versus ξ_0 are shown in Fig. 1.1. As follows from this figure, $\Im Z > 0$ and reaches its maximum at $\xi_0 = 0$. However, our theory is not valid for ξ_0 close to 0, as $|\Im\xi|$ in this case is of the same order of magnitude as or greater than $|\xi_0|$. In contrast to $\Im Z$, $\Re Z$ is an odd function of ξ_0: $\Re Z > 0$ at $\xi_0 < 0$ and $\Re Z < 0$ at $\xi_0 > 0$. $|\Re Z|$ reaches its maxima when $|\xi_0| = 0.924$. In the limit $|\xi| \gg 1$, $\Re Z(\xi_0)$ can be simplified to:

$$\Re Z(\xi_0) = -\frac{1}{\xi_0} - \frac{1}{2\xi_0^3} - \frac{3}{4\xi_0^5} - \dots . \qquad (1.22)$$

Taking the two first terms in this expansion we obtain, neglecting the contribution of the second and higher order terms, the solution of (1.16) in the form:

$$\omega_0 = \Pi(1 + \frac{3}{4}k^2\lambda_D^2). \qquad (1.23)$$

This equation for longitudinal waves was first obtained by Vlasov (1938) (the limiting case of this equation corresponding to $k = 0$ was first obtained by Tonks & Langmuir (1929)).

Using the same approximation as when deriving (1.23) we obtain from (1.17), (1.19) and (1.21) the following expression for γ:

$$\gamma = -\frac{\sqrt{\pi}\Pi}{|k^3\lambda_D^3|} \exp\left[-\left(\frac{1}{k^2\lambda_D^2} - \frac{3}{2}\right)\right]. \qquad (1.24)$$

As follows from (1.24), the waves propagating in a Maxwellian plasma are damped although no collisions are assumed. This collisionless damping of electrostatic waves was first theoretically predicted by Landau (1946) and so this phenomenon is known as Landau damping. A slight difference

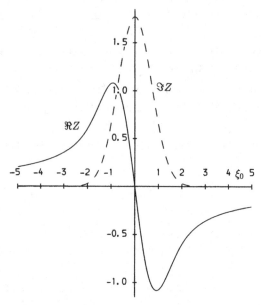

Fig. 1.1 Plots of $\Re Z(\xi_0) = -2 \int_0^{\xi_0} \exp\left(-\xi_0^2 + t^2\right) dt$ versus ξ_0 (solid) and $\Im Z(\xi_0) = \sqrt{\pi} \exp\left(-\xi_0^2\right)$ versus ξ_0 (dashed) (see equation (1.21)).

between equation (1.24) and the original expression for γ obtained by Landau lies in a different definition of electron thermal velocity w (see equation (1.18)): this velocity, as defined by Landau, is $\sqrt{2}$ times smaller than in our definition.

When $g_k(v)$ has its own poles, then the poles of the numerator of (1.10) should be added to the poles of the denominator determined by (1.12). This can result in undamped oscillations in a collisionless plasma (Akhiezer *et al.*, 1975). For more refined aspects of the mathematical analysis of the problem considered in this section readers are referred to the original papers, e.g. those of Backus (1960), Hayes (1961, 1963), Saenz (1965) and Trocheris (1965).

From equations (1.13) and (1.17) one can see that the electrostatic waves under consideration are unstable ($\gamma > 0$) when $\partial f_{0e}/\partial v > 0$ for $v \simeq \omega_0/k$. This and a number of other similar instabilities have been widely studied by many authors, although the theoretical foundation of their analysis has not yet been properly worked out. The Vlasov equation was derived from the chain of BBGKY equations for an isotropic Maxwellian plasma only, provided condition (1.1) was valid.

In the next section we derive the dispersion equation similar to (1.13), but for the general case of electromagnetic wave propagation in a magnetized plasma, with relativistic effects taken into account.

1.3 The dispersion equation

In the case of a relativistic magnetized plasma with electromagnetic disturbances the Vlasov equation (1.3) is generalized to:

$$\frac{\partial f_\alpha}{\partial t} + (\mathbf{v}\nabla)f_\alpha + q_\alpha(\mathbf{E} + \frac{1}{c}[\mathbf{v} \times \mathbf{B}])\frac{\partial}{\partial \mathbf{p}}f_\alpha = 0, \tag{1.25}$$

where ∇ is the spatial gradient, $\partial/\partial\mathbf{p}$ is the gradient in momentum space, and $f_\alpha = f_\alpha(\mathbf{r}, \mathbf{p})$ are the particle distribution functions which, as in the case of equation (1.3), will be asssumed to be normalized to unity, i.e.

$$\int d\mathbf{r} \int f_\alpha d\mathbf{p} = 1. \tag{1.26}$$

The electric ($\mathbf{E}$) and magnetic ($\mathbf{B}$) fields are determined by the Maxwell equations:

$$\nabla \times \mathbf{E} = -\frac{1}{c}\frac{\partial \mathbf{B}}{\partial t} \tag{1.27}$$

$$\nabla \times \mathbf{B} = \frac{4\pi}{c}\mathbf{j} + \frac{1}{c}\frac{\partial \mathbf{E}}{\partial t} \tag{1.28}$$

$$\nabla \cdot \mathbf{E} = 4\pi\rho \tag{1.29}$$

$$\nabla \cdot \mathbf{B} = 0, \tag{1.30}$$

where ρ and $\mathbf{j}$ are charge and current densities respectively. They can be presented in the form:

$$\rho = \Sigma_\alpha n_\alpha q_\alpha \int f_\alpha d\mathbf{p} + \rho_i \tag{1.31}$$

$$\mathbf{j} = \Sigma_\alpha n_\alpha q_\alpha \int \mathbf{v}f_\alpha d\mathbf{p} + \mathbf{j}_i, \tag{1.32}$$

where n_α are particle densities, and ρ_i and $\mathbf{j}_i$ are the densities of the external charges and currents respectively which will be assumed to be equal to zero. Integration in (1.31) and (1.32) is performed over all momenta $\mathbf{p}$. Equation (1.29) is a generalization of the Poisson equation (1.4) used in Section 1.2.

After taking the divergence of both sides of equations (1.27) and (1.28), and taking into account the continuity equation $\nabla \cdot \mathbf{j} = -\partial\rho/\partial t$, we have $\nabla \cdot \mathbf{B} = \text{const}$ and $\nabla \cdot \mathbf{E} = 4\pi\rho + \text{const}$. Thus equations (1.29) and (1.30)

can be considered as the boundary conditions for the equations (1.27) and
(1.28) determining the unknown constants. The continuity equation can be
obtained from (1.25); it also follows from equations (1.28) and (1.29).

Eliminating **B** from equations (1.27) and (1.28) we obtain:

$$\nabla \times \nabla \times \mathbf{E} + \frac{1}{c^2}\frac{\partial^2 \mathbf{E}}{\partial t^2} + \frac{4\pi}{c^2}\frac{\partial \mathbf{j}}{\partial t} = 0. \tag{1.33}$$

Equation (1.33) is a general one. In what follows it will be applied to the
problem of electromagnetic wave propagation in a magnetized plasma. We
assume that both **E** and **j** are proportional to $\exp i(\omega t - \mathbf{kr})$, i.e.

$$\mathbf{E} = \mathbf{E_1} \exp i(\omega t - \mathbf{kr}) \tag{1.34}$$

$$\mathbf{j} = \mathbf{j_1} \exp i(\omega t - \mathbf{kr}). \tag{1.35}$$

In a similar way we assume that **B** can be written as:

$$\mathbf{B} = \mathbf{B_0} + \mathbf{B_1} \exp i(\omega t - \mathbf{kr}), \tag{1.36}$$

where $\mathbf{B_0}$ is an external magnetic field, and the wave magnetic field $\mathbf{B_1}$ is
related to $\mathbf{E_1}$ and $\mathbf{j_1}$ via equations (1.27) and (1.28), $|\mathbf{B_1}| \ll |\mathbf{B_0}|$.

Also, we could base our theory on the assumptions that these fields are
proportional to $\exp i(\mathbf{kr} - \omega t)$ or $\exp i(\mathbf{kr} + \omega t)$. These assumptions would
lead us to slightly different intermediate equations but would not affect the
results.

Substituting (1.34) and (1.35) into (1.33) we obtain:

$$\mathbf{N} \times \mathbf{N} \times \mathbf{E} + \mathbf{E} - \frac{4\pi i}{\omega}\mathbf{j} = 0, \tag{1.37}$$

where $\mathbf{N} = \mathbf{k}c/\omega$ is the wave refractive index.

The value of **j** in (1.37) is determined by (1.32) and thus finally by (1.25).
Thus two mutually connected equations (1.25) and (1.37) can be considered as the basic equations for the analysis of the problem of plane wave
propagation in a plasma with an arbitrary distribution function (with the
reservation referring to the derivation of equation (1.25); see Section 1.2).

Introducing the plasma dielectric tensor $\hat{\epsilon}$:

$$\hat{\epsilon}\mathbf{E} = \mathbf{E} - \frac{4\pi i}{\omega}\mathbf{j} \tag{1.38}$$

we can write equation (1.37) as:

$$\mathbf{N} \times \mathbf{N} \times \mathbf{E} + \hat{\epsilon}\mathbf{E} = 0. \tag{1.39}$$

Assuming, without loss of generality, that the external magnetic field $\mathbf{B_0}$ is
parallel to the z axis of the right-handed system (x, y, z), $(\mathbf{e}_z = \mathbf{e}_x \times \mathbf{e}_y,$

where $\mathbf{e}_{x,y,z}$ are unit vectors in the corresponding directions) and $\mathbf{N}$ lies in the (x, z) plane, we rewrite (1.39) in matrix form:

$$\left\| \begin{matrix} -N^2 \cos^2 \theta + \epsilon_{xx} & \epsilon_{xy} & N^2 \cos \theta \sin \theta + \epsilon_{xz} \\ \epsilon_{yx} & -N^2 + \epsilon_{yy} & \epsilon_{yz} \\ N^2 \cos \theta \sin \theta + \epsilon_{zx} & \epsilon_{zy} & -N^2 \sin^2 \theta + \epsilon_{zz} \end{matrix} \right\| \left\| \begin{matrix} E_x \\ E_y \\ E_z \end{matrix} \right\| = 0. \quad (1.40)$$

The system of equations (1.40) has a non-trivial solution only when the determinant of the matrix is equal to zero, i.e. when

$$\left| \begin{matrix} -N^2 \cos^2 \theta + \epsilon_{xx} & \epsilon_{xy} & N^2 \cos \theta \sin \theta + \epsilon_{xz} \\ \epsilon_{yx} & -N^2 + \epsilon_{yy} & \epsilon_{yz} \\ N^2 \cos \theta \sin \theta + \epsilon_{zx} & \epsilon_{zy} & -N^2 \sin^2 \theta + \epsilon_{zz} \end{matrix} \right| = 0, \quad (1.41)$$

where θ is the wave normal angle, i.e. the angle between $\mathbf{N}$ and $\mathbf{B}_0$.

Equation (1.41) can be rewritten in a more convenient form:

$$D \equiv AN^4 + BN^2 + C = 0, \quad (1.42)$$

where:

$$A = \epsilon_{xx} \sin^2 \theta + 2\epsilon_{xz} \sin \theta \cos \theta + \epsilon_{zz} \cos^2 \theta \quad (1.43)$$

$$B = -\epsilon_{xx}\epsilon_{zz} - (\epsilon_{yy}\epsilon_{zz} + \epsilon_{yz}^2) \cos^2 \theta - (\epsilon_{xx}\epsilon_{yy} + \epsilon_{xy}^2) \sin^2 \theta$$

$$+ 2(\epsilon_{xy}\epsilon_{yz} - \epsilon_{yy}\epsilon_{xz}) \sin \theta \cos \theta + \epsilon_{xz}^2 \quad (1.44)$$

$$C = \epsilon_{zz}(\epsilon_{xx}\epsilon_{yy} + \epsilon_{xy}^2) + \epsilon_{xx}\epsilon_{yz}^2 + 2\epsilon_{xy}\epsilon_{xz}\epsilon_{yz} - \epsilon_{yy}\epsilon_{xz}^2. \quad (1.45)$$

Explicit expressions for the elements of the plasma dielectric tensor ϵ_{ij} are derived in the next section.

1.4 The plasma dielectric tensor (general expression)

To calculate the elements of the plasma dielectric tensor $\hat{\epsilon}$ defined by (1.38) we should first calculate the current $\mathbf{j}$ as a function of plasma and wave parameters, which can be done using the Vlasov equation (1.25). The analytical solution of this equation can be obtained if we assume, similarly to Section 1.2, that f_α can be represented as:

$$f_\alpha = f_{0\alpha} + f_{1\alpha}, \quad (1.46)$$

where $f_{0\alpha}$ is the unperturbed part of the particle distribution function which is given a priori, and $f_{\alpha 1}$ is the perturbation due to wave propagation,

$$f_{1\alpha} \ll f_{0\alpha}. \quad (1.47)$$

In a similar way to (1.34)–(1.36) we can write:

$$f_{1\alpha} = \tilde{f}_\alpha \exp i(\omega t - \mathbf{kr}). \qquad (1.48)$$

Remembering (1.34), (1.36), (1.47) and (1.48), and keeping only the first order terms with respect to the perturbations, we rewrite equation (1.25) as:

$$i\omega \left(1 - \frac{\mathbf{kv}}{\omega}\right)\tilde{f}_\alpha + \frac{q_\alpha}{c}(\mathbf{v} \times \mathbf{B}_0)\frac{\partial \tilde{f}_\alpha}{\partial \mathbf{p}} + q_\alpha \left[\mathbf{E}_1\left(1 - \frac{\mathbf{kv}}{\omega}\right) + \frac{\mathbf{k}(\mathbf{v}\mathbf{E}_1)}{\omega}\right]\frac{\partial f_{0\alpha}}{\partial \mathbf{p}} = 0. \qquad (1.49)$$

When deriving (1.49) we took into account that:

$$\frac{\mathbf{v} \times \mathbf{B}}{c} = \frac{\mathbf{v} \times \mathbf{B}_0}{c} + \frac{\mathbf{k}(\mathbf{v}\mathbf{E})}{\omega} - \frac{\mathbf{E}(\mathbf{k}\mathbf{v})}{\omega}, \qquad (1.50)$$

which follows from equation (1.27) and the definition of the vector product.

Similarly to Section 1.3 we assume that $\mathbf{B}_0$ is directed along the z axis of the right-handed coordinate system (x, y, z), and determine the angle between the projection of $\mathbf{p}$ on the (x, y) plane and the x axis as ϕ ($p_x = p\cos\phi$; $p_y = p\sin\phi$). Then we can write:

$$\frac{q_\alpha}{c}(\mathbf{v} \times \mathbf{B}_0)\frac{\partial \tilde{f}_\alpha}{\partial \mathbf{p}} = -\Omega_\alpha \frac{\partial \tilde{f}_\alpha}{\partial \phi}, \qquad (1.51)$$

where $\Omega_\alpha = q_\alpha B_0/(M_\alpha c)$ is the particle gyrofrequency, M_α is the relativistic particle mass:

$$M_\alpha = m_\alpha\sqrt{1 + p^2/m_\alpha^2 c^2} = m_\alpha / \sqrt{1 - \frac{v^2}{c^2}}, \qquad (1.52)$$

and m_α is the particle mass at rest.

Now we introduce the new parameters $a(\phi)$ and $b(\phi)$ determined by:

$$a(\phi) = -\frac{i\omega}{\Omega_\alpha}\left(1 - \frac{\mathbf{kv}}{\omega}\right) \qquad (1.53)$$

$$b(\phi) = \frac{q_\alpha}{\Omega_\alpha}\left[\mathbf{E}\left(1 - \frac{\mathbf{kv}}{\omega}\right) + \frac{\mathbf{k}(\mathbf{v}\mathbf{E})}{\omega}\right]\frac{\partial f_{0\alpha}}{\partial \mathbf{p}} \qquad (1.54)$$

and write equation (1.49) as:

$$\frac{d\tilde{f}_\alpha}{d\phi} + a(\phi)\tilde{f}_\alpha = b(\phi). \qquad (1.55)$$

The solution of equation (1.55) has the form (Korn & Korn, 1968):

$$\tilde{f}_\alpha(\phi) = \exp\left[-\int_0^\phi a(\phi')\mathrm{d}\phi'\right]\left[\int_0^\phi b(\phi')\exp\left(\int_0^{\phi'} a(\phi'')\mathrm{d}\phi''\right)\mathrm{d}\phi' + C_0\right],$$

(1.56)

where the constant C_0 is determined by the periodicity condition:

$$\tilde{f}_\alpha(\phi + 2\pi) = \tilde{f}_\alpha(\phi).$$

(1.57)

Remembering our assumption that $\mathbf{N}$ lies in the (x, z) plane, i.e. $\mathbf{k} = (k_x, 0, k_z) \equiv (k_\perp, 0, k_\parallel) = (k\sin\theta, 0, k\cos\theta)$, we obtain:

$$\int_0^{\phi'} a(\phi'')\mathrm{d}\phi'' = -\frac{\mathrm{i}\omega}{\Omega_\alpha}\left[\left(1 - \frac{k_\parallel v_\parallel}{\omega}\right)\phi' - \frac{k_\perp v_\perp}{\omega}\sin\phi'\right],$$

(1.58)

and

$$b(\phi') = \frac{q_\alpha}{\Omega_\alpha}\left\{E_x\left[\left(1 - \frac{k_\parallel v_\parallel}{\omega}\right)\frac{\partial f_{0\alpha}}{\partial p_\perp} + \frac{k_\parallel v_\perp}{\omega}\frac{\partial f_{0\alpha}}{\partial p_\parallel}\right]\cos\phi'\right.$$

$$+ E_y\left[\left(1 - \frac{k_\parallel v_\parallel}{\omega}\right)\frac{\partial f_{0\alpha}}{\partial p_\perp} + \frac{k_\parallel v_\perp}{\omega}\frac{\partial f_{0\alpha}}{\partial p_\parallel}\right]\sin\phi'$$

$$\left. + E_z\left[\frac{\partial f_{0\alpha}}{\partial p_\parallel} + \left(\frac{k_\perp v_\parallel}{\omega}\frac{\partial f_{0\alpha}}{\partial p_\perp} - \frac{k_\perp v_\perp}{\omega}\frac{\partial f_{0\alpha}}{\partial p_\parallel}\right)\cos\phi'\right]\right\}.$$

(1.59)

Substituting (1.58) and (1.59) into (1.56) and using the following expansions (Korn & Korn, 1968):

$$\exp(-\lambda_\alpha\sin\phi) = \sum_{n=-\infty}^{+\infty} J_n(\lambda_\alpha)\exp(-\mathrm{i}n\phi)$$

(1.60)

$$-\mathrm{i}\sin\phi\exp(-\lambda_\alpha\sin\phi) = \sum_{n=-\infty}^{+\infty} J_n'(\lambda_\alpha)\exp(-\mathrm{i}n\phi)$$

(1.61)

$$-\mathrm{i}\lambda_\alpha\cos\phi\exp(-\lambda_\alpha\sin\phi) = -\mathrm{i}\sum_{n=-\infty}^{+\infty} nJ_n(\lambda_\alpha)\exp(-\mathrm{i}n\phi),$$

(1.62)

where $\lambda_\alpha = -k_\perp v_\perp/\Omega_\alpha$, J_n are Bessel functions of order n, and $J_n'(\lambda_\alpha) =$

$dJ_n(\lambda_\alpha)/d\lambda_\alpha$ (equations (1.61) and (1.62) are obtained after taking the derivatives of (1.60) with respect to λ_α and ϕ respectively), we obtain:

$$\tilde{f}_\alpha(\phi) = \frac{iq_\alpha}{M_\alpha} \exp(i\lambda_\alpha \sin\phi)$$

$$\times \left[\sum_{n=-\infty}^{+\infty} \frac{\mathbf{a}(n)\mathbf{E}}{-k_\| v_\| + n\Omega_\alpha + \omega} \exp(-in\phi) + \frac{C_0}{i\Omega_\alpha} \exp\left[-\frac{i(k_\| v_\| - \omega)\phi}{\Omega_\alpha} \right] \right],$$

$$(1.63)$$

where

$$a_x(n) = \left[\frac{k_\| v_\perp}{\omega} \frac{\partial f_{0\alpha}}{\partial p_\|} + \left(1 - \frac{k_\| v_\|}{\omega} \right) \frac{\partial f_{0\alpha}}{\partial p_\perp} \right] \frac{n}{\lambda_\alpha} J_n(\lambda_\alpha), \tag{1.64}$$

$$a_y(n) = \left[\frac{k_\| v_\perp}{\omega} \frac{\partial f_{0\alpha}}{\partial p_\|} + \left(1 - \frac{k_\| v_\|}{\omega} \right) \frac{\partial f_{0\alpha}}{\partial p_\perp} \right] iJ_n'(\lambda_\alpha), \tag{1.65}$$

$$a_z(n) = \left\{ \frac{\omega + n\Omega_\alpha - k_\| v_\|}{\omega} \right.$$

$$\times \left(\frac{\partial f_{0\alpha}}{\partial p_\|} - \frac{p_\|}{p_\perp} \frac{\partial f_{0\alpha}}{\partial p_\perp} \right) + \frac{v_\|}{\omega} \left[k_\| \frac{\partial f_{0\alpha}}{\partial p_\|} + \left(\frac{\omega - k_\| v_\|}{v_\perp} \right) \frac{\partial f_{0\alpha}}{\partial p_\perp} \right] \right\} J_n(\lambda_\alpha).$$

$$(1.66)$$

The solution (1.63) predicts $|\tilde{f}_\alpha| \to \infty$ when

$$d \equiv k_\| v_\| - n\Omega_\alpha - \omega = 0 \tag{1.67}$$

(unless the numerator is equal to zero simultaneously) which would contradict our assumption (1.47). To resolve this apparent contradiction we should remember that ω may be complex, in which case d could be close to zero, but not equal to zero.

From (1.63) it follows that the periodicity condition (1.57) leads to $C_0 = 0$.

The perturbation $\tilde{f}_\alpha$ produces the electric current density $\mathbf{j}$ defined by the following expression (cf. equation (1.32)):

$$\mathbf{j} = \sum_\alpha q_\alpha \int_0^{+\infty} p_\perp dp_\perp \int_{-\infty}^{+\infty} dp_\| \int_0^{2\pi} \mathbf{v}\tilde{f}_\alpha d\phi, \tag{1.68}$$

where the summation is assumed over all particle species.

Substituting (1.63) with $C_0 = 0$ into (1.68) we obtain

$$
\begin{aligned}
\mathbf{j} &= \sum_\alpha \frac{iq_\alpha^2}{M_\alpha} \int_0^{+\infty} p_\perp dp_\perp \int_{-\infty}^{+\infty} dp_\parallel \int_0^{2\pi} d\phi \\
&\times \sum_{n=-\infty}^{+\infty} (\mathbf{a}(n)\mathbf{E}) \frac{\exp i(\lambda_\alpha \sin\phi - n\phi)}{-k_\parallel v_\parallel + n\Omega_\alpha + \omega} \mathbf{v},
\end{aligned} \tag{1.69}
$$

where $\mathbf{v} = (v_\perp \cos\phi;\ v_\perp \sin\phi;\ v_\parallel)$.

Similarly to Section 1.2 the singularity in the integral over $p_\parallel$ in (1.68) should be encircled from below, so that the whole integral is defined by equation (1.14). As has already been mentioned, this condition will be kept throughout the rest of our analysis.

Equation (1.69) can be simplified if we take into account expansions (1.60)–(1.62) and the following identities:

$$
\int_0^{2\pi} \exp[i(n-m)\phi]d\phi = 2\pi\delta_{nm}, \tag{1.70}
$$

where:

$$
\delta_{nm} = \begin{cases} 0 & \text{when } n \neq m \\ 1 & \text{when } n = m, \end{cases}
$$

$$
J_0^2 + 2\sum_{n=1}^{\infty} J_n^2 = 1, \tag{1.71}
$$

and

$$
J_{-n} = (-1)^n J_n. \tag{1.72}
$$

Then substituting this simplified expression for $\mathbf{j}$ into (1.38) we obtain the final expression for the elements of $\hat{\epsilon}$:

$$
\begin{aligned}
\epsilon_{ij} &= \delta_{ij} + \sum_\alpha \sum_{n=-\infty}^{+\infty} \frac{2\pi\Pi_\alpha^2 m_\alpha}{\omega^2} \\
&\times \int_0^{\infty} dp_\perp \int_{-\infty}^{+\infty} dp_\parallel \left\{ \left[(\omega M_\alpha - k_\parallel p_\parallel) \frac{\partial f_{0\alpha}}{\partial p_\perp} + k_\parallel p_\perp \frac{\partial f_{0\alpha}}{\partial p_\parallel} \right] \right. \\
&\times \left. \frac{\hat{\Pi}_{ij}^{(n,\alpha)} M_\alpha}{\omega M_\alpha - k_\parallel p_\parallel - n\Omega_{0\alpha} m_\alpha} + \frac{b_i b_j p_\parallel}{M_\alpha} \left(p_\perp \frac{\partial f_{0\alpha}}{\partial p_\parallel} - p_\parallel \frac{\partial f_{0\alpha}}{\partial p_\perp} \right) \right\},
\end{aligned} \tag{1.73}
$$

where

$$
\hat{\Pi}_{ij}^{(n,\alpha)} = \begin{vmatrix} \frac{n^2\Omega_\alpha^2}{k_\perp^2} J_n^2 & iv_\perp \frac{n\Omega_\alpha}{k_\perp} J_n J_n' & v_\parallel \frac{n\Omega_\alpha}{k_\perp} J_n^2 \\ -iv_\perp \frac{n\Omega_\alpha}{k_\perp} J_n J_n' & v_\perp^2 J_n'^2 & -iv_\parallel v_\perp J_n J_n' \\ v_\parallel \frac{n\Omega_\alpha}{k_\perp} J_n^2 & iv_\parallel v_\perp J_n J_n' & v_\parallel^2 J_n^2 \end{vmatrix}, \tag{1.74}
$$

$\mathbf{b} = \mathbf{B}_0/|\mathbf{B}_0|$; $J'_n \equiv dJ_n/d\lambda_\alpha$; $\Pi_\alpha = \sqrt{4\pi n_\alpha q_\alpha/m_\alpha}$ is the plasma frequency of the corresponding particles, and $\Omega_{0\alpha} = q_\alpha B_0/(m_\alpha c)$ is the particle gyrofrequency at rest. Note that all equations have so far been written in a full relativistic form and condition (1.2) has not been imposed.

In the non-relativistic limit, (1.73) reduces to that given by Akhiezer *et al.* (1975) provided that we take into account that λ_α in our definition has the sign opposite to that of Akhiezer *et al.* In contrast to our assumptions (1.34)–(1.36) Akhiezer *et al.* (1975) assumed that the field and current components are proportional to $\exp i(\mathbf{kr} - \omega t)$. Hence, the difference between our results. Note that there are some printing errors in the corresponding expression for ϵ_{ij} in the English translation of Akhiezer *et al.* In the Russian original this expression is correct except that the summation over n is omitted. Also, both editions contain printing errors in the intermediate expressions used when deriving the final expression for ϵ_{ij}.

The most important difference between expressions (1.73) and the corresponding non-relativistic expressions for ϵ_{ij} comes from the fact that the poles in (1.73) are two-dimensional and the particles which resonate with the wave should satisfy the following condition:

$$\sqrt{1 + p^2/(m_\alpha^2 c^2)} - p_\parallel/p_{ph} - nY_{0\alpha} = 0, \qquad (1.75)$$

where $p_{ph} = \omega m_\alpha/k_\parallel$ and $Y_{0\alpha} = \Omega_{0\alpha}/\omega$. Equation (1.75) determines either ellipses or hyperbolae in the $(p_\parallel, p_\perp)$ plane (Tsang, 1984), which reduce to circles if we assume that $\sqrt{1 + p^2/(m_\alpha^2 c^2)} \approx 1 + p^2/(2m_\alpha^2 c^2)$, i.e. in the weakly relativistic limit, and to the lines $p_\parallel = p_{ph}(1 - nY_{0\alpha})$ in the non-relativistic limit $p^2/(m_\alpha^2 c^2) \to 0$.

1.5 The plasma dielectric tensor (weakly relativistic approximation)

Expressions (1.73) for elements of the plasma dielectric tensor are very general ones and can be applied to a wide range of distribution functions $f_{0\alpha}$. As was mentioned in Section 1.1 we are interested mainly in the effects of electrons on wave propagation. Hence, unless stated otherwise, we restrict ourselves to considering the contribution of the electron distribution function $f_{0e} \equiv f_0$. Following Pritchett (1984) and Sazhin (1987a) we assume this distribution in the form:

$$f_0(p_\perp, p_\parallel) = (j!\pi^{3/2} p_{0\perp}^{2j+2} p_{0\parallel})^{-1} p_\perp^{2j} \exp(-p_\parallel^2/p_{0\parallel}^2 - p_\perp^2/p_{0\perp}^2), \qquad (1.76)$$

where $p_{0\|(\perp)}$ are relativistic analogues of the electron temperature in the direction parallel (perpendicular) to a magnetic field, $j = 0, 1, 2, \ldots$; when $j \neq 0$ function (1.76) describes a loss cone (reduced number of electrons with small $p_\perp$) distribution, the situation typical of magnetospheric conditions.

Having substituted (1.76) into (1.73) we obtain the elements of ϵ_{ij} expressed in terms of rather complicated integrals even for an isotropic plasma (see e.g. Yoon & Davidson, 1990). However, if we impose additional inequalities:

$$\left.\begin{array}{c} \max(p^2/2m_e^2c^2, |p_\||/p_{ph}|) \ll \min(|1 - nY|, 1) \\ |k_\perp v_\perp / \Omega| \ll 1 \end{array}\right\} \tag{1.77}$$

for $n = 0, \pm1, \pm2$, then the expressions for ϵ_{ij} can be written in a relatively simple form:

$$\epsilon_{ij} = \epsilon_{ij}^0 + \epsilon_{ij}^t + \epsilon_{ij}^r, \tag{1.78}$$

where ϵ_{ij}^0 are the elements of the cold plasma dielectric tensor (Holter & Kildal, 1973):

$$\left.\begin{array}{l} \epsilon_{xx}^0 = \epsilon_{yy}^0 = S \equiv \frac{1}{2}(R + L) \\ \epsilon_{xy}^0 = -\epsilon_{yx}^0 = iD \equiv \frac{1}{2}i(R - L) \\ \epsilon_{zz}^0 = P \\ \epsilon_{xz}^0 = \epsilon_{zx}^0 = \epsilon_{yz}^0 = \epsilon_{zy}^0 = 0 \end{array}\right\} \tag{1.79}$$

$(R = 1 + X/(Y - 1), L = 1 - X/(Y + 1), P = 1 - X)$; ϵ_{ij}^t are non-relativistic thermal corrections to ϵ_{ij}^0:

$$\begin{aligned} \epsilon_{xx}^t &= \beta_e N^2 Y^2 \left[\frac{\cos^2\theta \left[Y^2(3 + Y^2) + A_e(1 - Y^4)\right]}{(Y^2 - 1)^3} \right. \\ &\quad \left. - \frac{3A_e \sin^2\theta}{(4Y^2 - 1)(Y^2 - 1)} \right] \end{aligned} \tag{1.80}$$

$$\begin{aligned} \epsilon_{xy}^t &= -\epsilon_{yx}^t = i\beta_e N^2 Y^3 \left[\frac{\cos^2\theta \left[(3Y^2 + 1) + 2A_e(1 - Y^2)\right]}{(Y^2 - 1)^3} \right. \\ &\quad \left. - \frac{6A_e \sin^2\theta}{(4Y^2 - 1)(Y^2 - 1)} \right] \end{aligned} \tag{1.81}$$

$$\begin{aligned} \epsilon_{yy}^t &= \beta_e N^2 Y^2 \left[\frac{\cos^2\theta \left[Y^2(3 + Y^2) + A_e(1 - Y^4)\right]}{(Y^2 - 1)^3} \right. \\ &\quad \left. - \frac{A_e \sin^2\theta (8Y^2 + 1)}{(4Y^2 - 1)(Y^2 - 1)} \right] \end{aligned} \tag{1.82}$$

$$\epsilon_{xz}^t = \epsilon_{zx}^t = \beta_e \frac{N^2 Y^2 \cos\theta \sin\theta}{(Y^2-1)^2} \left[-\left(Y^2+1\right) + A_e\left(Y^2-1\right)\right] \quad (1.83)$$

$$\epsilon_{yz}^t = -\epsilon_{zy}^t = i\beta_e \frac{N^2 Y^3 \cos\theta \sin\theta}{(Y^2-1)^2} \left[2 - A_e\left(Y^2-1\right)\right] \quad (1.84)$$

$$\epsilon_{zz}^t = \beta_e \frac{N^2 Y^2 \left[1 + (2-3Y^2)\cos^2\theta\right]}{Y^2-1}, \quad (1.85)$$

ϵ_{ij}^r are relativistic thermal corrections to ϵ_{ij}^0:

$$\epsilon_{xx}^r = \epsilon_{yy}^r = \frac{\beta_e(1+Y^2)Y^2(1+4A_e)}{2(Y^2-1)^2} \quad (1.86)$$

$$\epsilon_{xy}^r = -\epsilon_{yx}^r = \frac{i\beta_e Y^3(1+4A_e)}{(Y^2-1)^2} \quad (1.87)$$

$$\epsilon_{zz}^r = \frac{\beta_e Y^2(3+2A_e)}{2} \quad (1.88)$$

$$\epsilon_{xz}^r = \epsilon_{zx}^r = \epsilon_{yz}^r = \epsilon_{zy}^r = 0, \quad (1.89)$$

where $Y = \Omega/\omega$ (Ω is the modulus of the electron gyrofrequency at rest: $\Omega = |\Omega_{0e}|$), $X = \Pi^2/\omega^2$, $\beta_e = \frac{1}{2}\left(\Pi p_{0\parallel}/(\Omega m_e c)\right)^2$, and $A_e = (j+1)p_{0\perp}^2/p_{0\parallel}^2$; as in Sections 1.3 and 1.4 we have written these equations in the right-handed coordinate system ($\mathbf{e}_z = \mathbf{e}_x \times \mathbf{e}_y$); in the left-handed system ($\mathbf{e}_z = -\mathbf{e}_x \times \mathbf{e}_y$) or when the sign in the power of the exponential terms in (1.34)–(1.36) and (1.48) is changed we should formally change the sign of Ω.

When writing the elements of the plasma dielectric tensor in the form (1.78) we neglected the contribution of residues in the integrals (1.73), which describe collisionless damping or amplification of the waves. In the case when the electron distribution can be represented as a sum of distributions of the type (1.76) with different j, $p_{0\parallel}$ and $p_{0\perp}$, then $\epsilon_{ij}^{t(r)}$ can be written as a sum of $\epsilon_{ij}^{t(r)}$ corresponding to each of these distributions; X in (1.79) would refer to the total electron plasma frequency.

As follows from (1.80)–(1.89) the contribution of relativistic thermal corrections to ϵ_{ij}^0 can be neglected when compared with ϵ_{ij}^t when $N^2 \gg 1$. However, in the opposite limiting case, $N^2 \ll 1$, relativistic corrections are the dominant ones. Thus, when N^2 is not specified we cannot neglect the contribution of relativistic corrections even if (1.2) is valid, in contrast to what used to be done in many papers where the thermal effects on wave propagation were considered (e.g. André, 1985). In fact, when considering wave propagation in the non-relativistic approximation we neglect the relativistic effects in the Vlasov equation, while the Maxwell equations are relativistic by their nature. Thus non-relativistic analysis of waves is in fact not self-consistent and can be justified only in the limiting case $N^2 \gg 1$ which will be assumed in the next three sections.

1.6 The plasma dielectric tensor (non-relativistic approximation)

Although a weakly relativistic approximation considered in the previous section allows us to present the elements of ϵ_{ij} in a rather simple form (1.78) when the conditions (1.77) are valid, a more general analysis of ϵ_{ij} in this approximation appears to be much more complicated. This makes it relevant to consider a simpler non-relativistic approximation. Using (1.78)–(1.89) we can see that when conditions (1.77) are valid this approximation is justified when $N^2 \gg 1$. As we will show in Chapter 3, within this limit the non-relativistic approximation is justified for parallel whistler-mode propagation even when conditions (1.77) are not valid. This allows us to assume that except for some special cases of whistler-mode propagation, the non-relativistic approximation is generally valid when $N^2 \gg 1$. This inequality will be assumed throughout this and the following two sections.

In the non-relativistic approximation we can put $p_{0\|(\perp)} = w_{\|(\perp)}m_e$, $p_{\|(\perp)} = v_{\|(\perp)}m_e$ and write the distribution function (1.76) with respect to velocities rather than momenta:

$$f(v_\perp, v_\|) = (j!\pi^{3/2}w_\perp^{2j+2}w_\|)^{-1}v_\perp^{2j}\exp(-v_\|^2/w_\|^2 - v_\perp^2/w_\perp^2). \qquad (1.90)$$

Substituting (1.90) into expressions (1.73) for the elements ϵ_{ij} taken in the limit $M_\alpha = m_\alpha$, and considering the contribution of electrons only, we can write:

$$\epsilon_{xx} = 1 + X\left[(A_e - 1) - \frac{4}{\alpha^2 j!w_\perp^{2j+2}}\sum_{n=-\infty}^{+\infty}n^2\mu_j^n Z_n\right] \qquad (1.91)$$

$$\epsilon_{xy} = -\epsilon_{yx} = -\frac{2iX}{\alpha j!w_\perp^{2j+2}}\frac{d}{d\alpha}\sum_{n=-\infty}^{+\infty}n\mu_j^n Z_n \qquad (1.92)$$

$$\epsilon_{yy} = 1 + X\left[(A_e - 1) - \frac{4}{j!w_\perp^{2j+2}}\sum_{n=-\infty}^{+\infty}\nu_j^n Z_n\right] \qquad (1.93)$$

$$\epsilon_{zx} = \epsilon_{xz} = \frac{2X\,w_\|}{\alpha j!w_\perp^{2j+2}}\sum_{n=-\infty}^{+\infty}n\mu_j^n Z_n' \qquad (1.94)$$

$$\epsilon_{yz} = -\epsilon_{zy} = -\frac{iXw_\|}{j!w_\perp^{2j+2}}\frac{d}{d\alpha}\sum_{n=-\infty}^{+\infty}\mu_j^n Z_n' \tag{1.95}$$

$$\epsilon_{zz} = 1 + \frac{2Xw_\|^2}{j!w_\perp^{2j+2}}\sum_{n=-\infty}^{+\infty}\xi_n\mu_j^n Z_n', \tag{1.96}$$

where:

$$\mu_j^n = M_j^n A_n,$$

$$\nu_j^n = M_j^n\left(\frac{n}{\alpha}\frac{d}{d\alpha} - \frac{n^2}{\alpha^2}\right)A_n + \tilde{M}_j^n A_{n+1},$$

$$M_j^n = \frac{n\Omega}{k_\| w_\|}\left[jL_{j-1} - \frac{1}{w_\perp^2}L_j + \frac{1}{w_\|^2}\frac{nY-1}{nY}L_j\right],$$

$$\tilde{M}_j^n = \frac{n\Omega}{k_\| w_\|}\left[jL_j - \frac{1}{w_\perp^2}L_{j+1} + \frac{1}{w_\|^2}\frac{nY-1}{nY}L_{j+1}\right],$$

$$L_j = (-1)^j\frac{d^j}{d\left(\frac{1}{w_\perp^2}\right)^j},$$

$$A_n = \frac{1}{2}w_\perp^2\exp\left(-\frac{1}{2}w_\perp^2\alpha^2\right)I_n\left(\frac{1}{2}w_\perp^2\alpha^2\right),$$

$A_e = (j+1)\,w_\perp^2/w_\|^2$, $\xi_n = (\omega - n\Omega)/(k_\| w_\|)$, $\alpha = k_\perp/\Omega$, $Z_n \equiv Z(\xi_n)$ is the plasma dispersion function (see equation (1.21)), $Z_n' \equiv dZ(\xi_n)/d\xi_n$, and I_n is the modified Bessel function.

In a similar way to (1.78)–(1.89), expressions (1.91)–(1.96) can be generalized to the case when electrons consist of different species with distributions (1.90) having different j and $w_{\|(\perp)}$ by summing the terms proportional to X over all these species. This will also be assumed for the expressions for ϵ_{ij} taken in different limiting cases.

Assuming that $x \equiv \frac{1}{2}w_\perp^2\alpha^2 \ll 1$ (cf. conditions (1.77)) we can expand $\exp(-x)$ and $I_n(x)$ in a Taylor series and rewrite (1.91)–(1.96) in the forms:

$$\epsilon_{xx} = 1 + X\left\{A_e - 1 + \Omega_\|\left[p_1 Z_1 - p_{-1}Z_{-1}\right.\right.$$
$$\left.\left. + \ x(-q_1 Z_1 + q_{-1}Z_{-1} + 2q_2 Z_2 - 2q_{-2}Z_{-2})\right]\right\} \tag{1.97}$$

$$\epsilon_{xy} = -\epsilon_{yx} = -iX\Omega_{\|}\Big[-p_1 Z_1 - p_{-1} Z_{-1}$$

$$+ \quad x(2q_1 Z_1 + 2q_{-1} Z_{-1} - 2q_2 Z_2 - 2q_{-2} Z_{-2})\Big] \qquad (1.98)$$

$$\epsilon_{yy} = 1 + X\Big\{ A_e - 1 + \Omega_{\|}\Big[p_1 Z_1 - p_{-1} Z_{-1} + x(2A_e(j+2)Z_0/Y$$

$$- \quad 3q_1 Z_1 + 3q_{-1} Z_{-1} + 2q_2 Z_2 - 2q_{-2} Z_{-2})\Big]\Big\} \qquad (1.99)$$

$$\epsilon_{xz} = \epsilon_{zx} = \frac{Xk_\perp}{4k_\|}\Big[-p_1 Z_1' - p_{-1} Z_{-1}'$$

$$+ \quad x(q_1 Z_1' + q_{-1} Z_{-1}' - q_2 Z_2' - q_{-2} Z_{-2}')\Big] \qquad (1.100)$$

$$\epsilon_{yz} = -\epsilon_{zy} = \frac{iXk_\perp}{4k_\|}\Big[-\frac{2A_e Z_0'}{Y} + p_1 Z_1' - p_{-1} Z_{-1}'$$

$$+ \quad x\Big(\frac{3A_e(j+2)Z_0'}{2Y} - 2q_1 Z_1' + 2q_{-1} Z_{-1}' + q_2 Z_2' - q_{-2} Z_{-2}'\Big)\Big]$$

$$(1.101)$$

$$\epsilon_{zz} = 1 + X\Big\{ \Big(-\xi_0^2 + \frac{A_e k_\perp^2}{2k_\|^2 Y^2}\Big) Z_0'$$

$$+ \quad \frac{k_\perp^2}{4k_\|^2 Y}\Big[(Y-1)p_1 Z_1' + (Y+1)p_{-1} Z_{-1}'\Big]$$

$$+ \quad \frac{k_\perp x}{8k_\|^2 Y^2}\Big[-\frac{3}{2}(j+2)A_e Z_0' - 2Y(Y-1)q_1 Z_1' - 2Y(Y+1)q_{-1} Z_{-1}'$$

$$+ \quad Y(2Y-1)q_2 Z_2' + Y(2Y+1)q_{-2} Z_{-2}'\Big]\Big\}, \qquad (1.102)$$

where $p_n = 1 - a_n$, $q_n = (j+1) - \frac{1}{2}(j+2)a_n$, $a_n = A_e(nY-1)/(nY)$, $\Omega_{\|} = \Omega/(2k_\| w_\|)$.

In deriving (1.97)–(1.102) we kept only linear terms with respect to x. These expressions can be further simplified if we assume that the electron temperature is so low that $|\xi_n| \gg 1$ for all n and the asymptotic expansion (1.22) of $\Re Z$ is valid. Remembering this expansion and keeping only zero- and first-order terms with respect to $w_{\|(\perp)}^2$ as well as the terms proportional to $\exp(-\xi_n)$, we can rewrite the elements of ϵ_{ij} in the form:

$$\epsilon_{ij} = \epsilon_{ij}^0 + \epsilon_{ij}^t + \epsilon_{ij}^I, \qquad (1.103)$$

where the elements ϵ_{ij}^0 and ϵ_{ij}^t have already been defined by (1.79)–(1.85) with $\beta_e = \frac{1}{2}(\Pi w_\|/\Omega c)^2$; the elements ϵ_{ij}^I contributing to the imaginary part of D are defined as (Sazhin, 1988a):

$$\epsilon_{xx}^I \equiv i\Im\epsilon_{xx} = \frac{iXY\sqrt{\pi}}{2N\tilde{w}_\parallel\cos\theta}[p_1 - xq_1]\exp(-\xi_1^2) \qquad (1.104)$$

$$\epsilon_{xy}^I \equiv \Re\epsilon_{xy} = \frac{XY\sqrt{\pi}}{2N\tilde{w}_\parallel\cos\theta}[-p_1 + 2xq_1]\exp(-\xi_1^2) \qquad (1.105)$$

$$\epsilon_{yy}^I \equiv i\Im\epsilon_{yy} = \frac{iXY\sqrt{\pi}}{2N\tilde{w}_\parallel\cos\theta}$$

$$\times\left\{[p_1 - 3xq_1]\exp(-\xi_1^2) + \frac{2xA_e(j+2)}{Y}\exp(-\xi_0^2)\right\} \qquad (1.106)$$

$$\epsilon_{xz}^I \equiv i\Im\epsilon_{xz} = \frac{iXY\sqrt{\pi}}{2N\tilde{w}_\parallel\cos\theta}\frac{(1-Y)\tan\theta}{Y}[p_1 - xq_1]\exp(-\xi_1^2) \qquad (1.107)$$

$$\epsilon_{yz}^I \equiv \Re\epsilon_{yz} = \frac{XY\sqrt{\pi}}{2N\tilde{w}_\parallel\cos\theta}\frac{(1-Y)\tan\theta}{Y}\left\{[p_1 - 2xq_1]\exp(-\xi_1^2)\right.$$

$$\left. +\frac{A_e}{Y(1-Y)}\left[-2 + \frac{3x(j+2)}{2}\right]\exp(-\xi_0^2)\right\} \qquad (1.108)$$

$$\epsilon_{zz}^I \equiv i\Im\epsilon_{yz} = \frac{iXY\sqrt{\pi}}{2N\tilde{w}_\parallel\cos\theta}\left\{\frac{(1-Y)^2\tan^2\theta}{Y^2}[p_1 - xq_1]\exp(-\xi_1^2)\right.$$

$$\left. +\frac{1}{Y^3}\left[\frac{4Y^2}{N^2\tilde{w}_\parallel^2\cos^2\theta} - 2A_e\tan^2\theta + \frac{3x}{4}\tan^2\theta(j+2)A_e\right]\exp(-\xi_0^2)\right\}, \qquad (1.109)$$

$\tilde{w}_\parallel = w/c$. When deriving (1.104)–(1.109) we took into account the future application of our theory to whistler-mode waves for which $Y > 1$, and assumed that ω was not very close to zero or Ω. (If this were not the case then we should have kept only the terms proportional to $\exp(-\xi_0^2)$, when ω is close to zero, or $\exp(-\xi_1^2)$, when ω is close to Ω, other terms being negligibly small). In this case we have the inequality:

$$\exp(-\xi_n^2) \ll \exp[-\max(\xi_1^2,\xi_0^2)], \qquad (1.110)$$

where $n \neq 0$ or 1, which allowed us to neglect the contribution of all the terms proportional to $\exp(-\xi_n^2)$ when $n \neq 0$ or 1.

Expression (1.103) will be widely used for the analysis of whistler-mode propagation, instabilities and damping in the following chapters. However, for completeness of our analysis we consider other possible simplifications of

(1.97)–(1.102) which will be relevant to some special cases of whistler-mode propagation.

1.7 The plasma dielectric tensor (special cases)

As was shown in Section 1.2, $\Re Z$ can be simplified for $|\xi_n| \gg 1$ and presented in the form of an asymptotic expansion (1.22). This expansion allowed us to further simplify (1.97)–(1.102) and write them in the form (1.103). However, expressions (1.97)–(1.102) could also be simplified for some other special values of ξ_n which are considered in this section following Sazhin & Sazhina (1988).

(a) $\xi_1 = -0.924$, $\xi_0 \gg 1$

One can easily show (see equation (1.21) and Fig. 1.1) that for this value of ξ_1:

$$2\xi_1 \int_0^{\xi_1} \exp(t^2)dt = \exp(\xi_1^2), \tag{1.111}$$

$$\Re Z(\xi_1) = -1/\xi_1, \tag{1.112}$$

and

$$\frac{d\Re Z(\xi_1)}{d\xi_1} = 0. \tag{1.113}$$

In view of (1.112) and (1.113) and using expansion (1.22) for $\Re Z$ when $\xi_n \neq \xi_1$ we simplify (1.97)–(1.102) to:

$$\epsilon_{ij} = \epsilon_{ij}^0 + \epsilon_{ij}^{t1} + \epsilon_{ij}^I, \tag{1.114}$$

where ϵ_{ij}^0 are the components of the same cold plasma dielectric tensor as in (1.79),

$$\epsilon_{xx}^{t1} = \beta_e N^2 Y^2 \left[\frac{\cos^2 \theta \left[Y - (Y+1) A_e \right]}{2(Y+1)^3} - \frac{3 A_e \sin^2 \theta}{(4Y^2 - 1)(Y^2 - 1)} \right] \tag{1.115}$$

$$\epsilon_{xy}^{t1} = -\epsilon_{yx}^{t1} = i\beta_e N^2 Y^2 \left[\frac{\cos^2 \theta \left[-Y + (Y+1) A_e \right]}{2(Y+1)^3} \right.$$

$$\left. - \frac{6 A_e Y \sin^2 \theta}{(4Y^2 - 1)(Y^2 - 1)} \right] \tag{1.116}$$

$$\epsilon_{yy}^{t1} = \beta_e N^2 Y^2 \left[\frac{\cos^2 \theta \left[Y - (Y+1) A_e \right]}{2 (Y+1)^3} \right.$$

$$\left. - \frac{A_e \sin^2 \theta \left(8 Y^2 + 1 \right)}{(4 Y^2 - 1)(Y^2 - 1)} \right] \tag{1.117}$$

$$\epsilon_{xz}^{t1} = \epsilon_{zx}^{t1} = \beta_e N^2 \frac{Y \cos \theta \sin \theta}{2 (Y+1)^2} \left[-Y + A_e (Y+1) \right] \tag{1.118}$$

$$\epsilon_{yz}^{t1} = -\epsilon_{zy}^{t1} = i \beta_e N^2 \frac{Y \cos \theta \sin \theta}{2 (Y+1)^2} \left[-Y - A_e \left(1 + 3Y + 2Y^2 \right) \right] \tag{1.119}$$

$$\epsilon_{zz}^{t1} = \beta_e N^2 \left[-3 Y^2 + \frac{Y (6 Y^2 + 6 Y + 1) \sin^2 \theta}{2 (Y+1)} + \frac{A_e \sin^2 \theta}{2} \right]; \tag{1.120}$$

the elements ϵ_{ij}^I are the same as in (1.104)–(1.109) although in this case the contribution of the terms proportional to $\exp(-\xi_0^2)$ can be neglected when compared with the terms proportional to $\exp(-\xi_1^2) = 0.426$.

In deriving (1.114), as well as all the subsequent expressions for ϵ_{ij} in this section, we have assumed that the contribution of the terms proportional to $w_{\parallel(\perp)}^n$ $(n > 2)$ is negligible, as was done when deriving (1.78) and (1.103).

(b) $\xi_0 = 0.924$, $|\xi_1| \gg 1$

In this case we can write equations (1.111)–(1.113) with ξ_1 replaced by ξ_0. As a result we obtain the expressions for ϵ_{ij} in the form:

$$\epsilon_{ij} = \epsilon_{ij}^{00} + \epsilon_{ij}^{t0} + \epsilon_{ij}^I, \tag{1.121}$$

where ϵ_{ij}^{00} coincide with ϵ_{ij}^0 except $\epsilon_{zz}^{00} = 1$, $\epsilon_{xx}^{t0} = \epsilon_{xx}^t$, $\epsilon_{yy}^{t0} = \epsilon_{yy}^t$, $\epsilon_{xy}^{t0} = -\epsilon_{yx}^{t0} = \epsilon_{xy}^t$, $\epsilon_{xz}^{t0} = \epsilon_{zx}^{t0} = \epsilon_{xz}^t$ (see equations (1.80)–(1.83)),

$$\epsilon_{yz}^{t0} = -\epsilon_{zy}^{t0} = i \beta_e \frac{N^2 Y \cos \theta \sin \theta}{(Y^2 - 1)^2} \left[2 Y^2 - A_e \left(Y^2 - 1 \right) \right], \tag{1.122}$$

$$\epsilon_{zz}^{t0} = \beta_e \frac{N^2 \sin^2 \theta}{Y^2 - 1} \left[Y^2 - A_e (Y^2 - 1) \right]. \tag{1.123}$$

The elements ϵ_{ij}^I are the same as in (1.104)–(1.109) although in this case the contribution of the terms proportional to $\exp(-\xi_1^2)$ can be neglected when compared with the terms proportional to $\exp(-\xi_0^2) = 0.426$.

The fact that ϵ_{zz}^{00} does not coincide with ϵ_{zz}^0 implies that the zero-order solution of (1.42) does not reduce to the cold plasma solution in general.

(c) $\xi_0 = -\xi_1 = 0.924$

This condition is satisfied only for $Y = 2$ and enables us to write (1.111)–(1.113) both for ξ_1 and ξ_0. Restricting ourselves to considering the case $\theta \ll 0$ (otherwise the waves are heavily damped for $A_e \lesssim 2$, typical values of A_e for magnetospheric conditions), we can write:

$$\epsilon_{ij} = \epsilon_{ij}^{00} + \epsilon_{ij}^{t2} + \epsilon_{ij}^{I}, \qquad (1.124)$$

where ϵ_{ij}^{00} are the same as in subsection (b), taken at $Y = 2$, while:

$$\epsilon_{xx}^{t2} = \beta_e N^2 \left[\frac{4}{27} - \frac{2}{9} A_e + \left(-\frac{4}{27} - \frac{2}{45} A_e \right) \theta^2 \right] \qquad (1.125)$$

$$\epsilon_{xy}^{t2} = -\epsilon_{yx}^{t2} = i\beta_e N^2 \left[-\frac{4}{27} + \frac{2}{9} A_e + \left(\frac{4}{27} - \frac{58}{45} A_e \right) \theta^2 \right] \qquad (1.126)$$

$$\epsilon_{yy}^{t2} = \beta_e N^2 \left[\frac{4}{27} - \frac{2}{9} A_e + \left(-\frac{4}{27} - \frac{122}{45} A_e \right) \theta^2 \right] \qquad (1.127)$$

$$\epsilon_{xz}^{t2} = \epsilon_{zx}^{t2} = -i\epsilon_{yz}^{t2} = i\epsilon_{zy}^{t2} = \beta_e N^2 \left(-\frac{2}{9} + \frac{1}{3} A_e \right) \theta \qquad (1.128)$$

$$\epsilon_{zz}^{t2} = \beta_e N^2 \left(\frac{1}{3} - \frac{1}{2} A_e \right) \theta^2. \qquad (1.129)$$

The elements ϵ_{ij}^{I} are the same as in (1.104)–(1.109) with $\exp(-\xi_0^2) = \exp(-\xi_1^2) = 0.426$.

1.8 The plasma dielectric tensor (effect of electron beams)

In this section we generalize the expressions for ϵ_{ij} derived in Section 1.6 to the case when there are beams of electrons in the directions parallel and/or antiparallel to the magnetic field, which are to be taken into account following Sazhin, Walker & Woolliscroft (1990b). This plasma model is, in particular, appropriate for the conditions in the vicinity of the magnetopause. We assume the electron distribution function in the form:

$$f_0 = \sum_i \kappa_i \left(j_i! \pi^{3/2} w_{\perp i}^{2j_i+2} w_{\parallel i} \right)^{-1} v_\perp^{2j_i} \exp \left[-\frac{\left(v_\parallel - v_{0_i} \right)^2}{w_{\parallel i}^2} - \frac{v_\perp^2}{w_{\perp i}^2} \right],$$
$$(1.130)$$

where $w_{\perp i(\parallel i)}$ are, as in (1.90), the electron thermal velocities in the direction perpendicular (parallel) to the external magnetic field, v_{0_i} are the beam velocities in the direction parallel to the external magnetic field, $j_i = 0, 1, 2, \ldots$

$(j_i = 0$ corresponds to a bi-Maxwellian distribution, $j_i \neq 0$ corresponds to a loss-cone distribution), and κ_i is the fractional abundance of the ith population of electrons present in the distribution,

$$\sum_i \kappa_i = 1. \tag{1.131}$$

The distribution function determined by (1.130) is an obvious generalization of the function (1.90) and reduces to the latter when we consider only one population of electrons and assume that $v_{0_i} = 0$. The main merit of the distribution function (1.130) is that it takes into account not only the effects of electron beams (finite v_{0_i}) but the effects of electron temperature and anisotropy as well.

Having substituted (1.130) into expressions (1.73) for ϵ_{ij} we can write:

$$\epsilon_{xx} = 1 + \sum_i X_i \left[(A_{e_i} - 1) - \frac{4 \sum_{n=-\infty}^{+\infty} n^2 \mu_{j_i}^n Z_{n_i}}{\alpha^2 j_i! w_{\perp_i}^{2j_i+2}} \right] \tag{1.132}$$

$$\epsilon_{xy} = -\epsilon_{yx} = -\frac{2\mathrm{i}}{\alpha} \sum_i \frac{X_i}{j_i! w_{\perp_i}^{2j_i+2}} \frac{\mathrm{d}}{\mathrm{d}\alpha} \sum_{n=-\infty}^{+\infty} n \mu_{j_i}^n Z_{n_i} \tag{1.133}$$

$$\epsilon_{zx} = \epsilon_{xz} = \frac{2}{\alpha} \sum_i \frac{X_i}{j_i! w_{\perp_i}^{2j_i+2}} \sum_{n=-\infty}^{+\infty} n \mu_{j_i}^n \left(w_{\|_i} Z_{n_i}' - 2v_{0_i} Z_{n_i} \right) \tag{1.134}$$

$$\epsilon_{yy} = 1 + \sum_i X_i \left[(A_{e_i} - 1) - \frac{4 \sum_{n=-\infty}^{+\infty} \nu_{j_i}^n Z_{n_i}}{j_i! w_{\perp_i}^{2j_i+2}} \right] \tag{1.135}$$

$$\epsilon_{yz} = -\epsilon_{zy} = -\mathrm{i} \sum_i \frac{X_i}{j_i! w_{\perp_i}^{2j_i+2}} \frac{\mathrm{d}}{\mathrm{d}\alpha} \sum_{n=-\infty}^{+\infty} \mu_{j_i}^n \left(w_{\|_i} Z_{n_i}' - 2v_{0_i} Z_{n_i} \right) \tag{1.136}$$

$$\epsilon_{zz} = 1 + 2 \sum_i X_i \left[\frac{v_{0_i}^2}{w_{\|_i}^2} + w_{\|_i}^2 \sum_{n=-\infty}^{+\infty} \frac{\tilde{\xi}_{n_i} \mu_{j_i}^n Z_{n_i}'}{j_i! w_{\perp_i}^{2j_i+2}} \right.$$

$$\left. + \frac{2v_{0_i} \sum_{n=-\infty}^{+\infty} \mu_{j_i}^n \left(w_{\|_i} Z_{n_i}' - v_{0_i} Z_{n_i} \right)}{j_i! w_{\perp_i}^{2j_i+2}} \right], \tag{1.137}$$

where:

$$\tilde{\xi}_{n_i} = \frac{\omega - k_\| v_{0_i} - n\Omega}{k_\| w_{\|_i}},$$

$\tilde{Y}_i = \Omega/(\omega - k_\parallel v_{0_i})$, other notations are the same as in (1.91)–(1.96) with Y replaced by $\tilde{Y}$, and index i refers to different electron populations. In the case of a one-component plasma and assuming $v_{0_i} = 0$, (1.132)–(1.137) reduce to (1.91)–(1.96). In the case where we consider one population of electrons with $w_\perp = w_\parallel$ and $j = 0$, (1.132)–(1.137) reduce to expressions (6.2.1.2) of Akhiezer *et al.* (1975).

Assuming that the conditions (1.77) are valid and making an additional assumption:

$$\frac{k_\parallel v_{0_i}}{\omega} \ll 1, \tag{1.138}$$

we can write the elements of ϵ_{ij} as:

$$\epsilon_{ij} = \epsilon^0_{ij} + \epsilon^t_{ij} + \Delta\epsilon_{ij} + \epsilon^I_{ij}, \tag{1.139}$$

where ϵ^0_{ij} and ϵ^t_{ij} are the same as defined by (1.79) and (1.80)–(1.85) in which the summation over X_i is to be performed, $\Delta\epsilon_{ij}$ are the corrections to ϵ^0_{ij} due to finite v_{0_i},

$$\Delta\epsilon_{xx} = \Delta\epsilon_{yy} = \frac{Y^2}{(Y^2-1)^3}\sum_i X_i \left[-2\left(Y^2-1\right)\delta_i + \left(Y^2+3\right)\delta_i^2\right] \tag{1.140}$$

$$\Delta\epsilon_{xy} = -\Delta\epsilon_{yx} = \frac{iY}{(Y^2-1)^3}\sum_i X_i \left[-\left(Y^4-1\right)\delta_i + \left(3Y^2+1\right)\delta_i^2\right] \tag{1.141}$$

$$\Delta\epsilon_{xz} = \Delta\epsilon_{zx} = \frac{\tan\theta}{(Y^2-1)^2}\sum_i X_i \left[\left(Y^2-1\right)\delta_i - \left(Y^2+1\right)\delta_i^2\right] \tag{1.142}$$

$$\Delta\epsilon_{yz} = -\Delta\epsilon_{zy} = \frac{iY\tan\theta}{(Y^2-1)^2}\sum_i X_i \left[-\left(Y^2-1\right)\delta_i + 2\delta_i^2\right] \tag{1.143}$$

$$\Delta\epsilon_{zz} = \frac{1}{(Y^2-1)}\sum_i X_i \left\{-2\left(Y^2-1\right)\delta_i + \left[3\left(-Y^2+1\right)+\tan^2\theta\right]\delta_i^2\right\} \tag{1.144}$$

with

$$\delta_i = k_\parallel v_{0_i}/\omega = N\cos\theta\tilde{v}_{0_i},$$

$$\tilde{v}_{0_i} = v_{0_i}/c,$$

and ϵ^I_{ij} are the same as in (1.104)–(1.109) with ω in the exponential terms replaced by $\omega - k_\parallel v_{0_i}$ and with the summation over i performed.

The smallest terms which we retained when deriving (1.139) were those proportional to β_i or δ_i^2, higher-order terms (e.g. proportional to $\beta_i\delta_i$) being neglected. Expressions (1.140)–(1.144) reduce to expressions (6.2.1.3) given by Akhiezer *et al.* (1975) provided the latter are taken in the limit $|\delta_i| \ll 1$.

1.9 The plasma dielectric tensor (cold plasma approximation)

The expressions for the elements of the dielectric tensor in a cold electron plasma have already been derived in Section 1.5 (see expressions (1.79)). In this section we (i) generalize these expressions so that the effects of protons and heavier ions (hereafter mentioned as ions) will be taken into account, (ii) rearrange them so that the effects of ions and finite electron density will be presented as perturbations to the corresponding elements of this tensor derived within the approximation of an infinitely dense electron plasma.

The terms proportional to X in the definitions of R, L and P in (1.79) result from the electron current, i.e. they describe the effect of the particles with mass m_e and the charge $q_e = -e$. The contribution of the currents due to other species of particles, e.g. ions, can be described in a straightforward way by adding to the terms ϵ_{ij}^0 the terms similar to those proportional to X but with m_e and q_e replaced by m_α and q_α, where m_α and q_α are the ion mass and charge respectively. As a result we will obtain the expressions for ϵ_{ij}^0 still defined by (1.79) but with R, L and P defined as:

$$R = 1 + \frac{X}{Y-1} - \sum_\alpha \frac{\Pi_\alpha^2}{\omega(\omega + \varepsilon_\alpha |\Omega_{0\alpha}|)} \tag{1.145}$$

$$L = 1 - \frac{X}{Y+1} - \sum_\alpha \frac{\Pi_\alpha^2}{\omega(\omega - \varepsilon_\alpha |\Omega_{0\alpha}|)} \tag{1.146}$$

$$P = 1 - X - \sum_\alpha \frac{\Pi_\alpha^2}{\omega^2}, \tag{1.147}$$

where Π_α and $\Omega_{0\alpha}$ are the plasma frequencies and the gyrofrequencies at rest of the corresponding ions, and $\varepsilon_\alpha = \mathrm{sign}(q_\alpha)$ (we assume that $\varepsilon_\alpha = +$).

Restricting ourselves to considering frequencies ω well above $|\Omega_{0\alpha}|$ we can neglect the contribution of the terms $\varepsilon_\alpha |\Omega_{0\alpha}|$ in (1.145) and (1.146). As a result the system of equations (1.145)–(1.147) can be simplified to:

$$\left. \begin{array}{l} R = \eta + X/(Y-1) \\ L = \eta - X/(Y+1) \\ P = \eta - X \end{array} \right\}, \tag{1.148}$$

where $\eta = 1 - \sum_\alpha \Pi_\alpha^2/\omega^2$.

In the limit of an infinitely dense plasma ($X \gg 1$) with the effects of ions neglected we assume $\eta = 0$ in (1.148) and write the elements ϵ_{ij}^0 with the upper index ^{0} replaced by ^{0d} as:

$$\left.\begin{array}{l} \epsilon_{xx}^{0d} = \epsilon_{yy}^{0d} = X/(Y^2 - 1) \\ \epsilon_{xy}^{0d} = -\epsilon_{yx}^{0d} = iXY/(Y^2 - 1) \\ \epsilon_{zz}^{0d} = -X \\ \epsilon_{xz}^{0d} = \epsilon_{zx}^{0d} = \epsilon_{yz}^{0d} = \epsilon_{zy}^{0d} = 0 \end{array}\right\}. \tag{1.149}$$

In view of (1.148) and (1.149) the elements ϵ_{ij}^0 with the effects of ions and finite electron density taken into account can be written as:

$$\epsilon_{ij}^0 = \epsilon_{ij}^{0d} + \epsilon_{ij}^d, \tag{1.150}$$

where ϵ_{ij}^{0d} are defined by (1.149),

$$\epsilon_{ij}^d = \eta\delta_{ij}, \tag{1.151}$$

$\delta_{ij} = 1$ when $i = j$ and $\delta_{ij} = 0$ when $i \neq j$.

Expressions (1.150) combined with (1.78) and (1.103) will allow us to consider additively different effects on whistler-mode propagation including the effects of finite electron density, the contribution of ions and thermal and relativistic effects. The first two effects will be considered in the next chapter, where basic properties of whistler-mode propagation in a cold plasma will be recalled. Meanwhile, in the next section we derive basic equations for wave polarization which will be applied later to whistler-mode waves.

1.10 Wave polarization

As was pointed out in Section 1.3, equation (1.39) has a non-trivial solution ($\mathbf{E} \neq 0$) only when the dispersion equation (1.42) is satisfied. The solution of the latter equation will be the primary interest of this book. However, sometimes it is essential to know not only whether the solution of equation (1.39) exists or not, but also the structure of the solution itself. This structure generally includes $|\mathbf{E}|$ and the orientation of $\mathbf{E}$ with respect to coordinate axes. Equation (1.39) cannot give us any information about the value of $|\mathbf{E}|$ but it allows us to predict the orientation of $\mathbf{E}$, namely, to calculate the parameters:

$$\left.\begin{array}{l} \alpha_1 = E_y/E_x \\ \alpha_2 = E_z/E_x \end{array}\right\} \tag{1.152}$$

describing wave polarization. Explicit expressions for α_1 and α_2 follow from equation (1.40) (which is a matrix form of equation (1.39)):

$$\alpha_1 = \frac{a_{xx}a_{zz} - a_{xz}a_{zx}}{a_{zy}a_{xz} - a_{xy}a_{zz}} \tag{1.153}$$

$$\alpha_2 = \frac{a_{xx}a_{yy} - a_{xy}a_{yx}}{a_{yz}a_{xy} - a_{xz}a_{yy}}, \tag{1.154}$$

where a_{ij} are the elements of the tensor $\hat{a}$ defined by:

$$\hat{a}\mathbf{E} = \mathbf{N} \times \mathbf{N} \times \mathbf{E} + \hat{\epsilon}\mathbf{E}. \tag{1.155}$$

We can write:

$$a_{ij} = \Lambda_{ij} + \epsilon_{ij}^{(1)}, \tag{1.156}$$

where $\Lambda_{xx} = S - N^2 \cos^2 \theta$, $\Lambda_{xy} = -\Lambda_{yx} = iD$, $\Lambda_{xz} = \Lambda_{zx} = N^2 \sin \theta \cos \theta$, $\Lambda_{yy} = S - N^2$, $\Lambda_{yz} = \Lambda_{zy} = 0$, $\Lambda_{zz} = P - N^2 \sin^2 \theta$; S, D and P are defined by expressions (1.145)–(1.147); $\epsilon_{ij}^{(1)}$ incorporates different corrections to the cold plasma dielectric tensor (e.g. those due to the effects of finite electron density, the contribution of ions, thermal and relativistic effects).

In view of (1.156) and assuming that the contribution of the terms proportional to $\epsilon_{ij}^{(1)}$ is small, expressions (1.153) and (1.154) can be written in a more explicit form (Sazhin, 1985):

$$\alpha_1 = \alpha_{10} + \sum_{j \geq 1} l_{ij}^{(1)} \epsilon_{ij}^{(1)}, \tag{1.157}$$

$$\alpha_2 = \alpha_{20} + \sum_{j \geq 1} l_{ij}^{(2)} \epsilon_{ij}^{(1)}, \tag{1.158}$$

where

$$\alpha_{10} = (\Lambda_{xz}^2 - \Lambda_{xx}\Lambda_{zz})/(\Lambda_{xy}\Lambda_{zz}) \tag{1.159}$$

and

$$\alpha_{20} = -(\Lambda_{xy}^2 + \Lambda_{xx}\Lambda_{yy})/(\Lambda_{xz}\Lambda_{yy}) \tag{1.160}$$

refer to the polarization in a cold plasma, and $l_{ij}^{(1)}$ and $l_{ij}^{(2)}$ are the elements of the tensors $\hat{l}^{(1)}$ and $\hat{l}^{(2)}$ defined by:

$$\hat{l}^{(1)} = \begin{pmatrix} -\Lambda_{xy}^{-1} & -(\Lambda_{xz}^2 - \Lambda_{xx}\Lambda_{zz})\Lambda_{xy}^{-2}\Lambda_{zz}^{-1} & 2\Lambda_{xz}\Lambda_{xy}^{-1}\Lambda_{zz}^{-1} \\ 0 & 0 & -(\Lambda_{xz}^2 - \Lambda_{xx}\Lambda_{zz})\Lambda_{xz}\Lambda_{xy}^{-2}\Lambda_{zz}^{-2} \\ 0 & 0 & -\Lambda_{xz}^2\Lambda_{xy}^{-1}\Lambda_{zz}^{-2} \end{pmatrix},$$

$$\tag{1.161}$$

$$\hat{I}^{(2)} = \begin{pmatrix} -\Lambda_{xz}^{-1} & -2\Lambda_{xy}\Lambda_{xz}^{-1}\Lambda_{yy}^{-1} & (\Lambda_{xy}^2 + \Lambda_{xx}\Lambda_{yy})\Lambda_{xz}^{-2}\Lambda_{yy}^{-1} \\ 0 & -\Lambda_{xy}^2\Lambda_{xz}^{-1}\Lambda_{yy}^{-2} & -(\Lambda_{xy}^2 + \Lambda_{xx}\Lambda_{yy})\Lambda_{xy}\Lambda_{xz}^{-2}\Lambda_{yy}^{-2} \\ 0 & 0 & 0 \end{pmatrix}.$$

$$(1.162)$$

The values of α_1 and α_2 also determine the polarization of the wave magnetic field. The corresponding expressions for B_y/B_x and B_z/B_x follow from (1.27) if we set $N_y = 0$:

$$B_y/B_x = (\alpha_2 \tan\theta - 1)/\alpha_1, \qquad (1.163)$$

$$B_z/B_x = -\tan\theta. \qquad (1.164)$$

Equations (1.157)–(1.164) are rather general and can be applied to different types of wave propagation. Later in this book they will be applied to the analysis of whistler-mode polarization in a cold and a hot anisotropic plasma.

Problems

Problem 1.1 Using the fact that the maximum of $|\Re Z|$ defined by (1.21) is achieved when $|\xi_0| = 0.924$, prove that the maximal value of $|\Re Z|$ is approximately equal to 1.082.

Problem 1.2 Obtain the fourth term in the asymptotic expansion (1.22).

Problem 1.3 Expression (1.40) was derived on the assumption that $\mathbf{N}$ lies in the (x, z) plane. Generalize this expression to the case when the projection of $\mathbf{N}$ on the (x, y) plane forms an angle ϕ with the x axis.

Problem 1.4 Expressions (1.73) were derived on the assumption that the disturbed parameters are proportional to $\exp[i(\omega t - \mathbf{kr})]$ and the wave propagates in the right-handed coordinate system $\mathbf{e}_z = \mathbf{e}_x \times \mathbf{e}_y$. How will these expressions change if the disturbed parameters are proportional to $\exp[i(\mathbf{kr} - \omega t)]$ or the wave propagates in the left-handed coordinate system $\mathbf{e}_z = -\mathbf{e}_x \times \mathbf{e}_y$?

Problem 1.5 Comparing expressions (1.80)–(1.85) and (1.86)–(1.88) we can see that both ϵ_{ij}^t and ϵ_{ij}^r have the same order of magnitude with respect to c^{-2}. What is the physical background of this?

2
Propagation in a cold plasma

2.1 The Appleton–Hartree equation

As mentioned in Section 1.1, for a really cold plasma ($T_\alpha \to 0$) the condition (1.1) is no longer valid and all the theory developed in Chapter 1 breaks down. Thus when speaking about cold plasma we will assume that its temperature is so low that the contributions of thermal and relativistic corrections to ϵ_{ij} (the terms ϵ_{ij}^t and ϵ_{ij}^r in (1.78)) to the process of wave propagation are small when compared with the contribution of ϵ_{ij}^0, but at the same time this temperature is high enough for condition (1.1) to remain valid. This definition of a cold plasma obviously depends on the type of waves under consideration. The cold plasma approximation allows us to write the dispersion equation for various waves in a particularly simple form and it has been widely used for the analysis of waves (in particular, whistler-mode) in the magnetosphere. Some results of plasma wave theory based on this approximation will be recalled below.

Neglecting the contribution of the terms ϵ_{ij}^t and ϵ_{ij}^r in (1.78) we can assume $\epsilon_{ij} = \epsilon_{ij}^0$ in the expressions for A, B and C defined by (1.43)–(1.45) and rewrite them as:

$$A = A_0 \equiv S \sin^2 \theta + P \cos^2 \theta, \tag{2.1}$$

$$B = B_0 \equiv RL \sin^2 \theta + PS(1 + \cos^2 \theta), \tag{2.2}$$

$$C = C_0 \equiv PRL, \tag{2.3}$$

where index $_0$ indicates that the corresponding coefficients refer to a cold plasma approximation; S, R, L and P are the same as in (1.79). When deriving (2.1)–(2.3) we took into account only the contribution of electrons to the process of wave propagation, as was done when deriving (1.78).

In view of (2.1)–(2.3) the cold plasma solution of (1.42) can be written as:

$$N_0^2 = (B_0 \pm F)/2A_0, \tag{2.4}$$

where

$$F^2 = (RL - PS)^2 \sin^4 \theta + 4P^2 D^2 \cos^2 \theta. \tag{2.5}$$

From (2.4) it follows that not more than two types of waves can propagate in a cold plasma in a fixed direction. When ω is real then N_0 can be either real ($N_0^2 > 0$) or imaginary (but not complex), which follows from the fact that $F^2 \geq 0$. If N_0 is imaginary then the wave cannot propagate; if N_0 is real then the wave propagates without damping or growth.

After some rearrangement equation (2.4) can be rewritten as (Ratcliffe, 1959; Holter & Kildal, 1973):

$$N_0^2 = 1 - \frac{2X(1-X)}{2(1-X) - Y^2 \sin^2 \theta \pm Y\sqrt{Y^2 \sin^4 \theta + 4(1-X)^2 \cos^2 \theta}}. \tag{2.6}$$

This equation is commonly known as the Appleton–Hartree equation (for a discussion about this name see Rawer & Suchy, 1976). It can be further simplified for some limiting values of θ, X or Y. For example, for $\theta = 0$ and $X > 1$ it reduces to:

$$N_0^2 = 1 - \frac{X}{1 \mp Y} \tag{2.7}$$

(for $X < 1$ the signs before Y in (2.7) are to be reversed; for $X = 1$ the cold plasma approximation is no longer valid unless θ is equal to zero: see equations (5.2)–(5.5)). In what follows we will be interested in the solution corresponding to the upper sign in (2.6) in the frequency range $Y > 1$. In this case (2.7) further reduces to:

$$N_0^2 = N_{0\|}^2 \equiv 1 + \frac{X}{Y-1}. \tag{2.8}$$

Waves described by equation (2.8) are known as whistler-mode waves (the origin of this name will be discussed later in this chapter) and they will be discussed in detail in the rest of the book.

Some attempts have been made to generalize (2.8) for finite θ avoiding, at the same time, returning to the full Appleton–Hartree equation. This was done by imposing the following conditions:

$$\left. \begin{array}{l} Y^2 \sin^2 \theta \ll 2|1 - X| \\ Y^2 \sin^4 \theta \ll 4(1 - X)^2 \cos^2 \theta \end{array} \right\} \tag{2.9}$$

known as the quasi-longitudinal approximation (e.g. Stix, 1962).

Conditions (2.9) seem to allow us to neglect the contribution of the terms proportional to $\sin^2\theta$ and $\sin^4\theta$ in equation (2.6) and simplify it to the equation similar to (2.8) but with Y replaced by $Y\cos\theta$ (e.g. Stix, 1962; Helliwell, 1965). However, as was first pointed out by Budden (1983), if we neglect the term proportional to $\sin^2\theta$ in equation (2.6) we should also assume $\cos^2\theta = 1$, as the terms proportional to $(1 - \cos^2\theta)$ have the same order of magnitude as those proportional to $\sin^2\theta$ unless

$$X \gg 1. \tag{2.10}$$

As a result, we can generalize (2.8) for finite θ only when condition (2.10) is valid. When both conditions (2.9) and (2.10) are valid, equation (2.6) for whistler-mode waves is simplified to:

$$N_0^2 = N_{0d}^2 \equiv \nu\tilde{N}_{0d}^2, \tag{2.11}$$

where $\nu = \Pi^2/\Omega^2$, $\tilde{N}_{0d} = Y/\sqrt{Y\cos\theta - 1}$. This equation is the generalization of equation (2.8) for finite θ provided conditions (2.9) and (2.10) are valid. The plots of $\tilde{N}_{0d}$ versus θ for $Y^{-1} = 0.2$, 0.4 and 0.6 are shown in Fig. 2.1. As follows from this figure, $\tilde{N}_{0d} > 2$ and increases when θ increases. Minimum $\tilde{N}_{0d} = 2$ is achieved when $\theta = 0$ and $Y^{-1} = 0.5$. This means that condition (2.10) can in fact be replaced by a less stringent condition:

$$\nu \gg 0.5. \tag{2.12}$$

In view of (2.10), conditions (2.9) are satisfied for a wide range of θ except in the immediate vicinity of $\pi/2$. However, equation (2.11) predicts wave propagation ($N_0^2 > 0$) only when

$$\theta < \arccos Y^{-1} \equiv \theta_{R0}. \tag{2.13}$$

Moreover, if θ is close to θ_{R0} then $N_0^2 \to \infty$ and ϵ_{ij}^t in (1.78) can no longer be neglected when compared with ϵ_{ij}^0. For these θ the cold plasma approximation breaks down altogether.

Also, equation (2.8) can be generalized for $\theta \neq 0$ when condition (2.10) (or (2.12)) is not necessarily valid but

$$|\theta| \ll 1. \tag{2.14}$$

In view of (2.14) we can expand $\sin\theta$ and $\cos\theta$ in (2.6) in a Taylor series with respect to θ and write this equation for whistler-mode waves as:

$$N_0^2 = N_{0\|}^2(1 + a_0\theta^2), \tag{2.15}$$

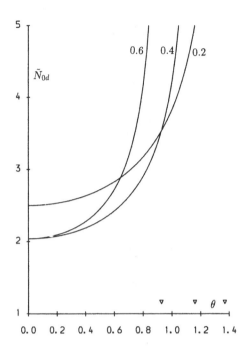

Fig. 2.1 Plots of $\tilde{N}_{0d} = Y/\sqrt{Y\cos\theta - 1}$ versus θ, the wave normal angle (in radians), for $Y^{-1} = 0.2, 0.4$ and 0.6 (figures near the curves). $\triangledown$ near the θ axis indicate $\theta = \theta_{R0} = \arccos Y^{-1}$ for these Y^{-1}.

where

$$a_0 = \frac{XY}{2(X-1)(Y-1)}, \tag{2.16}$$

and $N_{0\parallel}$ is defined by (2.8).

Equation (2.15) is valid when

$$|a_0\theta^2| \ll 1. \tag{2.17}$$

Condition (2.17) is hereafter considered as the quasi-longitudinal approximation for the plasma with arbitrary electron density, i.e. when condition (2.10) (or (2.12)) is not necessarily valid.

Remembering (2.16), condition (2.17) is always violated when X is close to 1 unless $\theta = 0$. Hence, one should be cautious when applying (2.8) to the interpretation of actual wave data at these X. When $X > 1$ then $a_0 > 0$ in the whistler-mode frequency range. Hence, N_0 determined by (2.15) increases with increasing θ as was the case with dense plasma (see equation (2.11) and Fig. 2.1).

Alternatively, if conditions (2.9) and (2.10) (or (2.12)) are valid but we are interested in retaining terms of the order of X^{-1} (or ν^{-1}), then equation (2.6) for whistler-mode waves is simplified to:

$$N_0^2 = N_{0d}^2(1 + \tilde{a}_c\nu^{-1}),$$ (2.18)

where

$$\tilde{a}_c = \frac{Y^2\cos^2\theta - 4Y\cos\theta + Y^2 + 2}{2Y^2(Y\cos\theta - 1)},$$ (2.19)

and N_{0d} is defined by (2.11); when deriving (2.18) we assumed that

$$|a_c| \equiv |\tilde{a}_c\nu^{-1}| \ll 1.$$ (2.20)

The term a_c describes the correction to N_{0d} due to the finite electron density. If $\theta = 0$ then $\tilde{a}_c$ simplifies to $(Y-1)/Y^2$. The same expression could be obtained from equation (2.8).

Remembering (1.150) and (1.151), equation (2.18) can be generalized so that the contribution of ions is taken into account as well (Sazhin, 1990a):

$$N_0^2 = N_{0d}^2(1 + \tilde{a}_c\nu^{-1} + \tilde{a}_r r),$$ (2.21)

where $\tilde{a}_r = -\tilde{a}_cY^2$, $r = \sum_\alpha \Pi_\alpha^2/\Pi^2$ (the summation is assumed over all ion species; in the case when only the contribution of protons is to be taken into account then $r = m_e/m_p$, m_p is the proton mass),

$$|\tilde{a}_r r| \ll 1.$$ (2.22)

Plots of $\tilde{a}_c$ versus θ and $\tilde{a}_r$ versus θ are shown in Fig. 2.2, for the same Y^{-1} as in Fig. 2.1. As follows from Fig. 2.2, $\tilde{a}_c > 0$ and increases with increasing θ. The maximal value of $\tilde{a}_c = 0.25$ is achieved at $\theta = 0$ and $Y^{-1} = 0.5$. In contrast to $\tilde{a}_c$, $\tilde{a}_r < 0$ and $|\tilde{a}_r|$ increases with increasing θ and/or Y^{-1}. In fact the contribution of ions tends to compensate for the contribution of finite electron density.

In a similar way to (2.21) we can generalize the expression for the resonance cone angle defined by (2.13) so that the contribution of finite electron density and ions could be taken into account (Sazhin, 1989a):

$$\theta_R = \theta_{R0} - \frac{\sqrt{Y^2-1}}{2Y^2}\nu^{-1} + \frac{\sqrt{Y^2-1}}{2}r.$$ (2.23)

N^2 for whistler-mode waves is positive when $\theta < \theta_R$, and $N \to \infty$ when $\theta \to \theta_R$.

As can be seen from (2.23) the contribution of ions tends to compensate for the contribution of finite electron density as was the case in (2.21).

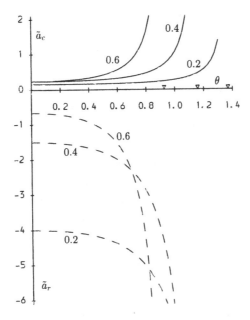

Fig. 2.2 Plots of $\tilde{a}_c$ (see equation (2.19)) versus θ (solid) and $\tilde{a}_r = -Y^2\tilde{a}_c$ versus θ (dashed) for $Y^{-1} = 0.2$, 0.4 and 0.6 (curves indicated). $\triangledown$ near the θ axis indicate $\theta = \theta_{R0} \equiv \arccos Y^{-1}$ for these Y^{-1}.

When deriving (2.23) it was assumed that the corrections to θ_{R0} due to finite ν^{-1} and r are small. In the general case of arbitrary ν but with the contribution of ions neglected, the expression for the resonance cone angle θ_R can be obtained by equating the denominator of (2.6) to zero. As a result we obtain:

$$\theta_R = \arcsin\sqrt{\frac{P(1-Y^2)}{\nu Y^4}}. \tag{2.24}$$

In the whistler-mode frequency range ($Y > 1$) expression (2.24) is defined when $P < 0$, i.e. when the wave frequency is below the electron plasma frequency.

Expressions (2.8), (2.11), (2.15), (2.21), (2.23) and (2.24) will be used for the analysis of different limiting cases of whistler-mode propagation throughout the whole book. Meanwhile we will consider another parameter which is also important for the analysis of the properties of whistler-mode waves, namely the group velocity.

2.2 Whistler-mode group velocity

The solution of a cold plasma dispersion equation referring to whistler-mode waves and discussed in the previous section allowed us to determine the wave refractive index N and, correspondingly, the wave phase velocity $v_{ph} = \omega/k = c/N$. However, this does not necessarily coincide with the velocity of energy flow. Without discussing details of the mathematical analysis of wave packet propagation (see Brillouin, 1960; Suchy, 1972; Anderson, Askne & Lisak, 1975, 1976; Tanaka, Fujiwara & Ikegami, 1986; Tanaka, 1989; Xu & Yeh, 1990, for details) we can refer only to the final result of this analysis. Namely, when the waves are not strongly damped or amplified and the dispersion is not large then the wave packet, or, in other words, cluster of wave energy, propagates with the so-called group velocity determined by the following equation:

$$\mathbf{v}_g = d\omega/d\mathbf{k}. \tag{2.25}$$

This result holds true for most cases of whistler-mode propagation in the magnetosphere and will be assumed throughout the whole book.

We begin our analysis with the simplest case of parallel whistler-mode propagation in a dense plasma when N_0 is determined by the equation (2.8) taken in the limit (2.10) (or (2.12)), or by equation (2.11) taken in the limit $\theta = 0$. After some straightforward algebra we obtain the following expression for v_g (which in this case is directed along $\mathbf{N}$, i.e. parallel to the magnetic field):

$$v_g = v_{g0} \equiv \frac{2c(Y-1)^{3/2}}{\sqrt{\nu}Y^2}, \tag{2.26}$$

where ν is the same as in (2.11).

As follows from (2.26), v_g approaches zero when either $\omega \to 0$ ($Y \to \infty$) or $\omega \to \Omega$ ($Y \to 1$). (Strictly speaking v_g never reaches zero in either case as the contribution of ions and finite electron temperature cannot be neglected when $\omega \to 0$ and $\omega \to \Omega$ respectively: see Section 4.3). Hence, we can expect v_g to be maximal for an intermediate frequency determined from the condition:

$$dv_g/dY = 0. \tag{2.27}$$

In view of (2.26) condition (2.27) can be rewritten as:

$$\frac{c(Y-1)^{1/2}}{2\sqrt{\nu}Y^3}(4-Y) = 0. \tag{2.28}$$

The obvious solution of (2.28) in the range $1 < Y < \infty$ is $Y = 4$ or $\omega = 0.25\Omega$. Having substituted this solution into (2.26) we obtain:

$$v_g = v_{g\max} \equiv \frac{3\sqrt{3}c}{8\sqrt{\nu}} \approx \frac{0.65c}{\sqrt{\nu}}. \tag{2.29}$$

This maximal whistler-mode group velocity is manifested in the minimal group delay time for whistler-mode waves propagating from one hemisphere to another, as can be seen from whistler dynamic spectra shown in Fig. I. This group delay time can be calculated from the equation:

$$t_g = \int_{-s_{\mathrm{ion}}}^{+s_{\mathrm{ion}}} \frac{\mathrm{d}s}{v_g}, \tag{2.30}$$

where the integration is assumed along the magnetic field line between the opposite hemispheres; in the simplest model of a cold dense plasma v_g is defined by (2.26). As follows from (2.30) and (2.26) the main contribution to the integral in equation (2.30) comes from the part of the integration path where Ω is minimal, i.e. from the vicinity of the magnetospheric equator. Calculation of this integral for realistic models of electron distribution in the magnetosphere leads us to the result that t_g is minimal when $\omega \approx 0.4\Omega_{\mathrm{eq}}$, where Ω_{eq} is the modulus of electron gyrofrequency at the magnetospheric equator (Carpenter & Smith, 1964; Park, 1972). Thus reading the frequency at which t_g is minimal, ω_n (nose frequency), from a whistler spectrogram (see e.g. Fig. I) we can determine the field line along which the whistler propagates. For given models of electron density distribution along the field lines and magnetic field (e.g. dipole field), t_g at the nose frequency depends on electron density at the magnetospheric equator (n_{eq}). Thus direct measurements of t_g allow us to estimate this density. Alternatively, measurements of ω_n at different moments of time allow us to get information about the motion of magnetic field tubes in the magnetosphere, and eventually about the large-scale electric field E_0 therein. Practical diagnostics of these parameters with the help of whistlers appears to be not so straightforward. Some particular problems related to this diagnostics will be discussed in Section 9.1. Meanwhile we shall consider other properties of v_g in some more detail.

Using (2.21), expression (2.26) for v_g can be generalized so that the contribution of finite electron density and ions can be taken into account (Sazhin, Smith & Sazhina, 1990):

$$v_g = v_{g0}(1 + \tilde{b}_c\nu^{-1} + \tilde{b}_r r), \tag{2.31}$$

where v_{g0} is determined by (2.26),

$$\tilde{b}_c = \frac{(4 - 3Y)(Y - 1)}{2Y^3},$$
(2.32)

$$\tilde{b}_r = \frac{1 - Y}{2}.$$
(2.33)

When deriving equation (2.31) it was assumed that:

$$\left.\begin{array}{l} |\tilde{b}_c \nu^{-1}| \ll 1 \\ |\tilde{b}_r r| \ll 1 \end{array}\right\}$$
(2.34)

(cf. conditions (2.20) and (2.22)).

As follows from (2.31) the contribution of ions cannot be neglected when Y is sufficiently large. Thus, this equation is not valid for these Y either (see the discussion following (2.26)). $\tilde{b}_c < 0$ when $Y^{-1} < 0.75$ and $\tilde{b}_c > 0$ when $0.75 < Y^{-1} < 1$. $|\tilde{b}_c|$ achieves its maximum $|\tilde{b}_c|_{\max} \approx 0.19$ when $Y^{-1} = 6/(14 + \sqrt{52}) \approx 0.28$. $|\tilde{b}_r|$ monotonically decreases when Y^{-1} increases.

When $\theta \neq 0$ the direction of $\mathbf{v}_g$ does not, in general, coincide with the direction of $\mathbf{k}$. In this case $\mathbf{v}_g$ can be presented in the form (Stix, 1962):

$$\mathbf{v}_g = \mathbf{e}_k \frac{\partial \omega}{\partial |\mathbf{k}|} + \mathbf{e}_\theta \frac{1}{|\mathbf{k}|} \frac{\partial \omega}{\partial \theta},$$
(2.35)

where $\mathbf{e}_k$ and $\mathbf{e}_\theta$ are the unit vectors in the directions parallel and perpendicular to $\mathbf{k}$ respectively, but coplanar with $\mathbf{v}_g$ and $\mathbf{k}$. The angles are hereafter assumed positive if measured in a clockwise direction.

Restricting our analysis to whistler-mode propagation in a cold dense plasma and neglecting the contribution of ions, we can write the dispersion equation in the form (2.11). This equation can be solved with respect to ω and written as:

$$\omega = -\frac{c^2 k^2 \Omega \cos \theta}{c^2 k^2 + \Pi^2}.$$
(2.36)

In view of (2.36), equation (2.35) can be rewritten as:

$$\mathbf{v}_g = v_{gk} \mathbf{e}_k + v_{g\theta} \mathbf{e}_\theta,$$
(2.37)

where

$$v_{gk} = \frac{2c(Y \cos \theta - 1)^{3/2}}{\sqrt{\nu} Y^2 \cos \theta},$$
(2.38)

$$v_{g\theta} = -\frac{c\sqrt{Y \cos \theta - 1}}{\sqrt{\nu} Y} \tan \theta.$$
(2.39)

Alternatively the expression for $\mathbf{v}_g$ can be presented as:

$$\mathbf{v}_g = v_{g\parallel}\mathbf{e}_\parallel + v_{g\perp}\mathbf{e}_\perp, \tag{2.40}$$

where

$$v_{g\parallel} = \frac{c\sqrt{Y\cos\theta - 1}(Y\cos^2\theta - 2\cos\theta + Y)}{\sqrt{\nu}Y^2\cos\theta}, \tag{2.41}$$

$$v_{g\perp} = -\frac{c\sqrt{Y\cos\theta - 1}\sin\theta(Y\cos\theta - 2)}{\sqrt{\nu}Y\cos\theta}, \tag{2.42}$$

$\mathbf{e}_\parallel$ is the unit vector in the direction parallel to the magnetic field (z axis), and $\mathbf{e}_\perp$ is the unit vector in the x direction (as in Chapter 1, $\mathbf{k}$ is supposed to lie in the (x, z) plane; $0 \leq \theta < \pi/2$).

From (2.38) and (2.39) or (2.41) and (2.42) we obtain the expression for the absolute value of v_g:

$$|v_g| = \frac{c\tilde{v}_g}{\sqrt{\nu}}, \tag{2.43}$$

where

$$\tilde{v}_g = \frac{\sqrt{Y\cos\theta - 1}}{Y^2\cos\theta}\sqrt{4(Y\cos\theta - 1)^2 + Y^2\sin^2\theta}. \tag{2.44}$$

Plots of $\tilde{v}_g$ versus θ are shown in Fig. 2.3 for the same Y^{-1} as in Figs. 2.1 and 2.2. As follows from this figure, $\tilde{v}_g$ decreases with increasing θ for any particular value of Y^{-1} until the cold plasma approximation breaks down for θ close to θ_{R0}.

As follows from (2.42), $v_{g\perp} = 0$ when

$$\theta = \theta_{G0} \equiv \arccos(2/Y), \tag{2.45}$$

which means that the whistler-mode group velocity for this particular θ is directed parallel to the magnetic field. This property of whistler-mode propagation seems to have been first noticed by Gendrin (1960) and so the angle θ_{G0} is known as the Gendrin angle. This angle is obviously defined only for $\omega < \Omega/2$. At $\omega > \Omega/2$, as well as at $\omega < \Omega/2$ and $\theta_{G0} < \theta < \theta_{R0}$ (see (2.13)), $v_{g\perp} < 0$ which means that the component of whistler-mode group velocity perpendicular to the external magnetic field is oppositely directed with respect to the corresponding component of $\mathbf{k}$ ($k_\perp$). When θ approaches θ_{R0}, then $|v_g| \to 0$. Whistler-mode waves in a cold dense electron plasma cannot propagate at $\theta > \theta_{R0}$. In the case when the contribution of finite electron density and ions is to be taken into account, θ_{R0} should be replaced by θ_R defined by (2.23) or (2.24). If $\theta < \theta_{R0}$ then $v_{g\parallel} > 0$ and so $v_{g\parallel}$ is parallel to $k_\parallel$.

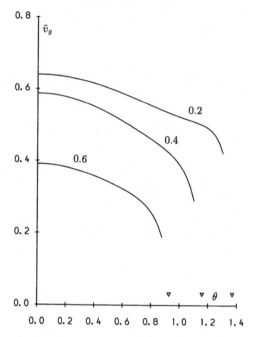

Fig. 2.3 Plots of $\tilde{v}_g$ (see equation (2.44)) versus θ for $Y^{-1} = 0.2, 0.4$ and 0.6 (curves indicated). $\triangledown$ near the θ axis indicate $\theta = \theta_{R0} \equiv \arccos Y^{-1}$ for these Y^{-1}.

In many practically important cases of whistler-mode propagation we are interested not only in the value of $|\mathbf{v}_g|$ but also in the direction of $\mathbf{v}_g$ with respect to $\mathbf{k}$ or $\mathbf{B}_0$. The corresponding angles between $\mathbf{k}$ and $\mathbf{v}_g$ (measured from $\mathbf{k}$ to $\mathbf{v}_g$) (θ_g) or between $\mathbf{B}_0$ and $\mathbf{v}_g$ $(\psi = \theta + \theta_g)$ can be determined from the relatively simple equations which will be considered below. In particular, from (2.35) it follows that:

$$\tan \theta_g = \frac{1}{|\mathbf{k}|} \frac{\partial \omega}{\partial \theta} \Big/ \frac{\partial \omega}{\partial |\mathbf{k}|} = -\frac{1}{N} \frac{\partial N}{\partial \theta}. \tag{2.46}$$

Expression (2.46) allows a rather simple geometrical interpretation: $\mathbf{v}_g$ is perpendicular to the surface $N(\theta)$. When deriving this expression we made no assumptions about the wave dispersion equation and so this expression can be applied to any type of plasma wave, not necessarily whistler-mode waves.

Having substituted (2.11) into (2.46) we obtain:

$$\theta_g = \theta_{g0} \equiv -\arctan\left[Y \sin\theta / (2(Y \cos\theta - 1))\right], \tag{2.47}$$

$$\psi = \psi_0 \equiv \theta - \arctan\left[Y \sin\theta / (2(Y \cos\theta - 1))\right]. \tag{2.48}$$

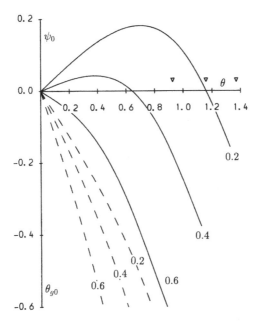

Fig. 2.4 Plots of θ_{g0} (see equation (2.47)) versus θ (dashed) and ψ_0 (see equation (2.48)) versus θ (solid) for $Y^{-1} = 0.2$, 0.4 and 0.6 (curves indicated). $\triangledown$ near the θ axis indicate $\theta = \theta_{R0} \equiv \arccos Y^{-1}$ for these Y^{-1}.

As follows from (2.48), $\psi_0 = \theta + \theta_{g0} = 0$ when $\theta = \theta_{G0}$ (see expression (2.45)) which agrees with the results of the analysis of (2.42). Also in agreement with the previous results we can see from (2.48) that $\psi_0 \leq 0$ when $1 < Y \leq 2$ or $Y > 2$ and $\theta \geq \theta_{G0}$, and $\psi_0 \geq 0$ when $Y > 2$ and $\theta \leq \theta_{G0}$. These properties of the angle ψ_0 follow from Fig. 2.4, where we show the plots ψ_0 versus θ and θ_{g0} versus θ. As follows from this figure, θ_{g0} is always negative and $|\theta_{g0}|$ increases with increasing θ and Y^{-1}. As to ψ_0, its behaviour appears to be different for $Y^{-1} > 0.5$ and $Y^{-1} < 0.5$. In the first case it is always negative and $|\psi_0|$ increases when θ and/or Y^{-1} increase. In the second case it is positive and first increases with increasing θ then reaches its maximum and then decreases with increasing θ reaching $\psi_0 = 0$ at $\theta = \theta_{G0}$. If $\theta > \theta_{G0}$ then $\psi_0 < 0$. If $\theta = 0$ then $\psi_0 = 0$. If θ approaches $\theta_{R0} = \arccos Y^{-1}$ then $\theta_{g0} \to \pi/2$ and $\psi_0 \to \theta - \pi/2$. Thus for the resonance cone whistler-mode wave normal angle $\mathbf{v}_g \perp \mathbf{N}$.

In the limit $Y \gg 2$ expression (2.46) is simplified to:

$$\theta_{g0} = -\arctan\left(0.5 \tan \theta\right). \tag{2.49}$$

In view of (2.48) and (2.49) we obtain:

$$\tan \psi_0 = \frac{\tan \theta + \tan \theta_{g0}}{1 - \tan \theta \tan \theta_{g0}} = \frac{0.5 \tan \theta}{1 + 0.5 \tan^2 \theta}. \tag{2.50}$$

As follows from (2.50), $\psi_0 \to 0$ when $\theta \to 0$ or when $\theta \to \pi/2$. Thus we can expect that ψ_0 should attain its maximal value for an intermediate value of θ. The value of θ at which ψ_0 is maximal follows from the equation:

$$\frac{\mathrm{d} \tan \psi_0}{\mathrm{d} \tan \theta} = 0. \tag{2.51}$$

In view of (2.50) this equation has a solution:

$$\theta = \theta_{s0} \equiv \arctan \sqrt{2} = 54.74° = 0.955 \, \mathrm{rad}. \tag{2.52}$$

Substituting (2.52) into (2.50) we obtain:

$$\psi(\theta_{s0}) = \psi_{s0} \equiv \arctan 0.25\sqrt{2} = 19.47° = 0.340 \, \mathrm{rad}. \tag{2.53}$$

That the whistler-mode group velocity at low frequencies does not deviate from the magnetic field by more than about 19.5° was first discovered by Storey (1953). Hence, the angle ψ_{s0} is known as the Storey angle.

The concept of the Storey angle can be generalized to the case when Y is above but not well above 2, when θ_{s0} and ψ_{s0} are determined by the following equations:

$$\theta_{s0} \equiv \arccos \left[Y^{-1} + \frac{\sqrt{1 - Y^{-2}}}{\sqrt{3}} \right], \tag{2.54}$$

$$\psi_{s0} \equiv \theta_{s0} - \arctan \left[\frac{Y \sin \theta_{s0}}{2(Y \cos \theta_{s0} - 1)} \right]. \tag{2.55}$$

The plots of θ_{s0} versus Y^{-1} and ψ_{s0} versus Y^{-1} are shown in Fig. 2.5. In the same figure we have shown for comparison the plot $-\psi_{R0} = -(\theta_{R0} - \pi/2) = -(\arccos Y^{-1} - \pi/2)$ versus Y^{-1}. The latter plot describes the direction of whistler-mode group velocity at $\theta = \theta_{R0}$. All the plots are shown only for $Y^{-1} \leq 0.5$ when θ_{s0} and ψ_{s0} are determined ($\cos \theta_{s0} \leq 1$).

As follows from Fig. 2.5, ψ_{s0} and θ_{s0} are maximal when $Y^{-1} \to 0$, while ψ_{R0} is close to zero for these Y^{-1}. An increase of Y^{-1} is accompanied by a decrease in both ψ_{s0} and θ_{s0} until they reach zero at $Y^{-1} = 0.5$. The values of $-\psi_{R0}$ increase almost linearly with increasing Y^{-1} in the same frequency range. At a certain Y^{-1} slightly below 0.2, $\psi_{s0} = -\psi_{R0}$. This means that the deviation of the direction of whistler-mode group velocity from the direction of magnetic field $\mathbf{B}_0$ is mainly controlled by ψ_{s0} at $Y^{-1} < 0.2$ and by ψ_{R0}

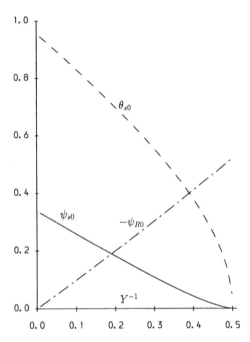

Fig. 2.5 Plots of θ_{s0} (see equation (2.54)) versus Y^{-1} (dashed), ψ_{s0} (see equation (2.55)) versus Y^{-1} (solid) and $-\psi_{R0} \equiv \pi/2 - \arccos Y^{-1}$ versus Y^{-1} (dashed–dotted).

at $Y^{-1} > 0.2$. Note that the values of ψ_{s0} determined by (2.55) correspond to the maxima of ψ_0 in Fig. 2.4.

In the case when finite electron density and the contribution of ions is taken into account, (2.48) can be generalized to:

$$\psi = \psi_0 + \tilde{\Delta}_c \nu^{-1} + \tilde{\Delta}_r r, \tag{2.56}$$

where

$$\tilde{\Delta}_c = \frac{(Y^2 - 2 + 2Y \cos\theta - Y^2 \cos^2\theta) \sin\theta}{Y(-Y^2 - 4 + 8Y \cos\theta - 3Y^2 \cos^2\theta)}, \tag{2.57}$$

$$\tilde{\Delta}_r = -Y^2 \tilde{\Delta}_c. \tag{2.58}$$

Plots of $\tilde{\Delta}_c$ versus θ and $\tilde{\Delta}_r$ versus θ are shown in Fig. 2.6. As follows from this figure, $\tilde{\Delta}_c$ is always negative and $|\tilde{\Delta}_c|$ increases with increasing θ and/or Y^{-1}. $\tilde{\Delta}_r$ is always positive and increases with increasing θ and/or Y^{-1}. Thus the effects of finite electron density and the contribution of ions tend to compensate for each other as was the case in evaluating N_0 (see equation (2.21) and Fig. 2.2).

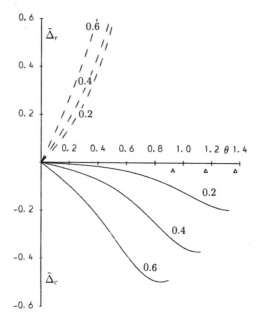

Fig. 2.6 Plots of $\tilde{\Delta}_c$ (see equation (2.57)) versus θ (solid) and $\tilde{\Delta}_r$ (see equation (2.58)) versus θ (dashed) for $Y^{-1} = 0.2$, 0.4 and 0.6 (curves indicated). $\triangle$ near the θ axis indicate $\theta = \theta_{R0} \equiv \arccos Y^{-1}$ for these Y^{-1}.

In view of (2.56) expressions (2.52) and (2.53) are generalized to (Sazhin, 1990a):

$$\theta_s = \theta_{s0} + \Delta\theta_s, \tag{2.59}$$

$$\psi_s = \psi_{s0} + \Delta\psi_s, \tag{2.60}$$

where

$$\Delta\theta_s = \Delta\tilde{\theta}_{cs}\nu^{-1} + \Delta\tilde{\theta}_{rs}r, \tag{2.61}$$

$$\Delta\tilde{\theta}_{cs} = \frac{(Y^2 - 1)(5Y\cos\theta_{s0} - 8)}{9Y^3\sin\theta_{s0}(1 - Y\cos\theta_{s0})}, \tag{2.62}$$

$$\Delta\tilde{\theta}_{rs} = -\Delta\tilde{\theta}_{cs}Y^2, \tag{2.63}$$

$$\Delta\psi_s = \tilde{\Delta}_{cs}\nu^{-1} + \tilde{\Delta}_{rs}r, \tag{2.64}$$

$$\tilde{\Delta}_{cs} = \frac{(Y^2 - 2 + 2Y\cos\theta_{s0} - Y^2\cos_{s0}^2\theta)\sin\theta_{s0}}{Y(-Y^2 - 4 + 8Y\cos\theta_{s0} - 3Y^2\cos^2\theta_{s0})}, \tag{2.65}$$

$$\tilde{\Delta}_{rs} = -Y^2\tilde{\Delta}_{cs}. \tag{2.66}$$

θ_{s0} and ψ_{s0} are the same as in (2.54) and (2.55).

In a similar way to (2.59), finite electron density and ion effects change the Gendrin angle defined by (2.45) so that the whistler-mode group velocity is directed along the magnetic field when (Sazhin, 1988b, 1990a):

$$\theta_G = \theta_{G0} + \Delta\theta_G, \tag{2.67}$$

where

$$\left.\begin{array}{l} \Delta\theta_G = \Delta\tilde{\theta}_{cG}\nu^{-1} + \Delta\tilde{\theta}_{rG}r \\ \Delta\tilde{\theta}_{cG} = (2 - Y^2)/(Y^2\sqrt{Y^2 - 4}) \\ \Delta\tilde{\theta}_{rG} = -(2 - Y^2)/\sqrt{Y^2 - 4} \end{array}\right\}. \tag{2.68}$$

As was the case with θ_R (see equation (2.23)) the corrections to θ_{G0} due to the contribution of finite electron density and the effect of ions tend to compensate each other.

The corrections to θ_R due to the contribution of finite electron density and ions determined by (2.23) result in the corresponding corrections to $\psi_R = \theta_R - \pi/2$. The corrections to the angles θ_{s0}, θ_{G0} and ψ_{s0} due to finite electron density and contribution of ions will be considered in more detail in Chapters 5 and 6 where we compare them with the corresponding corrections due to finite electron temperature and anisotropy.

2.3 Whistler-mode polarization

Whistler-mode dispersion and group velocity considered in the previous sections are important parameters for the study of the propagation of these waves in any realistic, and, in particular, magnetospheric plasma. However, they give us no information about the internal structure of the waves (polarization) and the physical background of their propagation. None of these aspects of whistler-mode theory seems to be of minor importance. Knowledge of whistler-mode polarization is essential for the study of the interaction of these waves with energetic electrons or for the determination of their wave normal angle based on the measurements of wave field components (see e.g. Sazhin, Walker & Woolliscroft, 1990a). Analysis of the physical background of whistler-mode propagation allows us to understand the process of wave propagation not in terms of the formal solution of the corresponding wave dispersion equation but rather in terms of the energy exchange between electric and magnetic fields of the wave and the electron current. A good feeling for this process would also contribute to a clearer understanding of the physical background of the process of whistler-mode electron interaction in general. Whistler-mode polarization in a cold plasma will be considered below while the physical background of whistler-mode propagation in a cold plasma will be considered in Section 2.4.

Initially, we restrict ourselves to whistler-mode propagation in relatively dense plasma when the dispersion equation of these waves is determined by equation (2.11). In this case the general expressions (1.159) and (1.160) of wave polarization in a cold plasma are simplified to:

$$\alpha_{10} \equiv E_y/E_x = \alpha_{10}^{(0)} \equiv -i(Y\cos\theta - 1)/(Y - \cos\theta), \qquad (2.69)$$

$$\alpha_{20} \equiv E_z/E_x = \alpha_{20}^{(0)} \equiv \sin\theta/(Y - \cos\theta). \qquad (2.70)$$

Plots of $\Im\alpha_{10}^{(0)}$ versus θ and $\alpha_{20}^{(0)}$ versus θ for the same Y^{-1} as in Fig. 2.1 are shown in Figs. 2.7 and 2.8. As follows from these figures and equations (2.69) and (2.70), if $\theta = 0$ then $E_y = -iE_x$, $E_z = 0$. Remembering our assumption (1.34), we can write for $\theta = 0$:

$$\left.\begin{array}{l} E_x = E_{1x}\exp(i\omega t - ikz) \\ E_y = E_{1x}\exp(i\omega t - ikz - i\pi/2) \end{array}\right\}. \qquad (2.71)$$

Equations (2.71) mean that the wave electric field rotates in a similar way to the electrons in a magnetic field. The amplitude of the wave remains the same during the process of rotation. This indicates a circular character of wave polarization for $\theta = 0$.

Also, it follows from Fig. 2.7 that $|E_y|$ decreases with respect to $|E_x|$ when θ increases and so wave polarization becomes elliptical in the (x, y) plane. $|E_y| = 0$ when $\theta = \theta_{R0} \equiv \arccos Y^{-1}$. In the latter case wave polarization becomes linear with the electric field vector directed along the wave vector **k**:

$$E_z = E_x \cot\theta_R. \qquad (2.72)$$

This means that an electromagnetic whistler-mode wave becomes an electrostatic one.

In the plane perpendicular to **k** the polarization of the whistler-mode electric field is always circular (see Problem 2.3).

From (1.163) and (1.164) we obtain the following expressions for the wave magnetic field polarization:

$$B_y/B_x = -i/\cos\theta, \qquad (2.73)$$

$$B_z/B_x = -\tan\theta. \qquad (2.74)$$

As follows from (2.73) and (2.74), at $\theta = 0$ the character of polarization of a wave magnetic field is exactly the same as that of a wave electric field.

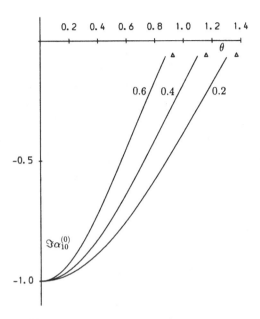

Fig. 2.7 Plots of $\Im\alpha_{10}^{(0)}$ (see equation (2.69)) versus θ for $Y^{-1} = 0.2$, 0.4 and 0.6 (curves indicated). $\triangle$ near the θ axis indicate $\theta = \theta_{R0} \equiv \arccos Y^{-1}$ for these Y^{-1}.

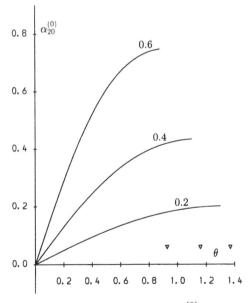

Fig. 2.8. The same as Fig. 2.7 but for $\alpha_{20}^{(0)}$ (see equation (2.70)).

However, in contrast to the wave electric field, $|B_y|/|B_x|$ increases when θ increases. Note that neither of the expressions (2.73) and (2.74) depends on the value of Y. These expressions have no physical meaning at $\theta = \arccos Y^{-1}$ as in this case $B_x = B_y = B_z = 0$.

Expressions (2.69) and (2.70) can be generalized so that the effects of finite electron density and the contribution of ions are taken into account. To do this we write the elements of a cold plasma dielectric tensor in the form (1.150) and the expression for the squared whistler-mode refractive index in the form (2.21). Then, after lengthy algebraic manipulations and neglecting the second order terms with respect to ν^{-1} and/or r, we can write the final expressions for α_{10} and α_{20} as (Sazhin, 1991a):

$$\alpha_{10} = \alpha_{10}^{(0)} (1 + \alpha_{11}^{(c)} \nu^{-1} + \alpha_{11}^{(r)} r), \tag{2.75}$$

$$\alpha_{20} = \alpha_{20}^{(0)} (1 + \alpha_{21}^{(c)} \nu^{-1} + \alpha_{21}^{(r)} r), \tag{2.76}$$

where

$$\alpha_{11}^{(c)} = -\frac{(Y^2 - 1) \sin^2 \theta}{2Y (Y \cos \theta - 1)(Y - \cos \theta)}, \tag{2.77}$$

$$\alpha_{21}^{(c)} = \frac{3Y^2 - 2 - 2Y^3 \cos \theta + Y^2 \cos^2 \theta}{2Y^2 \cos \theta (Y - \cos \theta)}, \tag{2.78}$$

$$\alpha_{11}^{(r)} = -\alpha_{11}^{(c)} Y^2, \tag{2.79}$$

$$\alpha_{21}^{(r)} = -\alpha_{21}^{(c)} Y^2, \tag{2.80}$$

$\alpha_{10}^{(0)}$ and $\alpha_{20}^{(0)}$ are the same as in (2.69) and (2.70), and ν and r are the same as in (2.21). When deriving (2.75) and (2.76) it was assumed that:

$$\left. \begin{array}{l} |\alpha_{11}^{(c)} \nu^{-1}| \ll 1 \\ |\alpha_{11}^{(r)} r| \ll 1 \\ |\alpha_{21}^{(c)} \nu^{-1}| \ll 1 \\ |\alpha_{21}^{(r)} r| \ll 1 \end{array} \right\}. \tag{2.81}$$

The plots of $\alpha_{11}^{(c)}$ versus θ, $\alpha_{11}^{(r)}$ versus θ, $\alpha_{21}^{(c)}$ versus θ and $\alpha_{21}^{(r)}$ versus θ for $Y^{-1} = 0.2$, 0.4 and 0.6 are shown in Figs. 2.9–2.12. As follows from (2.77) and (2.79) and Figs. 2.9–2.10, $\alpha_{11}^{(c)} \to 0$ and $\alpha_{11}^{(r)} \to 0$ when $\theta \to 0$ for all Y^{-1}. This contains a trivial result: the effects of finite electron density and the contribution of ions do not change the circular character of polarization of whistler-mode waves propagating parallel to the magnetic field. As follows from Fig. 2.9, the effects of finite electron density tend to decrease α_{10} when

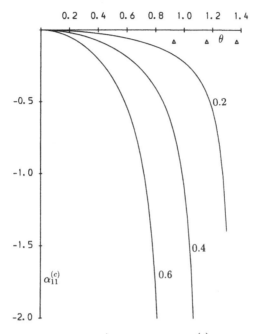

Fig. 2.9. The same as Fig. 2.7 but for $\alpha_{11}^{(c)}$ (see equation (2.77)).

$\theta \neq 0$ thus increasing the 'ellipticity' of whistler-mode polarization in the (x, y) plane. The degree of this increase is larger for larger Y^{-1}. In contrast to the effect of finite electron density, the contribution of ions tends to decrease the 'ellipticity' of whistler-mode polarization in this plane as $\alpha_{11}^{(r)} > 0$ (see Fig. 2.10). $\alpha_{11}^{(r)}$ increases when θ and/or Y^{-1} increase.

The character of the curves $\alpha_{21}^{(c)}$ versus θ and $\alpha_{21}^{(r)}$ versus θ appears to be more complicated as the values of $\alpha_{21}^{(c)}$ and $\alpha_{21}^{(r)}$ can have different signs for different θ. For some values of θ (e.g. $\theta \approx 0.3$ rad for $Y^{-1} = 0.6$, $\theta \approx 0.9$ rad for $Y^{-1} = 0.4$) $\alpha_{21}^{(c)} = \alpha_{21}^{(r)} = 0$. For these θ and Y^{-1} the values of α_{20} are not influenced by the effects of finite electron density and the contribution of ions provided (2.21) is valid. The fact that both $\alpha_{21}^{(c)}$ and $\alpha_{21}^{(r)}$ are real means that the effects of finite electron density and the contribution of ions do not change the linear character of whistler-mode polarization in the (x, y) plane.

Propagation in a cold plasma

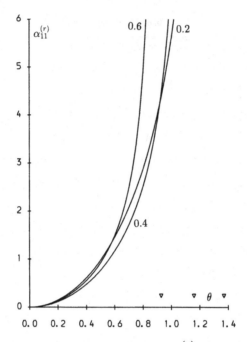

Fig. 2.10. The same as Fig. 2.7 but for $\alpha_{11}^{(r)}$ (see equation (2.79)).

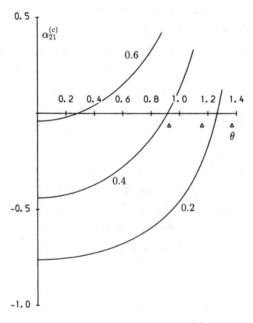

Fig. 2.11. The same as Fig. 2.7 but for $\alpha_{21}^{(c)}$ (see equation (2.78)).

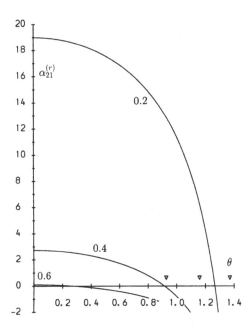

Fig. 2.12. The same as Fig. 2.7 but for $\alpha_{21}^{(r)}$ (see equation (2.80)).

2.4 A physical model of wave propagation

In this section we attempt to describe the process of whistler-mode propagation not in terms of the formal solution of the corresponding dispersion equation, but in terms of the periodic energy exchange between wave field and perturbed electron currents. We will restrict ourselves to two extreme cases of the propagation of these waves: their propagation strictly parallel to a magnetic field ($\theta = 0$) and their quasi-electrostatic propagation at $\theta = \theta_{R0}$. In these cases, which are important for particular applications, the process of the energy exchange between wave field and currents appears to be the simplest one. The understanding of this process could be a good basis for the study of the physical background of more sophisticated cases of wave propagation (Sazhin, 1987b, 1989b).

(a) Parallel propagation

As has been pointed out in Section 2.3, the polarization of whistler-mode waves is a circular one with the wave electric field components defined by equations (2.71). We can rewrite this system in the form more convenient for the analysis of the physical process of wave propagation:

$$\left.\begin{aligned} E_x &= E_0 \cos(\omega t - kz) \\ E_y &= E_0 \sin(\omega t - kz) \end{aligned}\right\}. \tag{2.82}$$

Having substituted (2.82) into equation (1.27) and remembering (1.36) we obtain the following expressions for the components of wave magnetic field ($\mathbf{B_1}$):

$$\left.\begin{aligned} B_x &= -N E_y \\ B_y &= N E_x \end{aligned}\right\}, \tag{2.83}$$

where $N = kc/\omega$ is the modulus of the wave refractive index.

In the absence of plasma, equations (1.27) and (1.28) are satisfied only when $N = 1$. Physically, the wave propagation in this case can be interpreted either as a sine disturbance of the magnetic field $\mathbf{B}_w$ moving with velocity ω/k and producing the induced field $\mathbf{E}$, or as a sine disturbance of $\mathbf{E}$ moving with the same velocity and producing the induced field $\mathbf{B}_w$. The stationarity is attained when these disturbances are equal, which takes place when $\omega/k = c$. In other words, this wave can be understood in terms of the coexistence of electric and magnetic fields oscillating in phase; no periodic energy transformation from one field to another takes place. Thus one cannot draw any analogy between this wave and a mathematical pendulum.

When the wave is propagating through plasma, the wave field $\mathbf{E}$ results not only in the appearance of the induced field $\mathbf{B}_w$ but also in the appearance of a conductivity current $\mathbf{j}$. Remembering (1.27), (2.82) and (2.83), we reduce (1.28) to the following system of equations:

$$E_y = N^2 E_y + \frac{4\pi}{\omega} j_x, \tag{2.84}$$

$$E_x = N^2 E_x - \frac{4\pi}{\omega} j_y. \tag{2.85}$$

The first terms on the right-hand sides of (2.84) and (2.85) describe the effect of the induced magnetic field. Neglecting the effects of protons and assuming electrons to be cold, we calculate the components of the current $\mathbf{j}$ related to the wave field $\mathbf{E}$ in a straightforward way:

$$j_x = -\frac{\Pi^2 E_y}{4\pi(\Omega - \omega)}, \tag{2.86}$$

$$j_y = \frac{\Pi^2 E_x}{4\pi(\Omega - \omega)}. \tag{2.87}$$

Remembering that $\omega < \Omega$ for whistler-mode waves, one can see from equations (2.84)–(2.87) that the current **j** tends to reduce the value of the induced magnetic field. The stationarity of the process is attained when this reduction of the induced magnetic field is compensated by its increase due to increasing N. Thus we can expect that N for whistler-mode waves in a cold plasma is always above 1.

During the process of wave propagation, the wave energy is periodically transported from **j** to **E** and $\mathbf{B}_w$ and back to **j**. This means that in contrast to the case of electromagnetic waves in vacuum, one can draw an analogy between parallel whistler-mode propagation in a cold plasma and a mathematical pendulum.

The value of N for parallel whistler-mode waves in a cold plasma (N_0) is obtained from (2.84)–(2.86) or (2.85)–(2.87) in a straightforward way:

$$N_0^2 = 1 + \frac{\Pi^2}{\omega(\Omega - \omega)}. \tag{2.88}$$

This expression has already been obtained in Section 2.1 (see equation (2.8)).

(b) Quasi-electrostatic propagation

As follows from (2.72) whistler-mode polarization at $\theta = \theta_R$ is linear with $\mathbf{E} \parallel \mathbf{N}$. In this case equations (1.27) and (1.28) reduce to:

$$\mathbf{E} - \frac{4\pi\mathrm{i}}{\omega}\mathbf{j} = 0. \tag{2.89}$$

Choosing as in Chapter 1 the coordinate system so that **N** lies in the (x, z) plane we obtain:

$$j_x = -\frac{\omega}{4\pi\mathrm{i}}\frac{\Pi^2 E_x}{(\Omega^2 - \omega^2)}, \tag{2.90}$$

$$j_y = \frac{\mathrm{i}\Omega}{4\pi\mathrm{i}}\frac{\Pi^2 E_y}{(\Omega^2 - \omega^2)}, \tag{2.91}$$

$$j_z = \frac{\omega}{4\pi\mathrm{i}}\frac{\Pi^2 E_z}{\omega^2}. \tag{2.92}$$

In view of (2.90)–(2.92) and remembering our assumption $\mathbf{E} \parallel \mathbf{N}$, equation (2.89) is satisfied when

$$1 + \frac{\Pi^2(\omega^2 - \Omega^2 \cos^2\theta)}{\omega^2(\Omega^2 - \omega^2)} = 0. \tag{2.93}$$

Equation (2.93) is a condition determining the resonance cone angle $\theta = \theta_R$

in a plasma with arbitrary electron density. The explicit expression for the latter angle is simplest in the dense plasma limit $\Pi \gg \Omega$, when

$$\cos\theta_R \approx \omega/\Omega. \tag{2.94}$$

This expression was also derived in Section 2.1 (see equation (2.13)).

From the physical point of view we do not have wave propagation, but rather a periodic exchange of energy between the electric field and particles, similarly to Langmuir oscillations in a cold plasma described by equation (1.23) in the limit $w = 0$.

Problems

Problem 2.1 Generalize expression (2.26) for v_g to a plasma with arbitrary density, the contribution of ions being neglected.

Problem 2.2 Show that $\mathbf{v}_g(\theta = \theta_{G0})$ does not depend on wave frequency in the limit $\nu \gg 1$.

Problem 2.3 Prove that the polarization of whistler-mode electric and magnetic fields in a cold dense plasma and in a plane perpendicular to $\mathbf{k}$ is always circular.

3

Parallel propagation (weakly relativistic approximation)

3.1 Simplification of the general dispersion equation

In contrast to Chapter 2, in this chapter we restrict ourselves to considering the parallel whistler-mode propagation, but impose no restrictions on magnetospheric electron temperature except the condition (1.2) of weakly relativistic approximation. We begin with some general simplifications of the dispersion equation (1.42) with the elements of the plasma dielectric tensor defined by (1.73) in the limiting case $\theta \to 0$. In this limit we can assume that $|\lambda_\alpha| \equiv |k_\perp v_\perp/\Omega_\alpha| \ll 1$ and expand the Bessel functions in (1.74) using the following formula (Abramovitz & Stegun, 1964):

$$J_n(\lambda_\alpha) = \left(\frac{\lambda_\alpha}{2}\right)^n \sum_{k=0}^{\infty} \frac{(-\lambda_\alpha^2/4)^k}{k!\,\Gamma(n+k+1)}. \tag{3.1}$$

Remembering (3.1) and keeping only zero-order terms with respect to λ_α we obtain the following expressions for the non-zero elements of the tensor $\Pi_{ij}^{(n,\alpha)}$ defined by (1.74):

$$\left.\begin{array}{l} \Pi_{11}^{(\pm 1,\alpha)} = \Pi_{22}^{(\pm 1,\alpha)} = v_\perp^2/4 \\[4pt] \Pi_{12}^{(\pm 1,\alpha)} = -\Pi_{21}^{(\pm 1,\alpha)} = \pm i v_\perp^2/4 \\[4pt] \Pi_{33}^{(0,\alpha)} = v_\parallel^2 \end{array}\right\}. \tag{3.2}$$

From (1.73) and (3.2) it follows that:

$$\left.\begin{array}{l} \epsilon_{13} = \epsilon_{23} = \epsilon_{31} = \epsilon_{32} = 0 \\[4pt] \epsilon_{21} = -\epsilon_{12} \\[4pt] \epsilon_{11} = \epsilon_{22} \end{array}\right\}. \tag{3.3}$$

In view of (3.3) and remembering our assumption that $\theta = 0$ we can simplify the dispersion equation (1.42) to

$$\epsilon_{33}(N^4 - 2\epsilon_{11}N^2 + \epsilon_{11}^2 + \epsilon_{12}^2) = 0. \tag{3.4}$$

This equation can be further simplified if we take into account the contribution of electrons only (which is justified for whistler-mode waves) and present $\epsilon_{11} = \epsilon_{22}$ as $\epsilon_{11} = 1 + \epsilon_+ + \epsilon_-$ and $\epsilon_{12} = -\epsilon_{21}$ as $\epsilon_{12} = i(\epsilon_+ - \epsilon_-)$, where ϵ_+ and ϵ_- are the contributions of the electron currents corresponding to $n = 1$ and $n = -1$ respectively in the term ϵ_{11} in (1.73). In this case equation (3.4) can be written as

$$\epsilon_{33}(N^2 - 1 - 2\epsilon_-)(N^2 - 1 - 2\epsilon_+) = 0. \qquad (3.5)$$

In order to satisfy (3.5) at least one of three factors in the left-hand side of this equation must be equal to zero. The equation $\epsilon_{33} = 0$ is the dispersion equation for electrostatic Langmuir waves, propagating along the magnetic field (see equation (1.13)). These waves are similar to those considered in Section 1.2. The equation $N^2 - 1 - 2\epsilon_- = 0$ describes the propagation of the so-called left-hand polarized waves. In a cold plasma limit this equation reduces to (2.7) taken with the '+' sign. Finally, the equation:

$$N^2 - 1 - 2\epsilon_+ = 0 \qquad (3.6)$$

is just the dispersion equation for the whistler-mode waves in which we are primarily interested in this book. In the cold plasma limit it reduces to (2.8). Keeping in mind the definition of ϵ_+, (3.6) can be written in a more explicit form:

$$N^2 = 1 + \pi X m_e \int_0^\infty p_\perp^2 \, dp_\perp \int_{-\infty}^{+\infty} \frac{(\omega M_e - k p_\parallel)\frac{\partial f_{0e}}{\partial p_\perp} + k p_\perp \frac{\partial f_{0e}}{\partial p_\parallel}}{M_e(\omega M_e - k p_\parallel - \Omega_0 m_e)} dp_\parallel, \qquad (3.7)$$

where $k \equiv k_\parallel$.

Alternatively, equation (3.7) can be written as (Sazhin, 1989c):

$$N^2 = 1 - i\pi X \int_0^\infty dt \int_{-\infty}^{+\infty} dp_\parallel \int_0^\infty p_\perp^2 M_e^{-1}$$

$$\times \exp[i(\omega m_e/M_e - \Omega_0 - k p_\parallel/m_e)t] \left[(\omega M_e - k p_\parallel)\frac{\partial f_{0e}}{\partial p_\perp} + k p_\perp \frac{\partial f_{0e}}{\partial p_\parallel} \right] dp_\perp.$$
$$(3.8)$$

The integral with respect to t converges when $\Im \omega > 0$. However, it can be analytically continued to the region $\Im \omega \leq 0$ (cf. Section 1.2).

Considering the weakly relativistic limit ($p = \sqrt{p_\parallel^2 + p_\perp^2} \ll m_e c$; $M_e = m_e(1 + p^2/2m_e^2 c^2)$) and substituting (1.76) taken for $j = 0$ into (3.8) we obtain after some rearrangement:

$$N^2 = 1 + \frac{2iX}{\tilde{r}} \int_0^\infty d\tau \exp\left[-\frac{h^2\tau^2}{1-i\tau} + \frac{2i(1-Y)\tau}{\tilde{r}}\right] \frac{1}{(1-i\tau)^{1/2}(1-iA_{e0}\tau)^2}$$

$$\times \left\{1 + \frac{i(A_{e0}-1)N^2\tau}{1-i\tau}\left[-1 + \frac{A_{e0}\tilde{r}}{1-iA_{e0}\tau} + \frac{3\tilde{r}}{4(1-i\tau)} - \frac{N^2\tau^2}{2(1-i\tau)^2}\right]\right\},$$

$$(3.9)$$

where $\tau = p_{0\parallel}\omega t/2p_c^2$; $h = p_c^2/p_{ph}p_{0\parallel}$; $p_{ph} = m_e\omega/k$; $p_c = m_ec$; $A_{e0} = p_{0\perp}^2/p_{0\parallel}^2$; $\tilde{r} = p_\parallel^2/p_c^2$. When deriving (3.9) we assumed that the integral in (3.9) taken from $\tau_{cr} \equiv \min(1, A_{e0}^{-1})$ to infinity is negligibly small.

Equation (3.9) gives us a slight generalization of the corresponding equation which could be obtained from equation (14) of Tsai *et al.* (1981). The latter was derived under the assumption that relativistic effects could be neglected everywhere except in the exponential term.

Neglecting the contribution of the second-order terms with respect to $\tilde{r}$ and/or τ in (3.9) (see the discussion in Section 3.2) we can rewrite this equation in a more compact form:

$$N^2 = 1 - \frac{2X}{\tilde{r}}\left[\mathcal{F}_{1/2,2} - \frac{d\mathcal{F}_{3/2,2}}{dz}(A_{e0}-1)N^2\right], \qquad (3.10)$$

where the function $\mathcal{F}_{q,p}$ is defined as:

$$\mathcal{F}_{q,p} \equiv \mathcal{F}_{q,p}(z,a,b) = -i\int_0^\infty e^{izt-at^2/(1-it)}(1-it)^{-q}(1-ibt)^{-p}dt, \quad (3.11)$$

and $z = 2(1-Y)/\tilde{r}$; $a = N^2/\tilde{r}$; $b = A_{e0}$.

Equation (3.10) can be considerably simplified in the limit of a weakly anisotropic plasma:

$$|\Delta A_{e0}| = |A_{e0} - 1| \ll 1, \qquad (3.12)$$

and rewritten as (Sazhin, 1989c):

$$D \equiv N^2 - 1 + \frac{2X}{\tilde{r}}\left[\mathcal{F}_{5/2} + \Delta A_{e0}(2-N^2)\mathcal{F}'_{7/2}\right] = 0, \qquad (3.13)$$

where

$$\mathcal{F}_q \equiv \mathcal{F}_q(z,a) = -i\int_0^\infty e^{izt-at^2/(1-it)}(1-it)^{-q}dt \qquad (3.14)$$

is the Shkarofsky function (Shkarofsky, 1966, 1986; Airoldi & Orefice, 1982; Krivenski & Orefice, 1983; Robinson, 1986, 1987a), and

$$\mathcal{F}'_q \equiv d\mathcal{F}_q/dz = \mathcal{F}_q - \mathcal{F}_{q-1}. \qquad (3.15)$$

Comparing (3.11) and (3.14) we can see that $\mathcal{F}_{q,p}$ reduces to $\mathcal{F}_q$ when $p = 0$. Hence, we will call $\mathcal{F}_{q,p}$ a generalized Shkarofsky function (Sazhin

& Temme, 1990; Temme, Sumner & Sazhin, 1992). Note that exactly the same function was introduced at about the same time by Bornatici, Ghiozzi & de Chiara (1990) and was called the generalization to the anisotropic-temperature case of the weakly relativistic Shkarofsky dispersion function.

For isotropic plasma ($\Delta A_{e0} = 0$) equation (3.13) is further simplified to

$$D \equiv N^2 - 1 + \frac{2X}{\tilde{r}} \mathcal{F}_{5/2} = 0. \tag{3.16}$$

Equation (3.16) has been used by many authors for numerical analysis of whistler-mode propagation in a hot plasma (see e.g. Robinson, 1987b). From the formal point of view the Shkarofsky function in (3.13) and (3.16) is defined on two sheets of the Riemann surface corresponding to $\Re[(1-\mathrm{i}t)^{5/2}] > 0$ and $\Re[(1-\mathrm{i}t)^{5/2}] < 0$. However, in deriving (3.13) and (3.16), we implicitly assumed that $\Re[(1-\mathrm{i}t)^{5/2}] > 0$ and thus restricted our analysis to the first sheet.

In the nonrelativistic limit, we should set $c \to \infty$ and write equation (3.9) as:

$$N^2 = 1 + \frac{2\mathrm{i}X}{\tilde{r}} \int_0^\infty \exp\left[-h^2\tau^2 + \frac{2\mathrm{i}(1-Y)\tau}{\tilde{r}}\right] \times [1 - \mathrm{i}(A_{e0} - 1)N^2\tau]\mathrm{d}\tau. \tag{3.17}$$

The integral on the right-hand side of (3.17) can be solved in terms of the following general formula:

$$\int_0^\infty \exp\left(-\frac{x^2}{4\beta} - \gamma x\right)\mathrm{d}x = \sqrt{\pi\beta}\exp(\beta\gamma^2)[1 - \Phi(\gamma\sqrt{\beta})], \tag{3.18}$$

where:

$$\Phi(x) = \frac{2}{\sqrt{\pi}}\int_0^x \exp(-t^2)\mathrm{d}t, \tag{3.19}$$

and $\Re\beta > 0$.

In view of (3.18), equation (3.17) can be reduced to a more conventional form:

$$N^2 = 1 + (A_{e0} - 1)X + \frac{X\omega}{kw_\parallel}[A_{e0} + (1 - A_{e0})Y]Z(\xi), \tag{3.20}$$

where $w_\parallel = p_{0\parallel}/m_e$, $\xi = (1-Y)/(\tilde{r}h)$, and $Z(\xi)$ is the plasma dispersion function (see equation (1.20)). Equation (3.20) will be studied in detail in Chapters 4 and 7.

Equations (3.10), (3.13) and (3.20) can be generalized in a straightforward way for the case when $j \neq 0$ using the following identity (Tsai *et al.*, 1981):

$$N^2(j \neq 0) = 1 + \frac{(-1)^j}{j!}\frac{1}{p_{0\perp}^{2j+2}}\frac{\mathrm{d}^j}{\mathrm{d}(p_{0\perp}^{-2})^j}[p_{0\perp}^2(N^2(j=0) - 1)]. \tag{3.21}$$

In the case of (3.20) this would require a formal replacement of A_{e0} by $A_e = (j+1)p_{0\perp}^2/p_{0\|}^2$.

3.2 Low-temperature limit

In this section we apply the general analysis of Section 3.1 to the problem of parallel whistler-mode propagation in a low-temperature plasma. As in Section 1.2 we assume that whistler-mode amplification or damping does not influence wave propagation and the latter is determined by an equation of the type (1.16), where D is the difference between the left-hand and right-hand sides of equation (3.9). Remembering our assumption of a weakly relativistic plasma (see condition (1.2)) and assuming that

$$h^2 \gg 1, \tag{3.22}$$

we can expect that the main contribution to the integral (3.9) comes from $\tau \ll 1$. For these τ we can write the following expansions:

$$\frac{1}{(1-i\tau)^{1/2}(1-iA_{e0}\tau)^2} = 1 + \frac{i(4A_{e0}+1)\tau}{2}, \tag{3.23}$$

$$\exp\left(-\frac{h^2\tau^2}{1-i\tau}\right) = (1-ih^2\tau^3)\exp(-h^2\tau^2), \tag{3.24}$$

$$\frac{1}{1-i\tau}\left[-1 + \frac{A_{e0}\tilde{r}}{1-iA_{e0}\tau} + \frac{3\tilde{r}}{4(1-i\tau)} - \frac{N^2\tau^2}{2(1-i\tau)^2}\right]$$

$$= -1 - i\tau + (A_{e0}+0.75)\tilde{r} - N^2\tau^2/2. \tag{3.25}$$

Remembering (3.18) and (3.23)–(3.25), and restricting ourselves to the analysis of the real part of (3.9), we can simplify the latter equation to:

$$N^2 = 1 + \frac{X}{\tilde{r}h}\Re Z + \frac{X\Re Z'}{2}\left[(1-A_{e0}) + \frac{4A_{e0}+1}{2N^2} + (A_{e0}+0.75)(A_{e0}-1)\tilde{r}\right]$$

$$+ \frac{X(1-A_{e0})\Re Z''}{4h} + \frac{X\Re Z'''}{16N^2}\left[2 + N^2(A_{e0}-1)\tilde{r}\right], \tag{3.26}$$

where N, ω, Y and X hereafter in this chapter are assumed to be real, the argument of the Z-function is $\xi_0 = \Re\xi$, and the superscript "'" indicates differentiation with respect to ξ_0, e.g. $\Re Z''' = d^3\Re Z/d\xi_0^3$. Note that:

$$Z''' = (8 - 8\xi^2) + (12\xi - 8\xi^3)Z. \tag{3.27}$$

In equation (3.26) we should further neglect the terms proportional to $\tilde{r}$ and h^{-1} (but not $(\tilde{r}h)^{-1}$) as they have a smaller order of magnitude than the retained terms. (Moreover, if these terms were kept, then more terms in the expansions (3.23)–(3.24) should have been kept.) As a result equation (3.26) simplifies to:

$$N^2 = 1 + (A_{e0} - 1)X + \frac{X\omega}{kw_{\|}}[A_{e0} + (1 - A_{e0})Y]\Re Z$$

$$+ \frac{X}{2N^2}\left[-(4A_{e0} + 1) + \frac{(Y-1)\omega}{kw_{\|}}(4A_{e0} + 1)\Re Z + \frac{\Re Z'''}{4}\right]. \qquad (3.28)$$

Using (3.21), equation (3.28) is generalized in a straightforward way to the case $j \neq 0$ by replacing A_{e0} by $A_e = (j+1)p_{0\perp}^2/p_{0\|}^2$. In the asymptotic limit $N^2 \to \infty$, equation (3.28) reduces to (3.20). In the limit $|\xi_0| \gg 1$ it has a solution:

$$N^2 = N_{0\|}^2(1 + \tilde{a}_{\beta\|}\beta_e), \qquad (3.29)$$

where $N_{0\|}$ is the same as in (2.8), β_e is the same as in (1.80)–(1.88),

$$\tilde{a}_{\beta\|} = \tilde{a}_{\|t} + \tilde{a}_{\|R}, \qquad (3.30)$$

$$\tilde{a}_{\|t} = \frac{Y^2[Y + A_e(1 - Y)]}{(Y - 1)^3}, \qquad (3.31)$$

$$\tilde{a}_{\|R} = \frac{Y^2(1 + 4A_e)}{2(Y - 1 + \nu Y^2)(Y - 1)}. \qquad (3.32)$$

When deriving (3.29) it was assumed that:

$$|a_{\beta\|}| \equiv |\tilde{a}_{\beta\|}\beta_e| \ll 1. \qquad (3.33)$$

In the limit $\beta_e \to 0$ equation (3.29) reduces to (2.8).

The term $\tilde{a}_{\|t}$ describes the contribution of the non-relativistic thermal correction to N_0. It could be obtained from equation (3.20). However, the term $\tilde{a}_{\|R}$ could not be obtained from equation (3.20) and describes the relativistic thermal correction to $N_{0\|}$. In Figs. 3.1 and 3.2 we present the curves $\tilde{a}_{\|t}$, $\tilde{a}_{\|R}$ and $\tilde{a}_{\beta\|}$ versus Y^{-1} for $\nu = 4, 1$ and 0.25 and $A_e = 1$ and 2.

As follows from Fig. 3.1, for $A_e = 1$ and $\nu = 4$ the term $\tilde{a}_{\|t}$ brings the main contribution to $\tilde{a}_{\beta\|}$ so that the relativistic effects (the term $\tilde{a}_{\|R}$) on parallel whistler-mode propagation could be neglected altogether. However, for $\nu = 1$ and especially for $\nu = 0.25$ the relativistic term $\tilde{a}_{\|R}$ dominates the non-relativistic correction $\tilde{a}_{\|t}$ at least for relatively small Y^{-1}. However, for

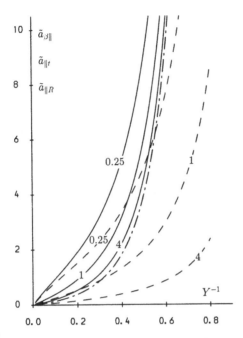

Fig. 3.1 Plots of $\tilde{a}_{\|R}$ (see equation (3.32)) versus Y^{-1} (dashed), $\tilde{a}_{\|t}$ (see equation (3.31)) versus Y^{-1} (dashed–dotted) and $\tilde{a}_{\beta\|}$ (see equation (3.30)) versus Y^{-1} (solid) for $\nu = 0.25$, 1 and 4 (curves indicated; $\tilde{a}_{\|t}$ does not depend on ν) and $A_e = 1$.

Y^{-1} approaching 1 the term $\tilde{a}_{\|t}$ is always dominant. Both terms $\tilde{a}_{\|t}$ and $\tilde{a}_{\|R}$ are positive and increase with increasing Y^{-1}.

In contrast to the case $A_e = 1$, for $A_e = 2$ and relatively small Y^{-1}, $\tilde{a}_{\|t}$ and $\tilde{a}_{\beta\|}$ are negative so that the effect of finite anisotropy tends to compensate for the influence of thermal effects on parallel whistler-mode propagation (see Fig. 3.2). At the same time, comparing Figs. 3.1 and 3.2 we can see that an increase in A_e tends to increase the contribution of the term $\tilde{a}_{\|R}$. In a similar way to the case $A_e = 1$, the contribution of the relativistic term for $A_e = 2$ is smallest for $\nu = 4$. Hence, we can conclude that the non-relativistic analysis of the thermal effects on whistler-mode propagation is justified only in a relatively dense plasma $\nu \gg 1$ or at ω close to Ω_0 provided our approximations are valid. Non-relativistic analysis of whistler-mode propagation in a rarefied plasma ($\nu \lesssim 1$) even by rigorous numerical methods (e.g. André, 1985) can be misleading.

Equation (3.29) could also be obtained from the general equation (1.42) taken in the limit $\theta = 0$ and $|N^2 - N_0^2| \ll N_0^2$, provided the elements of the plasma dielectric tensor can be presented in the form (1.78) (see Section

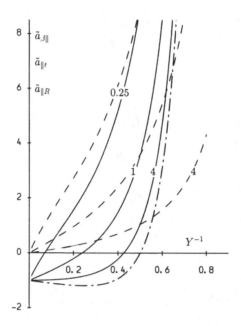

Fig. 3.2. The same as Fig. 3.1 but for $A_e = 2$.

5.1). Alternatively we could expand the integrand (3.8) for small $p_\parallel$ and $p_\perp$ keeping the first order terms with respect to $p_\parallel$ and $p_\perp$. Then imposing the same condition $|N^2 - N_0^2| \ll N_0^2$ and after a straightforward integration we obtain (3.29) (see e.g. Jaquinot & Leloup, 1971). Note that when deriving (3.29) we imposed no restrictions on the value of electron density. If this density is not too low then N_0^2 can be considered as a perturbation of N_{0d}^2 (see equation (2.18) taken at $\theta = 0$). Assuming this is true and taking into account the contribution of ions we can write N_0^2 in the form of (2.21). Remembering this equation we can rewrite (3.29) as:

$$N^2 = N_{0d\parallel}^2 (1 + \tilde{a}_{c\parallel}\nu^{-1} + \tilde{a}_{r\parallel}r + \tilde{a}_{\beta\parallel}\beta_e), \qquad (3.34)$$

where $N_{0d\parallel} = N_{0d}(\theta = 0)$;

$$\tilde{a}_{c\parallel} = (Y - 1)/Y^2 \qquad (3.35)$$

and

$$\tilde{a}_{r\parallel} = 1 - Y \qquad (3.36)$$

are the values of $\tilde{a}_c$ and $\tilde{a}_r$ (see equations (2.19) and (2.21)) taken in the limit $\theta = 0$.

Considering another limiting case $\xi_0 = -0.924$ when $Z'(\xi) = 0$ (see equation (1.113)), we can reduce (3.28) to (Sazhin, 1989c):

$$N^2 = N_{0\parallel}^2 - X/(2N^2). \tag{3.37}$$

The solution of equation (3.37):

$$N_{1,2}^2 = \frac{2N_{0\parallel}^2 \pm \sqrt{4N_{0\parallel}^4 - 8X}}{4} \tag{3.38}$$

should satisfy the condition:

$$\frac{Y-1}{N\sqrt{\tilde{r}}} = 0.924. \tag{3.39}$$

The same equation (3.37) has also been derived by Sazhin & Temme (1990) using an asymptotic expansion of the generalized Shkarofsky function (3.11). Although the case $\xi_0 = -0.924$ is the exception rather than the rule for whistler-mode propagation in magnetospheric conditions, our solution (3.38) can be used as an independent test of the validity of a numerical analysis of the whistler-mode dispersion equation.

3.3 Physical interpretation

Equation (3.29) in the previous section was obtained from a formal analysis of the corresponding dispersion equation. In this section we interpret it in terms of a periodic energy exchange between wave electric and magnetic fields and perturbed electron currents in a similar way to that in Section 2.4, where we considered a physical model of whistler-mode propagation in a cold plasma. In order to do this we should first consider the dynamics of weakly relativistic electrons in the wave field and the corresponding dynamics of perturbed currents, following Sazhin (1987b).

(a) Dynamics of weakly relativistic electrons

The acceleration of relativistic particles under the influence of electric ($\mathbf{E}$) and magnetic ($\mathbf{B}$) fields is described by:

$$\frac{d\mathbf{v}}{dt} = \frac{q_\alpha}{m_\alpha}\sqrt{1 - \frac{v^2}{c^2}}\left[\mathbf{E} + \frac{1}{c}\mathbf{v} \times \mathbf{B} - \frac{1}{c^2}\mathbf{v}(\mathbf{v}.\mathbf{E})\right], \tag{3.40}$$

where q_α is the particle charge, m_α its mass at rest, and $\mathbf{v}$ its velocity. As in Section 2.4 we restrict ourselves to considering the motion of electrons in the wave electric ($\mathbf{E}_w$) and magnetic ($\mathbf{B}_w$) fields. We assume that $\mathbf{E}_w = \mathbf{E}$

and $\mathbf{B}_w = \mathbf{B} - \mathbf{B}_0$, where $\mathbf{B}_0$ is the external homogeneous magnetic field. Also we assume the fields $\mathbf{E}_w$ and $\mathbf{B}_w$ to be small so that we can look for the solution of (3.40) in the form:

$$\mathbf{v} = \mathbf{v}_0 + \mathbf{v}_1, \tag{3.41}$$

where $|\mathbf{v}_1| \ll |\mathbf{v}_0|$, and $\mathbf{v}_0$ is the solution of (3.41) corresponding to $|\mathbf{E}_w| = |\mathbf{B}_w| = 0$:

$$\left.\begin{array}{l} v_{0x} = v_{0t}\cos(\Omega t + \varsigma_e) \\ v_{0y} = v_{0t}\sin(\Omega t + \varsigma_e) \\ v_{0z} = \mathrm{const} \end{array}\right\}, \tag{3.42}$$

$\Omega = \Omega_0\sqrt{1 - v_0^2/c^2}$ is the relativistic electron gyrofrequency, ς_e is the initial phase of the electrons, and $v_0^2 = v_{0t}^2 + v_{0z}^2$.

When looking for the first-order solution of (3.40), we first keep only linear terms in this equation. These terms are of the order of v_1/c, E_w/B_0 or B_w/B_0 (they are assumed to be of the same order of magnitude). Then we assume $|\mathbf{v}_0| \ll \min(v_{ph}, c)$ and neglect the terms proportional to $(v_0/v_{ph})^n$ and $(v_0/c)^n$, where $n > 2$, in the coefficients before v_1, E_w and B_w. Remembering these assumptions as well as (2.82), (2.83), (3.41) and (3.42), we obtain from (3.40):

$$\frac{dv_{1x}}{dt} = -\frac{|e|E_0}{m_e}\left\{\left(1 - \frac{v_{0z}}{v_{ph}} - \frac{v_{0z}^2}{2c^2} - \frac{v_{ot}^2}{c^2}\right)\cos(\omega t - kz + \emptyset)\right.$$

$$\left. - \frac{v_{0t}^2}{2c^2}\cos[(2\Omega - w)t + kz + (2\varsigma_e - \emptyset)]\right\}$$

$$- \Omega_0\left[\left(1 - \frac{v_0^2}{2c^2} - \frac{v_{0t}^2\sin^2(\Omega t + \varsigma_e)}{c^2}\right)v_{1y} - \frac{v_{0t}^2\sin(2\Omega t + 2\varsigma_e)}{2c^2}v_{1x}\right], \tag{3.43}$$

$$\frac{dv_{1y}}{dt} = -\frac{|e|E_0}{m_e}\left\{\left(1 - \frac{v_{0z}}{v_{ph}} - \frac{v_{0z}^2}{2c^2} - \frac{v_{ot}^2}{c^2}\right)\sin(\omega t - kz + \emptyset)\right.$$

$$\left. - \frac{v_{0t}^2}{2c^2}\sin[(2\Omega - w)t + kz + (2\varsigma_e - \emptyset)]\right\}$$

$$+ \Omega_0\left[\left(1 - \frac{v_0^2}{2c^2} - \frac{v_{0t}^2\cos^2(\Omega t + \varsigma_e)}{c^2}\right)v_{1x} - \frac{v_{0t}^2\sin(2\Omega t + 2\varsigma_e)}{2c^2}v_{1y}\right], \tag{3.44}$$

$$\frac{dv_{1z}}{dt} = -\frac{|e|E_0}{m_e}\left(\frac{v_{0t}}{v_{ph}} - \frac{v_{0z}v_{0t}}{c^2}\right)\cos[(\Omega - \omega)t + kz + (\varsigma_e - \emptyset)], \quad (3.45)$$

where E_0 is the amplitude of the wave electric field (cf. equation (2.82)). The solution of (3.43)–(3.45) can be written as:

$$v_{1x} = -v_1\sin(\omega t - kz + \emptyset) - v_2\sin[(2\Omega - \omega)t + kz + (2\varsigma_e - \emptyset)], \quad (3.46)$$

$$v_{1y} = v_1\cos(\omega t - kz + \emptyset) + v_2\cos[(2\Omega - \omega)t + kz + (2\varsigma_e - \emptyset)], (3.47)$$

$$v_{1z} = v_3\sin[(\Omega - \omega)t + kz + (\varsigma_e - \emptyset)], \quad (3.48)$$

where:

$$v_1 = \frac{|e|E_0}{m_e(\omega - \Omega_0)}\left[1 + \frac{N\Omega_0 v_{0z}}{c(\omega - \Omega_0)} + \frac{N^2\omega\Omega_0 v_{0z}^2}{c^2(\omega - \Omega_0)^2} - \frac{\omega(v_{0z}^2 + 2v_{0t}^2)}{2c^2(\omega - \Omega_0)}\right], \quad (3.49)$$

$$v_2 = \frac{|e|E_0(\omega - 2\Omega_0)}{m_e(\omega - \Omega_0)^2}\frac{v_{0t}^2}{c^2}, \quad (3.50)$$

$$v_3 = \frac{|e|E_0}{m_e(\omega - \Omega_0)}\left[\frac{Nv_{0t}}{c} + \frac{N^2\omega v_{0t}v_{0z}}{c^2(\omega - \Omega_0)} - \frac{v_{0z}v_{0t}}{c^2}\right]. \quad (3.51)$$

It follows from (3.46) and (3.47) that the wave field $\mathbf{E}_w$ gives rise not only to oscillations of the electrons with frequency ω but also to oscillations with frequency $(2\Omega - \omega)$. The latter is an essentially relativistic effect which could not be predicted by non-relativistic theory. However, the currents caused by electron oscillations with frequency $(2\Omega - \omega)$, being non-coherent with the wave field, do not contribute to the process of whistler-mode propagation and will be disregarded. The currents caused by other perturbations of electron velocities described by (3.46)–(3.48) are calculated in the next subsection.

(b) Perturbed currents

The general equation for the perturbed currents can be presented as:

$$\mathbf{j} = -|e|n_0\int_0^{2\pi}d\emptyset_p\int_{-\infty}^{+\infty}dp_\|\int_0^\infty p_\perp f_0\left(\mathbf{v}_1 + \mathbf{v}_0\frac{d\mathbf{p}_1}{d\mathbf{p}}\right)dp_\perp, \quad (3.52)$$

where $\mathbf{v}_1$ is the perturbation of the electron velocity (see equations (3.46)–(3.48)), $\mathbf{v}_0$ is the unperturbed electron velocity, and $d\mathbf{p}_1/d\mathbf{p}$ is the derivative of the perturbation of the electron momentum $\mathbf{p}_1$ with respect to $\mathbf{p}$ in the direction of $\mathbf{p}$.

Remembering that the element of volume in the phase space is conserved we can see that $d\mathbf{p}_1/d\mathbf{p} = -\delta n_e/n_0$, where δn_e is the perturbation of electron density. We denote the part of the electron current (3.52) caused by the perturbation of $\mathbf{v}$ by $\mathbf{j}_v$, while the part of the electron current related to the perturbation of n_0 will be denoted by $\mathbf{j}_d$.

In order to calculate perturbed currents caused by the perturbations of electron velocities (3.46)–(3.48) ($\mathbf{j}_v$), we must choose the corresponding unperturbed electron distribution function. As in Section 1.5 we take this function in the form (1.76). Remembering (1.76) we can present the equation for this perturbed electron current as

$$\mathbf{j}_v = -|e|n_0 \int_0^{2\pi} d\emptyset_p \int_{-\infty}^{+\infty} dp_{\|} \int_0^\infty p_\perp f_0(v_{1x}\mathbf{i}_x + v_{1y}\mathbf{i}_y + v_{1z}\mathbf{i}_z)dp_\perp, \quad (3.53)$$

where $\emptyset_p$ is the phase of the momentum which is equal to $\emptyset = \Omega t + \varsigma_e$, the corresponding phase of the velocity (see equation (3.42)); $v_{1x,y,z}$ are defined by (3.46)–(3.48). When substituting the latter expressions into (3.52), we must represent v_{0t} and v_{0z} in (3.42) as functions of $p_\perp$ and $p_{\|}$ according to the equation $\mathbf{p} = m_e\mathbf{v}_0/\sqrt{1 - v_0^2/c^2}$. However, the relativistic corrections of v_0 give higher-order terms with respect to v_0^2/c^2 and can be neglected. Thus we can set

$$\mathbf{p} = m_e\mathbf{v}_0.$$

There is an apparent contradiction between (3.53), which contains an integration over infinitely large momenta, and our assumption $|\mathbf{v}_0| \ll \min(v_{ph}, c)$. However, the contribution of high energy electrons in (3.53) falls exponentially to a small value and does not influence the results.

Having substituted (3.46)–(3.48) into (3.53), remembering our assumption that $\mathbf{p} = m_e\mathbf{v}_0$ and neglecting the terms proportional to v_2, we have

$$j_x = -\frac{\omega}{4\pi}\left[\frac{\Pi^2}{\omega(\Omega_0 - \omega)} + \frac{N^2\beta_e\Omega_0^3}{(\Omega_0 - \omega)^3} + \frac{\beta_e(1 + 4A_e)\Omega_0^2}{2(\Omega_0 - \omega)^2}\right]E_y, \quad (3.54)$$

$$j_y = \frac{\omega}{4\pi}\left[\frac{\Pi^2}{\omega(\Omega_0 - \omega)} + \frac{N^2\beta_e\Omega_0^3}{(\Omega_0 - \omega)^3} + \frac{\beta_e(1 + 4A_e)\Omega_0^2}{2(\Omega_0 - \omega)^2}\right]E_x, \quad (3.55)$$

where β_e and A_e are the same as in Section 3.2.

The first terms on the right-hand sides of (3.54) and (3.55) correspond to the perturbed currents in a cold plasma $j_{cx(cy)}$ (see equations (2.86) and (2.87)). The second terms correspond to the perturbed currents related to non-relativistic thermal motion of electrons (hereafter referred to as $j_{nx(ny)}$). The last terms correspond to relativistic contributions to the perturbed currents (hereafter referred to as $j_{rx(ry)}$). It follows from (3.54) and (3.55) that all three currents have the same sign, which means that both currents $j_{nx(ny)}$ and $j_{rx(ry)}$ tend to increase the effect of $j_{cx(cy)}$.

The z component of the perturbed current $\mathbf{j}_v$ does not contribute directly to the process of wave propagation. However, it causes a perturbation of electron density which in its turn contributes to the perturbed current $\mathbf{j}_d$, and thus influences the process of wave propagation. In order to estimate this effect quantitatively, we introduce the perturbed current density

$$\delta j_z = -|e|n_0 f_0 v_{1z}, \tag{3.56}$$

which refers to the electron current density in unit volume of phase space. The corresponding electron density δn_e can be obtained from the Poisson equation:

$$|e|\frac{\partial \delta n_e}{\partial t} = \nabla \delta \mathbf{j} = \frac{\partial \delta j_z}{\partial z}. \tag{3.57}$$

Remembering (3.48), (3.51) and (3.56), the solution of (3.57) can be presented as

$$\delta n_e = \frac{n_0|e|E_0 N^2\omega}{(\Omega_0 - \omega)^2}\frac{p_\perp}{m_e^2 c^2}f_0\sin\left[(\Omega - \omega)t + kz + (\varsigma_e - \emptyset)\right]. \tag{3.58}$$

In deriving (3.58), we neglected the contribution of the higher-order terms.

The perturbed current $\mathbf{j}_d$ due to δn_e is given by:

$$\mathbf{j}_d = -|e|\int_0^{2\pi}d\emptyset\int_{-\infty}^{\infty}dp_\|\int_0^{\infty}p_\perp dp_\perp v_{0t}\delta n_e[\cos(\Omega t + \varsigma_e)\mathbf{i}_x + \sin(\Omega t + \varsigma_e)\mathbf{i}_y]. \tag{3.59}$$

After substituting (3.58) into (3.59) and neglecting the terms which are not coherent with the wave field, we have

$$j_{dx} = \frac{\omega}{4\pi}\frac{N^2\beta_e A_e\Omega_0^2}{(\Omega_0 - \omega)^2}E_y, \tag{3.60}$$

$$j_{dy} = -\frac{\omega}{4\pi}\frac{N^2\beta_e A_e\Omega_0^2}{(\Omega_0 - \omega)^2}E_x. \tag{3.61}$$

The sign of $j_{dx(dy)}$ is opposite to that of $j_{cx(cy)}$, $j_{nx(ny)}$ and $j_{rx(ry)}$. Hence, the current $\mathbf{j}_d$, in contrast to $\mathbf{j}_n$ and $\mathbf{j}_r$, tends to counteract the effect of $\mathbf{j}_c$. It is important to notice that the current $\mathbf{j}_d$ determined by (3.60) and (3.61) is essentially non-relativistic. The relativistic terms would have higher order with respect to v_0^2/c^2 and have been neglected in our analysis.

From (2.84), (3.54), (3.60) or (2.85), (3.55), (3.61), there follows the dispersion equation for parallel whistler-mode propagation in a weakly relativistic plasma:

$$N^2 = N_{0\|}^2\left[1 + \frac{\beta_e\Omega_0^3}{(\Omega_0 - \omega)^3} + \frac{\beta_e(1 + 4A_e)\Omega_0^2}{2(\Omega_0 - \omega)^2 N_{0\|}^2} - \frac{\beta_e A_e\Omega_0^2}{(\Omega_0 - \omega)^2}\right], \tag{3.62}$$

where $N_{0\parallel}$ is determined by (2.8). Equation(3.62) reduces in a straightforward way to (3.29).

The physical meaning of the terms in (3.62) is self-evident: the first term on the right-hand side of (3.62) is controlled by the perturbed current in a cold plasma $\mathbf{j}_c$ (see Section 2.4), the second describes the increase of $\mathbf{j}_c$ due to the non-relativistic current $\mathbf{j}_n$ (second terms in (3.54) and (3.55)), the third describes the increase of $\mathbf{j}_c$ due to relativistic currents $\mathbf{j}_r$ (third terms in the (3.54) and (3.55)), and the fourth gives the decrease of $\mathbf{j}_c$ due to the non-relativistic current $\mathbf{j}_d$ (equations (3.60) and (3.61)). From (3.62), it follows that the current $\mathbf{j}_n$ dominates the currents $\mathbf{j}_r$ and $\mathbf{j}_d$ when ω is close to Ω_0 (but not too close to Ω_0 as in this case the influence of damping on whistler-mode propagation cannot be neglected and our approximation fails). For whistler-mode waves in a dense plasma ($N_0 \gg 1$), the contribution of the current $\mathbf{j}_r$ is negligibly small. The contribution of the current $\mathbf{j}_d$ increases with increasing A_e.

The total process of wave propagation is accompanied by periodic energy exchange between the currents $\mathbf{j}_c$, $\mathbf{j}_n$, $\mathbf{j}_r$, $\mathbf{j}_d$ and the electric and magnetic fields of the wave. From this point of view, the basic physics of parallel whistler-mode propagation in a weakly relativistic plasma is essentially the same as the physics of propagation in a cold plasma. The difference comes only from the different values of $j_{x(y)}$ for these two cases.

The compatibility of the expressions (3.62) and (3.29) confirms the validity of both approaches to this problem.

Problems

Problem 3.1   Using the equation $N^2 - 1 - 2\epsilon_- = 0$ (see equation (3.5)) obtain the dispersion equation for a left-hand polarized wave with $\theta = 0$ in a weakly relativistic plasma (analogue of expression (3.29) for a whistler-mode wave).

Problem 3.2   Writing the expression for $\mathcal{F}_{q,p}$ (see equation (3.11)) in the form:

$$\mathcal{F}_{q,p} = \mathcal{F}_{q,p}(\zeta, a) = \int_0^\infty e^{-\zeta s - as/(1+s)} f(s) \, ds, \qquad (3.63)$$

where $\zeta = z - a$, and $f(s) = 1/((1+s)^q(1+bs)^p)$, prove the following recursions:

$$(b-1)\mathcal{F}_{q,p} = b\mathcal{F}_{q-1,p} - \mathcal{F}_{q,p-1},$$

$$\zeta\mathcal{F}_{q,p} = 1 - a\mathcal{F}_{q+2,p} - q\mathcal{F}_{q+1,p} - pb\mathcal{F}_{q,p+1},$$

$$b\zeta\mathcal{F}_{q-1,p} = 1 + a(b-1)\mathcal{F}_{q+2,p} + [q(b-1) - ab]\mathcal{F}_{q+1,p}$$
$$+ [\zeta(b-1) - qb - (p-1)b]\mathcal{F}_{q,p}.$$

Problem 3.3 Show details of the derivation of equation (3.13) from equation (3.10) in the limit (3.12).

4

Parallel propagation (non-relativistic approximation)

4.1 Analytical solutions

The general dispersion equation for whistler-mode propagation, instability or damping in a non-relativistic plasma with the electron distribution function in the form (1.90) has already been derived in Section 3.1 (see equation (3.20)). Assuming, as in Sections 1.2 and 3.2, that whistler-mode growth or damping does not influence wave propagation we can simplify equation (3.20) to:

$$\Re D \equiv N^2 - 1 - (A_e - 1)\nu Y^2 - \frac{\nu^{3/2}Y^2}{N\sqrt{2\beta_e}}[A_e + (1 - A_e)Y]\,\Re Z(\xi_1) = 0, \quad (4.1)$$

where N, ω and Y hereafter in this chapter are assumed to be real, the argument of the Z function is $\xi_1 = \Re\xi = (1 - Y)/N\tilde{w}_{\parallel}$, $\tilde{w}_{\parallel} = w_{\parallel}/c$ (cf. similar assumptions in Section 3.2),

$$\Re Z(\xi_1) = -2\int_0^{\xi_1} \exp(-\xi_1^2 + t^2)\mathrm{d}t \qquad (4.2)$$

(cf. the definition of the Z function by equation (1.21)), and $A_e = (j + 1)w_{\perp}^2/w_{\parallel}^2$ (when deriving (4.1) we have generalized equation (3.20) for arbitrary integer j).

As follows from the analysis of Chapter 3, the non-relativistic approximation and, in particular, equation (4.1) is valid in a relatively dense plasma when $\nu \gg 1$ and $N^2 \gg 1$, in general. Hence, the second term '1' in equation (4.1) will either be neglected altogether or taken into account when calculating the perturbation of N^2 due to non-zero ν^{-1} (cf. equation (3.34)).

Although equation (4.1) is much simpler than the corresponding weakly relativistic dispersion equation (cf. equation (3.10)), it still has no analytical solution in general. This solution is possible only in some limiting cases which are considered below following Sazhin (1986a, 1989d, 1990b).

(a) *Low-temperature limit (first-order approximation)*

In the case when $\tilde{w}_{\parallel}$ is low enough we can assume $|\xi_1| \gg 1$ and use the asymptotic expansion of $\Re Z(\xi_1)$ in the form (1.22). As a result, using only the first two terms in this expansion:

$$\Re Z(\xi_1) = Z_1 \equiv -\frac{1}{\xi_1} - \frac{1}{2\xi_1^3}, \tag{4.3}$$

and assuming $\nu \gg 1$ we obtain:

$$N^2 = \frac{\nu Y^2}{(Y-1)\left[1 - \frac{\beta_e Y^2}{(Y-1)^2}\left(\frac{1}{Y-1} + 1 - A_e\right)\right]} \equiv \nu \tilde{N}_1^2. \tag{4.4}$$

Remembering that our assumption $|\xi_1| \gg 1$ is satisfied when $\beta_e \ll 1$ and Y is not close to 1, solution (4.4) can be written in a more conventional form:

$$N^2 = \frac{\nu Y^2}{Y-1}\left[1 + \frac{\beta_e Y^2}{(Y-1)^2}\left(\frac{1}{Y-1} + 1 - A_e\right)\right] \equiv \nu \tilde{N}_t^2. \tag{4.5}$$

For finite ν and taking into account the contribution of ions, equation (4.5) can be generalized to equation (3.34) with $\tilde{a}_{\beta\parallel} = \tilde{a}_{\parallel t}$. This equation has already been considered in Section 3.2 where it was pointed out that the term $\tilde{a}_{\parallel R}$ (relativistic correction) can in fact be neglected when compared with $\tilde{a}_{\parallel t}$ when $\nu \gg 1$.

(b) *Metastable propagation*

As follows from (3.20) taken for $A_{e0} = A_e$, both parts of this equation are real for real ω when

$$A_e = A_{ph} \equiv \frac{Y}{Y-1}, \tag{4.6}$$

although the waves can grow or be damped if (4.6) is not strictly valid. Hence, the condition (4.6) is known as the condition of metastable propagation.

Having substituted (4.6) into (4.1) we obtain equation (2.8) for whistler-mode propagation in a cold plasma. Hence, in the case of metastable propagation the non-zero electron temperature does not influence either wave amplitude or wave dispersion.

(c) *Propagation at the frequency corresponding to $\xi_1 = -0.924$*

As was pointed out in Section 1.7, at $\xi_1 = -0.924$ we have

$$\Re Z(\xi_1) = Z_0 \equiv -\frac{1}{\xi_1}. \tag{4.7}$$

Having substituted (4.7) into (4.1) we again obtain equation (2.8) for whistler-mode propagation in a cold plasma. In the dense plasma limit this equation can be written as:

$$N^2 = \frac{\nu Y^2}{Y - 1} \equiv \nu \tilde{N}_0^2. \tag{4.8}$$

Equation (4.8) could be anticipated since equation (4.7) corresponds to the zero-order approximation of the low-temperature limit (cf. equation (4.3)).

In view of (4.8) we can see that the condition $\xi_1 = -0.924 \equiv \xi_{10}$ is satisfied when:

$$Y^{-1} = Y_x^{-1} \equiv 1 - (\beta_e \xi_{10}^2)^{1/3} \left[\left(1 - \sqrt{8\beta_e \xi_{10}^2/27 + 1}\right)^{1/3} \right.$$

$$\left. + \left(1 + \sqrt{8\beta_e \xi_{10}^2/27 + 1}\right)^{1/3} \right]. \tag{4.9}$$

If we consider the parameter β_e to be small enough ($\beta_e \xi_{10}^2 \ll 27/8$) then (4.9) can be simplified to

$$Y^{-1} = Y_{xs}^{-1} \equiv 1 - (2\beta_e \xi_{10}^2)^{1/3}. \tag{4.10}$$

The plots of Y_x^{-1} versus β_e and Y_{xs}^{-1} versus β_e are shown in Fig. 4.1. As follows from this figure, both equations (4.9) and (4.10) predict a decrease of Y_x^{-1} and Y_{xs}^{-1} respectively with increasing β_e. The values of Y_x^{-1} are always above those of Y_{xs}^{-1}. The divergence between the curves increases with increasing β_e. For $\beta_e = 0.005$ the approximation Y_{xs}^{-1} introduces an error of the order of 2%; for $\beta_e = 0.05$ this error increases up to 10%; for $\beta_e = 0.1$ it reaches 18%.

(d) Low-temperature limit (second-order approximation)

Keeping three terms in the expansion (1.22) of the Z function we can write:

$$\Re Z(\xi_1) = Z_2 \equiv -\frac{1}{\xi_1} - \frac{1}{2\xi_1^3} - \frac{3}{4\xi_1^5}. \tag{4.11}$$

Having substituted (4.11) into (4.1) and again assuming $\nu \gg 1$ we can write the solution of the latter equation in the form:

$$N^2 = \frac{\nu \left[1 - \frac{Y^2[A_e + (1 - A_e)Y]\beta_e}{(Y-1)^3} - \sqrt{x}\right](Y - 1)^5}{6Y^2[A_e + (1 - A_e)Y]\beta_e^2} \equiv \nu \tilde{N}_2^2, \tag{4.12}$$

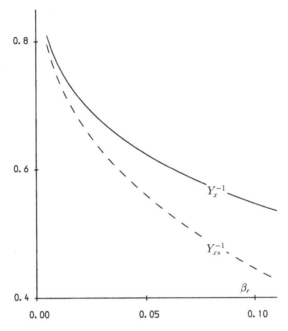

Fig. 4.1 Plots of Y_x^{-1} versus β_e (solid) and Y_{xs}^{-1} versus β_e (dashed) (see equations (4.9) and (4.10)).

where

$$\bar{\chi} = 1 - \frac{2Y^2[A_e + (1 - A_e)Y]\beta_e}{(Y - 1)^3}$$
$$+ \frac{Y^4[A_e + (1 - A_e)Y][-12 + A_e + (1 - A_e)Y]\beta_e^2}{(Y - 1)^6}.$$

Were we to expand the radical in (4.12) and keep only the terms proportional to β_e^2 then we would obtain the cold plasma solution (4.8). Note that when deriving (4.12) we kept only the solution referring to whistler-mode waves (sign '−' before the radical in (4.12)).

(e) *Approximation of* $\Re Z(\xi_1)$ *by* $Z_A = 2\xi_1/(1 - 2\xi_1^2)$

The approximation

$$Z_A \equiv \frac{2\xi_1}{1 - 2\xi_1^2} \tag{4.13}$$

gives the exact expression for $\Re Z(\xi_1)$ for $\xi_1 = -1.502$, which corresponds to the case when $\mathrm{d}^2 \Re Z/\mathrm{d}\xi_1^2 = 0$. For $|\xi_1| \gg 1$ the approximation (4.13) reduces to (4.3).

Having substituted (4.13) into (4.1) we can write the solution of the latter equation in the limit $\nu \gg 1$ in the form:

$$N^2 =$$
$$\nu \frac{(Y-1)^2 - \beta_e Y^2(1-A_e) - \sqrt{[(Y-1)^2 - \beta_e Y^2(1-A_e)]^2 - 4(Y-1)\beta_e Y^2}}{2\beta_e}$$
$$\equiv \nu \tilde{N}_A^2. \tag{4.14}$$

Solution (4.14) reduces to (4.5) in the low temperature limit $|\xi_1| \gg 1$ and gives the exact solution of (4.1) for $\xi_1 = -1.502$.

(f) Approximation of $\Re Z$ by $Z_B = -2\xi_1/(1+2\xi_1^2)$

The main merit of the approximation

$$Z_B = -\frac{2\xi_1}{1+2\xi_1^2} \tag{4.15}$$

is that it can reasonably well approximate to $\Re Z$ in the whole range of ξ_1 from zero to infinity. The divergence between the actual values of $\Re Z$ and Z_B is maximal near $\xi_1 = -1$ where $Z_B/\Re Z \approx 0.6$. For $|\xi_1| \gg 1$ it gives the correct first term in the expansion of $\Re Z$ (see equation (4.7)). For $|\xi_1| \ll 1$ it also gives the correct first term in the expansion of $\Re Z(\xi_1)$ for small $|\xi_1|$. Note that one should be cautious when applying this approximation for $|\xi_1| \ll 1$ as in this case our assumption $|\Im \xi_1| \ll |\Re \xi_1|$ (wave growth or damping does not influence wave propagation) may be no longer valid.

Again assuming $\nu \gg 1$ and substituting (4.15) into (4.1) we obtain a solution of the latter equation in the form:

$$N^2 = 2\nu Y^2(Y-1)\left\{(Y-1)^2 + (1-A_e)Y^2\beta_e + \left[(Y-1)^4 + 2(Y-1)\right.\right.$$
$$\left.\left.\times(Y+1-A_e Y+A_e)Y^2\beta_e + (1-A_e)^2 Y^4\beta_e^2\right]^{1/2}\right\}^{-1} \equiv \nu \tilde{N}_B^2. \tag{4.16}$$

Although the accuracy of the solution (4.16) can be worse than about 40% for $|\xi_1|$ close to 1, this solution can be of some use for interpreting wave phenomena in the magnetospheric plasma where the error of determination of the input parameters can be 100% or even more.

(g) Comparison of the results

All the solutions to equation (4.1) considered in this section except that referring to metastable propagation (subsection (b)) are based on different approximations of $\Re Z$. In this subsection we compare all these approximations as well as the corresponding solutions to equation (4.1).

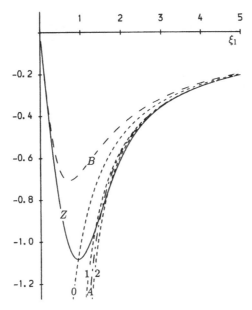

Fig. 4.2 The plot of $\Re Z(\xi_1)$ (see equation (4.2)) versus ξ_1 (solid curve Z), the plot of $Z_0 \equiv -1/\xi_1$ versus ξ_1 (dashed curve 0), the plot of $Z_1 \equiv -1/\xi_1 - 1/2\xi_1^3$ versus ξ_1 (dashed curve 1), the plot of $Z_2 \equiv -1/\xi_1 - 1/2\xi_1^3 - 3/4\xi_1^5$ versus ξ_1 (dashed curve 2), the plot of $Z_A \equiv 2\xi_1/(1-2\xi_1^2)$ versus ξ_1 (dashed–dotted curve A between curves 1 and 2), and the plot of $Z_B \equiv -2\xi_1/(1+2\xi_1^2)$ versus ξ_1 (dashed curve B). All these functions are odd with respect to ξ_1. The intersection points of the curve Z with the other curves are given in Table 4.1.

The curves Z_0 versus ξ_1, Z_1 versus ξ_1, Z_2 versus ξ_1, Z_A versus ξ_1 and Z_B versus ξ_1 as well as the curve $Z \equiv \Re Z \xi_1$ versus ξ_1 (not based on any approximation) are shown in Fig. 4.2. All the curves, except those of $\Re Z$, Z_B and Z_0 are shown for $\xi_1 > 1$. For smaller ξ_1 these approximations are not relevant for $\Re Z$. Note that $\Re Z$ as well as all approximations Z_i, where $i = 0, 1, 2, A, B$, are odd functions of ξ_1. $\xi_1 < 0$ for whistler-mode waves.

As follows from Fig. 4.2, when $|\xi_1|$ is close to 5 all the approximations to $\Re Z$ appear to be very close to $\Re Z$. In fact, the approximations Z_A, Z_1 and Z_2 all coincide with $\Re Z$ within the accuracy of plotting at $|\xi_1| \gtrsim 3$. At smaller $|\xi_1|$ these approximations deviate slightly from $\Re Z$. However, at certain $|\xi_1| < 2$ these approximations give exact values of $\Re Z$, as can be seen better in Fig. 4.3 which differs from Fig. 4.2 by the scale of the axes. The values of ξ_1 at which this happens for different approximations and the corresponding values of $\Re Z$ for these ξ_1 are shown in Table 4.1. As follows from Figs. 4.2–4.3 and Table 4.1, the second-order approximation

Table 4.1 *The values of ξ_{1i} and Z_i ($i = 0, 1, A, 2$), where $Z_0 \equiv -1/\xi_1 = \Re Z(\xi_1)$ (1st column), $Z_1 \equiv -1/\xi_1 - 1/2\xi_1^3 = \Re Z(\xi_1)$ (2nd column), $Z_A \equiv 2\xi_1/(1-2\xi_1^2) = \Re Z(\xi_1)$ (3rd column), $Z_2 \equiv -1/\xi_1 - 1/2\xi_1^3 - 3/4\xi_1^5 = \Re Z(\xi_1)$ (4th column). Note that $\xi_{1i} < 0$ and $Z_i > 0$ for whistler-mode waves.*

i	0	1	A	2
ξ_{1i}	± 0.924	± 1.358	± 1.502	± 1.685
Z_i	∓ 1.082	∓ 0.936	∓ 0.855	∓ 0.753

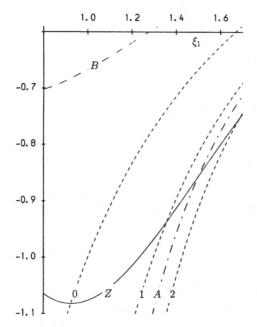

Fig. 4.3. The same as Fig. 4.2 but for different scales of the axes.

to $\Re Z$, given by (4.11), is the best one for $|\xi_1| \gtrsim 1.685$ and is exact for $|\xi_1| = \xi_{12} \equiv 1.685$. The approximation Z_A is best for $1.502 \lesssim |\xi_1| \lesssim 1.685$ and is exact for $|\xi_1| = \xi_{1A} \equiv 1.502$. The first-order approximation to $\Re Z$ given by (4.3) is suitable for $|\xi_1|$ close to $\xi_{11} \equiv 1.358$, while Z_0 is relevant for $|\xi_1|$ close to $\xi_{10} \equiv 0.924$. Z_B differs from $\Re Z$ more than other approximations at $|\xi_1| > 1$, but is the closest to $\Re Z$ at $|\xi_1| < 1$ where other approximations are not valid at all.

As follows from expressions (4.4), (4.5), (4.8), (4.12), (4.14) and (4.16), all these solutions to equation (4.1) are proportional to ν. Hence, when comparing these solutions it is sufficient to compare the corresponding expressions

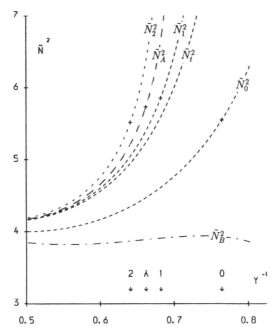

Fig. 4.4 The plot of $\tilde{N}_0^2$ (see equation (4.8)) versus Y^{-1}, the plot of $\tilde{N}_1^2$ (see equation (4.4)) versus Y^{-1}, the plot of $\tilde{N}_t^2$ (see equation (4.5)) versus Y^{-1}, the plot of $\tilde{N}_2^2$ (see equation (4.12)) versus Y^{-1}, the plot of $\tilde{N}_A^2$ (see equation (4.14)) versus Y^{-1} and the plot of $\tilde{N}_B^2$ (see equation (4.16)) versus Y^{-1} for $\beta_e = 0.01$ and $A_e = 1$. Arrows indicate the values of Y^{-1} at which $\xi_{12} = (Y-1)/(\tilde{N}_2\sqrt{2\beta_e}) = 1.685$ (arrow 2), $\xi_{1A} = (Y-1)/(\tilde{N}_A\sqrt{2\beta}) = 1.502$ (arrow A), $\xi_{11} = (Y-1)/(\tilde{N}_1\sqrt{2\beta}) = 1.358$ (arrow 1) and $\xi_{10} = (Y-1)/(\tilde{N}_0\sqrt{2\beta}) = 0.924$ (arrow 0). The corresponding values of $\tilde{N}_2^2$, $\tilde{N}_A^2$, $\tilde{N}_1^2$ and $\tilde{N}_0^2$ are shown by crosses on the curves. These values of $\tilde{N}^2$ coincide with the exact solution to equation (4.1).

for $\tilde{N}_0^2$, $\tilde{N}_1^2$, $\tilde{N}_t^2$, $\tilde{N}_2^2$, $\tilde{N}_A^2$ and $\tilde{N}_B^2$. Plots of $\tilde{N}_i^2$ versus Y^{-1} ($i = 0, 1, t, 2, A, B$) for $\beta_e = 0.01$ and $A_e = 1$ and $A_e = 2$ are shown in Figs. 4.4 and 4.5 respectively. These values of β_e and A_e are relevant for magnetospheric conditions, although β_e in the plasma sheet and magnetosheath can be at least an order of magnitude larger, while in the plasmasphere β_e can be an order of magnitude smaller.

We have indicated by arrows those values of $Y^{-1} = Y_i^{-1}$ ($i = 2, A, 1, 0$) for which the corresponding parameters ξ_1 are equal to ξ_{12}, ξ_{1A}, ξ_{11} and ξ_{10} respectively. The corresponding values of $\tilde{N}^2$ are shown by crosses on the curves. These values coincide with exact numerical solutions to (4.1). For $Y^{-1} \lesssim 0.5$ all the curves in Fig. 4.5 coincide within the accuracy of plotting with the exact numerical solution to (4.1). For Y^{-1} close to 1, $\Im Z$ can no

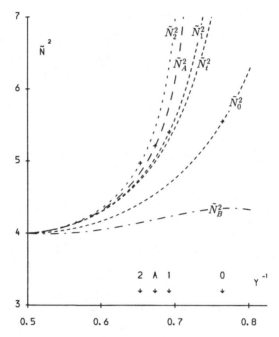

Fig. 4.5. The same as Fig. 4.4 but for $A_e = 2$.

longer be neglected and different methods for the solution of the original equation (3.20) must be used (Sazhin, 1983a).

Also, it follows from Figs. 4.4 and 4.5 that the solution (4.14) gives better results for $Y^{-1} \lesssim Y_A^{-1}$ than any of the solutions (4.4), (4.5), (4.8), (4.12) and (4.16). The solution (4.12) gives even better results for $Y^{-1} \lesssim Y_2^{-1}$. However, the form of the latter solution seems to be too complicated for practical applications.

Although the solution $\tilde{N}_B^2$ deviates most of all from the exact solution to (4.1) at $Y^{-1} \lesssim 0.8$, it appears to be the only possible approximation in the wide range of Y^{-1} from zero to that Y^{-1} at which the contribution of wave growth and amplification to whistler-mode propagation can no longer be neglected. Comparing Figs. 4.4 and 4.5 we can see that a decrease of A_e results in greater separation between the curves $\tilde{N}_1^2$, $\tilde{N}_t^2$, $\tilde{N}_2^2$, $\tilde{N}_A^2$ and $\tilde{N}_B^2$ versus Y^{-1} from that of $\tilde{N}_0^2$ versus Y^{-1}.

For lower values of β_e, say $\beta_e = 0.001$, all the curves except $\tilde{N}_0^2$ versus Y^{-1} shift towards larger Y^{-1} so that they approach closer to the curve $\tilde{N}_0^2$ versus Y^{-1}. The values of Y_i^{-1} and the corresponding values of $\tilde{N}_i^2$, where $i = 0, 1, 2, A$ for $\beta_e = 0.01, 0.001$ and $A_e = 1, 2$ are shown in Table 4.2.

Were we to extrapolate all these solutions to the vicinity of Y^{-1} corre-

Table 4.2 *The values of Y_i^{-1} and $\tilde{N}_i^2$ ($i = 0, 1, A, 2$), corresponding to the intersection points (ξ_{1i}, Z_i) given in Table 4.1, for different β_e and A_e.*

		Z_0		Z_1		Z_A		Z_2	
		Y_0^{-1}	$\tilde{N}_0^2$	Y_1^{-1}	$\tilde{N}_1^2$	Y_A^{-1}	$\tilde{N}_A^2$	Y_2^{-1}	$\tilde{N}_2^2$
$\beta_e = 0.01$;	$A_e = 1$	0.764	5.55	0.682	5.85	0.662	5.73	0.641	5.51
$\beta_e = 0.01$;	$A_e = 2$	0.764	5.55	0.692	5.40	0.673	5.21	0.653	4.97
$\beta_e = 0.001$;	$A_e = 1$	0.885	9.83	0.842	9.55	0.831	9.14	0.820	8.55
$\beta_e = 0.001$;	$A_e = 2$	0.885	9.83	0.844	9.27	0.834	8.85	0.822	8.28

sponding to $\xi_1 = \xi_{12}, \xi_{1A}, \xi_{11}$ and ξ_{10}, we would have a smooth approximate solution to (4.1) in the whole range of Y^{-1} from 0 up to $Y^{-1} = Y_0^{-1}$, corresponding to $|\xi_1| = \xi_{10}$. The formal theory of this extrapolation is discussed e.g. by Sazhin (1988c). Meanwhile we consider another approach to the solution of (4.1) based on a graphical technique.

4.2 A graphical solution

In this section we suggest an alternative approach to the solution of (4.1) based on simple graphical constructions. This graphical method is based on an alternative presentation of equation (4.1) in the form (Sazhin, 1989d):

$$a\tau + b = Q(\tau), \qquad (4.17)$$

where

$$\tau = \xi_1^{-2} = N^2 \tilde{w}_\parallel^2 / (Y - 1)^2, \qquad (4.18)$$

$$a = \frac{(Y - 1)^3}{2\beta_e Y^2 [A_e + (1 - A_e)Y]}, \qquad (4.19)$$

$$b = \frac{-(Y - 1) + (1 - A_e)(Y - 1)\nu Y^2}{\nu Y^2 [A_e + (1 - A_e)Y]}, \qquad (4.20)$$

$$Q(\tau) = \tau^{-1/2} \Re Z(-\tau^{-1/2}). \qquad (4.21)$$

When deriving (4.17) we assumed that $A_e + (1 - A_e)Y \neq 0$, i.e. the whistler-mode propagation is not metastable. If this is not true then the solution to (4.1) is straightforward (see subsection 4.1(b)).

In order to solve equation (4.17) we should first plot the curve $Q(\tau)$ versus $\tau > 0$ (see Fig. 4.6). Then we calculate the coefficients a and b according to

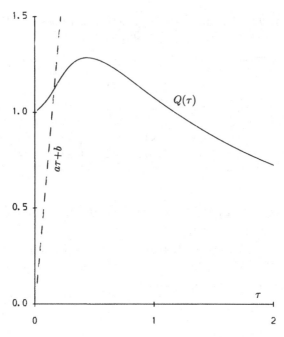

Fig. 4.6 Plots of $Q(\tau) = \tau^{-1/2}\Re Z(-\tau^{-1/2})$ versus τ (solid) and $a\tau+b = 6.94\tau-0.03$ versus τ (dashed); these plots intersect when $\tau \approx 0.16$ (see equation (4.17)).

equations (4.19) and (4.20) for given values of the parameters and plot the line $a\tau+b$. The abscissa of the intersection point (or points) of this line with the curve $Q(\tau)$ gives the solution (or solutions) $\tau = \tau_0$ to equation (4.1). (If there are no such points then equation (4.1) has no solutions.) The value N^2 is calculated from equation (4.18) as:

$$N^2 = \tau_0(Y-1)^2/\tilde{w}_\parallel^2. \qquad (4.22)$$

In order to illustrate our method let us assume, following Sazhin (1989d), $\nu = 9$, $\beta_e = 0.018$, $Y = 2$ and $A_e = 1$. From equations (4.19) and (4.20) we obtain $a \approx 6.94$ and $b \approx -0.03$. The line $a\tau + b = 6.94\tau - 0.03$ is plotted in Fig. 4.6. It follows from this figure that the line intersects the curve $Q(\tau)$ when $\tau = \tau_0 \approx 0.16$. Substituting this value of τ_0 into (4.22) we obtain $N^2 \approx 40$. This value coincides with that which follows from equation (3.29), relativistic effects (the term $\tilde{a}_{\parallel R}$) being neglected.

For a plasma consisting of a mixture of hot and cold electrons with plasma frequencies Π_h and Π_c, respectively, b in equation (4.17) is to be redefined as:

$$b = \hat{b} \equiv \frac{-(Y-1) + (1 - A_e)(Y-1)\nu_h Y^2 - \nu_c Y^2}{\nu_h Y^2 [A_e + (1 - A_e)Y]}, \qquad (4.23)$$

where $\nu_h = \Pi_h^2/\Omega_0^2$, $\nu_c = \Pi_c^2/\Omega_0^2$.

Once we have obtained the solution to (4.1) it is essential to calculate the increment of instability or decrement of damping (γ), so that we can be sure that the equation (4.1) itself is valid. This will be done in Chapter 7.

Another graphical technique for the solution of (3.20) for Y close to 1 was suggested by Sazhin (1983a). The main merit of his approach is that it has not imposed any specific restrictions on the value of $|\gamma|$. The main conclusion of this paper is that the whistler-mode refractive index at frequencies close to the electron gyrofrequency is close to 1. But in this case the whole non-relativistic approach to the problem becomes doubtful (cf. Winglee, 1983; Robinson, 1987b).

4.3 Group velocity

As was already mentioned in Section 2.2, the whistler-mode group velocity defined by (2.25) is an important parameter describing the propagation of wave energy. In this section we generalize the results of Section 2.2 to the case when whistler-mode waves propagate in a hot anisotropic (but non-relativistic) plasma, but restrict ourselves to considering waves propagating strictly parallel to a magnetic field. Similarly to Sections 4.1 and 4.2 we assume that wave growth or damping does not influence wave propagation which is determined by the dispersion equation in the form (4.1). In view of this equation we can redefine whistler-mode group velocity as:

$$v_g = -(\partial \Re D/\partial k)/(\partial \Re D/\partial \omega_0). \qquad (4.24)$$

Having substituted $\Re D$ defined by (4.1) into (4.24) we obtain the expression for whistler-mode group velocity in the form:

$$v_g = \frac{c \left\{ 2N^2 + X A_e \xi_1 \tilde{\varepsilon} \left[-2\xi_1 + \Re Z(\xi_1)(1 - 2\xi_1^2) \right] \right\}}{N \left\{ 2 - X A_e \xi_1 (Y-1)^{-1} \left[-2\xi_1 \tilde{\varepsilon} + \Re Z(\xi_1)(1 - 2\xi_1^2 \tilde{\varepsilon}) \right] \right\}}, \qquad (4.25)$$

where:

$$\tilde{\varepsilon} = 1 + \frac{Y}{A_e(1-Y)}.$$

Equation (4.25) gives us a general expression for whistler-mode group velocity provided we know the explicit expression for N. In particular, using the solutions for N obtained in Section 4.1 we can automatically obtain the corresponding expressions for v_g. Some of these expressions will be obtained and discussed below.

(a) Low-temperature limit

As in Section 4.1 we can assume that in the low temperature limit $|\xi_1| \gg 1$ and use the asymptotic expansion of $\Re Z(\xi_1)$ in the form (1.22). However, in contrast to Section 4.1 we should use three terms in this expansion, i.e. equation (4.11), when calculating the first order thermal corrections to v_{g0}, as equation (4.25) contains the terms $\Re Z$ and $\xi_1^2 \Re Z$ simultaneously. Restricting ourselves to the dense plasma limit, where the non-relativistic approximation is justified, we obtain after having substituted (4.11) into (4.25) and keeping only the terms proportional to β_e:

$$v_g = v_{g0}(1 + \tilde{b}_\beta \beta_e), \tag{4.26}$$

where v_{g0} was defined by (2.26),

$$\tilde{b}_\beta = -\frac{3Y^2 + 0.5Y^3 + A_e(2Y - 1.5Y^2 - 0.5Y^3)}{(Y-1)^3}. \tag{4.27}$$

The plots of $\tilde{b}_\beta$ versus Y^{-1} for different A_e are shown in Fig. 4.7. As follows from this figure, for $A_e = 1$ we have $\tilde{b}_\beta < 0$ and $|\tilde{b}_\beta|$ increases with increasing Y^{-1}. The increase of A_e tends to compensate for the influence of finite electron temperature, so that for some A_e this influence disappears altogether ($\tilde{b}_\beta = 0$). The values of A_e for which this happens,

$$A_e = A_g \equiv \frac{6Y^2 + Y^3}{Y^3 + 3Y^2 - 4Y}, \tag{4.28}$$

versus Y^{-1} are shown in Fig. 4.8. Also we have shown in the same figure for comparison the plots of $A_e = A_{ph}$ (see equation (4.6)) versus Y^{-1} when the electron finite temperature does not influence whistler-mode refractive index or phase velocity. As follows from this figure, both values A_g and A_{ph} increase with increasing Y^{-1}, A_{ph} being always below A_g. Keeping in mind that for magnetospheric conditions A_e is almost always below 2 (see e.g. Burton, 1976; Tsurutani *et al.*, 1982), we can expect that the complete compensation for the influences of anisotropy and temperature on the value of v_g can occur only for $Y^{-1} \lesssim 0.35$. A similar compensation for the value of N or v_{ph} occurs for $Y^{-1} \lesssim 0.5$. This means that when analysing temperature effects on N or v_{ph} for a given frequency, one cannot make any definite conclusion about the corresponding effects on the value of group velocity and vice versa.

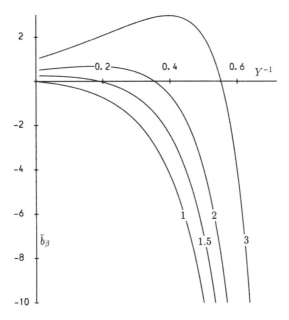

Fig. 4.7 Plots of $\tilde{b}_\beta$ (see equation (4.27)) versus Y^{-1} for different A_e (curves indicated).

Equation (4.26) can be generalized in a straightforward way so that the contribution of finite electron density and ions are taken into account as well:

$$v_g = v_{g0}(1 + \tilde{b}_c \nu^{-1} + \tilde{b}_r r + \tilde{b}_\beta \beta_e), \qquad (4.29)$$

where $\tilde{b}_c$ and $\tilde{b}_r$ are defined by (2.32) and (2.33) respectively. As follows from these definitions of $\tilde{b}_c$ and $\tilde{b}_r$, finite electron density effects at $Y^{-1} < 0.75$ and ion effects tend to reduce v_g in a similar way as the effects of finite electron temperature. However, at $Y^{-1} > 0.75$ finite density effects tend to increase v_g acting in a similar way to the effects of finite anisotropy of the electron distribution function.

(b) Metastable propagation

As was mentioned in subsection 4.1(*b*), a non-zero electron temperature does not change the whistler-mode dispersion equation during metastable propagation when (4.6) is valid. However, this influence exists for the whistler-

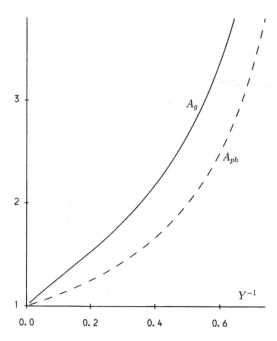

Fig. 4.8 Plots of A_g (solid) (see equation (4.28)) versus Y^{-1} and A_{ph} (dashed) (see equation (4.6)) versus Y^{-1}.

mode group velocity v_g. The value of v_g corresponding to this propagation can be obtained from (4.25) if we set in the latter equation $\tilde{\varepsilon} = 0$:

$$v_g = \frac{2N_{0\parallel}c}{2 + A_e \nu Y^2 \Re Z(\xi_1)/(N_{0\parallel}\tilde{w}_\parallel)}, \tag{4.30}$$

where $N_{0\parallel}$ is determined by the cold plasma dispersion equation (2.8).

Equation (4.30) can be further simplified if we restrict our analysis to considering dense plasma ($\nu \gg 1$):

$$v_g = \frac{2(A_e - 1)w_\parallel}{A_e \Re Z(\xi_1)}, \tag{4.31}$$

where:

$$\xi_1 = -\left(\sqrt{2\beta_e} A_e \sqrt{A_e - 1}\right)^{-1}. \tag{4.32}$$

When $\xi_1 \gg 1$ equation (4.31) reduces to (2.26).

The plots of $v_g/w_\parallel$ versus β_e based on equation (4.31) are shown in Fig. 4.9 for different values of A_e. As follows from this figure, this ratio tends to decrease with increasing β_e for $A_e = 1.5$ and $A_e = 2$. However, the curve

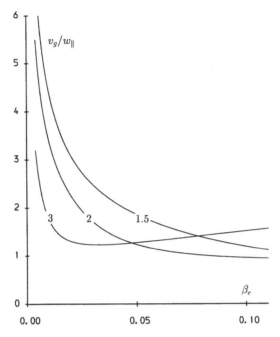

Fig. 4.9 Plots of $v_g/w_\|$, determined by equation (4.30), versus β_e for different values of A_e (curves indicated).

$A_e = 3$ has a hook-like form. The minimum of $v_g/w_\|$ in the latter case corresponds to $\xi_1 = -0.924$. The corresponding minima for other curves occur for the values of β_e which are outside the scale of the figure. The decrease of $v_g/w_\|$ for small β_e agrees with equation (4.26). However, the subsequent increase of this ratio could never be predicted within the perturbation theory considered in the previous subsection. β_e being fixed, the value of v_g increases when $w_\|$ increases. It is important to note that (4.30) remains valid up to frequencies very close to the electron gyrofrequency as $|\gamma| = 0$ for the metastable propagation.

(c) *Propagation at the frequency corresponding to* $\xi_1 = -0.924$

Having substituted $\xi_1 = -0.924 \equiv \xi_{10}$ and $\Re Z(\xi_1) = -1/\xi_1 \equiv 1/\xi_{10}$ into (4.25) and assuming that N is determined by (2.11) (dense plasma approximation) we obtain v_g at the frequency corresponding to $\xi_1 = -\xi_{10}$ in the form:

$$v_g = \frac{\xi_{10}w_\| \nu \left[(Y_x - 1)^2/(\xi_{10}^2 \beta_e) + Y_x^3/(Y_x - 1) - A_e Y_x^2 \right]}{2(Y_x - 1) + \nu A_e Y_x^2}, \qquad (4.33)$$

where Y_x is determined by (4.9).

Remembering our assumption $\nu \gg 1$, equation (4.33) can be simplified to:

$$v_g = \frac{\xi_{10}w_\parallel}{A_e}\left[\frac{(1-Y_x^{-1})^2}{\xi_{10}^2\beta_e} + \frac{1}{1-Y_x^{-1}} - A_e\right]. \tag{4.34}$$

Assuming that β_e is so small that we can approximate Y_x by Y_{xs} (see equation (4.10)) then (4.34) can be further simplified to:

$$v_g = \xi_{10}w_\parallel\left[\frac{3}{A_e(2\beta_e\xi_{10}^2)^{1/3}} - 1\right]. \tag{4.35}$$

The plots of $v_g/w_\parallel$ versus β_e based on equations (4.33) and (4.35) are shown in Fig. 4.10 for different values of A_e. The solid and dashed–dotted curves refer to equation (4.33): the first corresponds to $\nu = 16$, the second to $\nu = 4$. These curves become closer when A_e increases and practically coincide for $A_e = 3$. The dashed curves refer to equation (4.35). As follows from this figure, both equations predict a decrease in $v_g/w_\parallel$ when β_e or A_e increases. The value of this ratio predicted by (4.35) appears to be about 25% larger than that predicted by (4.33). However, in many cases this inaccuracy may not be very important for a qualitative estimate of the value of $v_g/w_\parallel$.

A decrease in $v_g/w_\parallel$ with increasing A_e, which follows from Fig. 4.10, agrees with the results of numerical computations presented by Sazhin, Ponyavin & Varshavski (1979). When $A_e = 3(2\beta_e\xi_{10})^{-1/3}$ the values of v_g appear to be close to zero. When $A_e > 3(2\beta_e\xi_{10})^{-1/3}$ equation (4.35) predicts anomalous dispersion of the waves $(v_g < 0)$.

When applying (4.33)–(4.35) to the interpretation of particular wave phenomena in magnetospheric or laboratory plasmas one should check whether the corresponding value of $|\gamma|$ is well below $\min(\omega_0, \Omega - \omega_0)$. Different approximations for calculations of γ will be considered in Chapter 7.

The analysis of the whistler-mode group velocity similar to that given in subsections (a)–(c) could be done for other approximate solutions discussed in Section 4.1 (subsections (d), (e) and (f)) as well. However, the corresponding expressions for v_g in these cases appear to be too complicated for practical applications and we will not discuss them.

Problems

Problem 4.1   As mentioned in Section 4.2, equation (4.17) was derived on the assumption that $\tilde{D} \equiv A_e + (1 - A_e)Y \neq 0$ (whistler-mode propagation is not metastable). Prove, however, that in the limit $\tilde{D} \to 0$ its solution is that of a cold plasma, i.e. it reduces to equation (2.8).

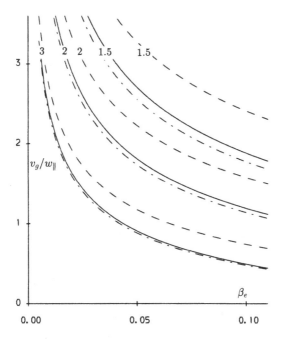

Fig. 4.10 Plots of $v_g/w_{\parallel}$, determined by equation (4.33), versus β_e with $\nu \equiv \Pi^2/\Omega^2 = 16$ (solid), and $\nu = 4$ (dashed–dotted); and plot of $v_g/w_{\parallel}$, determined by equation (4.35), (dashed) for $A_e = 1.5$, 2, 3 (curves marked).

Problem 4.2 In the cold plasma limit, $|\xi_1|$ tends to infinity and $Z(\xi_1)$ can be approximated by $-1/\xi_1$ (see equation (4.3)). However, if we substitute this approximation for $Z(\xi_1)$ into expression (4.25) for v_g, we obtain expression (4.34), which does not reduce to that which follows from the theory of wave propagation in a cold plasma (see e.g. expression (2.26)). Explain the paradox.

5

Quasi-longitudinal approximation

5.1 Almost parallel propagation (low-temperature limit)

In this chapter we generalize the results of Chapters 3 and 4 to the case of quasi-longitudinal whistler-mode propagation. As in Section 2.1 we consider whistler-mode propagation as quasi-longitudinal if either $|\theta| \ll 1$ (see inequalities (2.14) and (2.17)) in the plasma with arbitrary electron density (inequalities (2.10) and (2.12) are not necessarily valid), or inequalities (2.9) and (2.10) (or (2.12)) are valid simultaneously provided the whistler-mode wave normal angle θ is not close to the resonance cone angle θ_{R0} in a cold plasma (see equation (2.13)). First we consider the case $|\theta| \ll 1$ when whistler-mode waves propagate almost parallel to the magnetic field.

As was shown in Chapter 1, when we impose no restrictions on the electron density we should use the general relativistic expressions for the elements of the plasma dielectric tensor in the form (1.73). Assuming that the waves propagate through plasma with the electron distribution function (1.76) and imposing conditions (1.77) we write these expressions in a much simpler form (1.78). Also, we assume that the electron temperature is so low that it can only slightly perturb the corresponding whistler-mode dispersion equation in a cold plasma, i.e.

$$|N^2 - N_0^2| \ll N_0^2, \tag{5.1}$$

where N is the whistler-mode refractive index in a hot plasma, and N_0 is the whistler-mode refractive index in a cold plasma defined by (2.15).

Inequality (5.1) is always violated when θ is close to θ_{R0} even for a very low-temperature plasma. Whistler-mode propagation at θ in the immediate vicinity of θ_{R0} (quasi-electrostatic propagation) will be considered in Chapter 6.

Having substituted (1.78) into the general dispersion equation (1.42) and

94

remembering our conditions (2.14) and (5.1) we obtain (after very lengthy but straightforward algebra) the following expression for N^2 (Sazhin, 1987a):

$$N^2 = N^2_{0\|}[1 + \tilde{a}_{\beta\|}\beta_e + (a_0 + \tilde{a}_\theta\beta_e)\theta^2], \qquad (5.2)$$

where $N_{0\|} \equiv N^2_0(\theta = 0)$ and is defined by (2.8), $\tilde{a}_{\beta\|}$ is the same as in (3.30), a_0 is the same as in (2.16),

$$\tilde{a}_\theta = \tilde{a}_{\theta t} + \tilde{a}_{\theta R}, \qquad (5.3)$$

$$\tilde{a}_{\theta t} = \frac{Y^2 \sum_{i=0}^5 \tilde{\xi}_i Y^i}{2P^2(2Y-1)(Y-1)^4}, \qquad (5.4)$$

$$\tilde{a}_{\theta R} = \frac{Y^2 \sum_{i=0}^3 \tilde{\eta}_i Y^i}{4P^2(Y-P)(Y-1)^2}, \qquad (5.5)$$

$$\tilde{\xi}_0 = P^2(1 - A_e),$$
$$\tilde{\xi}_1 = -4P - 9P^2 + A_e(5P + 5P^2),$$
$$\tilde{\xi}_2 = 3 + 21P + 17P^2 + A_e(-17P - 4P^2),$$
$$\tilde{\xi}_3 = -12 - 32P - 6P^2 + 16A_eP,$$
$$\tilde{\xi}_4 = 15 + 12P - 4A_eP,$$
$$\tilde{\xi}_5 = -6,$$
$$\tilde{\eta}_0 = 2P^2(A_e - 1),$$
$$\tilde{\eta}_1 = 2P[-2(A_e - 1) + (2 + 3A_e)P],$$
$$\tilde{\eta}_2 = -(3 + 2A_e)(1 + 2P),$$
$$\tilde{\eta}_3 = 3 + 2A_e,$$

and $P = 1 - \nu Y^2$ is the same as in (1.79).

In the limit $\theta \to 0$ equation (5.2) reduces to (3.29), which has already been analysed in Section 3.2 (see Figs. 3.1 and 3.2). The additional term $\tilde{a}_\theta\beta_e\theta^2$ in (5.2) which appears for $\theta \neq 0$ can be considered as thermal ($\tilde{a}_{\theta t}$) and relativistic ($\tilde{a}_{\theta R}$) corrections to $a_0\theta^2$. The plots of $\tilde{a}_\theta$, $\tilde{a}_{\theta t}$ and $\tilde{a}_{\theta R}$ versus Y^{-1} for the same ν and A_e as in Figs. 3.1 and 3.2 are shown in Figs. 5.1 and 5.2. As follows from these figures, relativistic corrections to a_0 are most important for a relatively rarefied plasma corresponding to $\nu = 0.25$ and 1, while in a relatively dense plasma with $\nu = 4$ relativistic and non-relativistic curves are close to one another. The difference between relativistic and non-relativistic terms looks especially important for the curves corresponding to $\nu = 0.25$ and $A_e = 2$ at $Y^{-1} \approx 0.25$ (see Fig. 5.2), when relativistic analysis predicts $\tilde{a}_\theta$ close to zero, while non-relativistic analysis predicts $\tilde{a}_\theta \approx -15$ (cf. the difference between the corresponding curves $\tilde{a}_{\|t}$ and $\tilde{a}_{\beta\|}$ for $\nu = 0.25$ and $A_e = 2$ in Fig. 3.2). Comparing Figs. 5.1 and 5.2 we can see that the

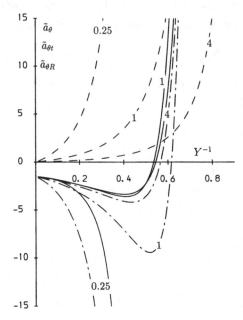

Fig. 5.1 Plots of $\tilde{a}_{\theta R}$ (see equation (5.5)) versus Y^{-1} (dashed), $\tilde{a}_{\theta t}$ (see equation (5.4)) versus Y^{-1} (dashed–dotted) and $\tilde{a}_{\theta}$ (see equation (5.3)) versus Y^{-1} (solid) for $\nu = 0.25$, 1 and 4 (curves indicated) and $A_e = 1$.

increase in A_e for a dense plasma with $\nu = 4$ leads to a decrease in $\tilde{a}_{\theta}$. This can be predicted analytically if we consider $\tilde{a}_{\theta}$ for $\nu \gg 1$:

$$\tilde{a}_{\theta} = \tilde{a}_{\theta d} \equiv \frac{Y^2 \left[1 - 9Y + 17Y^2 - 6Y^3 + A_e(-1 + 5Y - 4Y^2) \right]}{2(Y-1)^4(2Y-1)}. \qquad (5.6)$$

As follows from (5.6), the expression in brackets after A_e is always negative as $-1 + 5Y - 4Y^2 = -(2Y-1)^2 + Y$ and $(2Y-1)^2 > 2Y - 1 > Y$ for $Y > 1$ (whistler-mode waves). Hence, the decrease in A_e leads to the decrease in $\tilde{a}_{\theta}$.

Comparing expressions (5.4) and (5.5) we can see that in the limit $Y \gg 1$ (low-frequency waves) the contribution of $\tilde{a}_{\theta R}$ can be neglected when compared with $\tilde{a}_{\theta t}$. In this case the expression (5.3) can be simplified to:

$$\tilde{a}_{\theta} = \tilde{a}_{\theta l} \equiv -1.5. \qquad (5.7)$$

Note that the expression for $\tilde{a}_{\theta l}$ depends on neither Y nor A_e nor ν. Expression (5.7) is consistent with the forms of the curves $\tilde{a}_{\theta t}$ and $\tilde{a}_{\theta}$ shown in Figs. 5.1 and 5.2 in the limit $Y^{-1} \to 0$.

As follows from equations (5.4) and (5.5), both $\tilde{a}_{\theta t}$ and $\tilde{a}_{\theta R}$ become infinitely large as $P \to 0$, which corresponds to the case when wave frequency

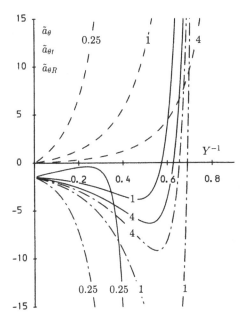

Fig. 5.2. The same as Fig. 5.1 but for $A_e = 2$.

approaches electron plasma frequency. In this case (5.1) and correspondingly (5.2) become invalid and we need a quite different approach to the problem, which will be considered later in this section. Meanwhile, the validity of (5.2) will be assumed.

As follows from equation (5.2), for small β_e and θ the term $\tilde{a}_\theta \beta_e \theta^2$ is usually very small and does not significantly influence the value of N^2. However, its contribution can be important when we consider some general characteristics of whistler-mode propagation at $|\theta| \ll 1$. In what follows we illustrate this with the example of focusing of whistler-mode energy along the magnetic field at $|\theta| \ll 1$, i.e. when the whistler-mode group velocity is directed along the magnetic field, while $\mathbf{N}$ forms an angle $|\theta| \ll 1$ with this field. The condition for this focusing is satisfied when:

$$N \cos \theta = \text{const},\qquad (5.8)$$

i.e. $N \cos \theta$ does not depend on θ provided $|\theta| \ll 1$.

Having substituted (5.2) into (5.8) and restricting ourselves to $|\theta| \ll 1$ we can rewrite condition (5.8) as

$$- \tilde{a}_{\beta\|}\beta_e - 1 + a_0 + \tilde{a}_\theta \beta_e = 0.\qquad (5.9)$$

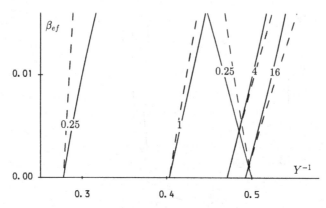

Fig. 5.3 Plots of β_{ef} (see equation (5.11)) versus Y^{-1} for $\nu = 0.25, 1, 4$ and 16 (curves indicated), $A_e = 1$ (solid) and $A_e = 2$ (dashed).

For a cold plasma ($\beta_e = 0$) condition (5.9) is satisfied when

$$a_0 = 1. \tag{5.10}$$

In the general case when β_e is not necessarily equal to zero (5.9) is satisfied when

$$\beta_e = \beta_{ef} \equiv \frac{1 - a_0}{\tilde{a}_\theta - \tilde{a}_{\beta\|}}. \tag{5.11}$$

Plots of β_{ef} versus Y^{-1} for the same values of ν and A_e as in Figs. 5.1 and 5.2 along with the plots for $\nu = 16$ and $A_e = 1, 2$ (representing dense plasma) are shown in Fig. 5.3. As follows from this figure, when $\nu = 1, 4$ and 16, an increase of β_{ef} from 0 to about 0.015 results in an increase of the corresponding values of Y^{-1} (when whistler-mode focusing takes place) by about 10%. This value is large enough to be noticed in experiments. For $\nu = 4$ and 16 the increase in A_e increases the effect of finite β_{ef} on the condition of focusing, while for $\nu = 1$ the increase in A_e leads to an opposite effect. The behaviour of the curves β_{ef} versus Y^{-1} for $\nu = 0.25$ appears to be more complicated. At frequencies close to $0.28\Omega_0$, where a_0 is close to 1, the forms of the curves are similar to those for $\nu = 1$. However, at frequencies close to $0.5\Omega_0$, where X is close to 1, an increase in β_e results in a decrease in Y^{-1}.

We believe that the focusing considered in this section could explain a maximum in the intensity of natural whistler-mode emissions of the chorus type at frequencies slightly above 0.5 electron gyrofrequency when the plasma density is relatively high ($\nu \gg 1$) (Tsurutani & Smith, 1974; Burtis & Helliwell, 1976).

As we have already mentioned, at frequencies close to electron plasma frequency when $P \to 0$, both $\tilde{a}_{\theta t}$ and $\tilde{a}_{\theta R}$ become infinitely large and equation (5.2) becomes invalid for any non-zero β_e and θ. Hence, a different approach is needed for the analysis of the waves in this frequency range. This will be considered below.

Replacing condition (5.1) by the condition

$$|P| \ll 1 \qquad (5.12)$$

and keeping our assumptions $|\theta| \ll 1$ and $\beta_e \ll 1$ we can simplify the general dispersion equation (1.42) to (Sazhin, 1988d):

$$-6\beta_e(Y^2 - 1)N^6$$

$$+ \left\{ 2\theta^2 + 2P(Y^2 - 1)Y^{-2} + \beta_e[(Y^2 - 1)(3 + 2A_e) + 12Y^2] \right\} N^4$$

$$+ \left\{ -2\theta^2 - 4P + \beta_e[-2Y^2(3 + 2A_e) - 6Y^2] \right\} N^2 + [2P + \beta_e Y^2(3 + 2A_e)] = 0. \qquad (5.13)$$

When $P = -\beta_e Y^2(3 + 2A_e)/2$ one of the solutions of equation (5.13) corresponds to a plasma cutoff for all θ ($\mathbf{N} = \mathbf{0}$). This result agrees with that of Sazhin (1987a) (see his Section 3). For $\theta = 0$ equation (5.13) further simplifies to:

$$\left[N^2 - \frac{2P + \beta_e Y^2(3 + 2A_e)}{6\beta_e Y^2} \right] \left[N^2 - \frac{Y}{Y - 1} \right] \left[N^2 - \frac{Y}{Y + 1} \right] = 0. \qquad (5.14)$$

Equation (5.14) has three solutions, of which the first is an analogue to Langmuir waves in a non-relativistic plasma (see Section 1.2), the second refers to whistler-mode waves in a cold plasma (see equation (2.8) taken for $X = 1$), and the third refers to left-hand polarized waves in a cold plasma (see equation (2.7) taken for $X = 1$ and a '+' sign in the denominator of its right-hand side). At first sight it seems paradoxical that equation (5.14) does not contain thermal and relativistic corrections to N^2 as well as the corrections due to finite P. However, these corrections would appear in the second-order terms in equations (5.13) and (5.14) which were neglected therein.

As to the dispersion equation for Langmuir waves, it has different forms in the vicinity of $P \to 0$ and $\beta_e \to 0$ depending on how we approach the point $(P, \beta_e) = (0, 0)$. If we assume that $|P|$ is small but finite then in the limit $\beta_e \to 0$ this equation reduces to the non-relativistic dispersion equation already derived in Section 1.2 (see equation (1.23)). However, if we assume

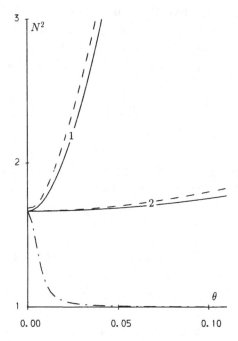

Fig. 5.4 Plots of N^2 versus θ for $P = 10^{-4}$, $Y^{-1} = 0.4$, $\beta_e = 0$ (dashed–dotted), $\beta_e = 10^{-4}$ (curves 1), $\beta_e = 10^{-2}$ (curves 2), $A_e = 1$ (solid) and $A_e = 2$ (dashed).

that β_e is small but finite, then in the limit $|P| \to 0$ this equation reduces to:

$$N^2 = (3 + 2A_e)/6. \tag{5.15}$$

This result is drastically different from that predicted by non-relativistic theory where N^2 is equal to zero for $P = 0$ (see equation (1.23)). This provides good justification for a relativistic formulation of the problem.

In the general case $\theta \neq 0$ equation (5.13) can also be resolved analytically with respect to N^2. However, the corresponding solutions appear to be so complicated that in practice only their numerical presentation could be of any use. Restricting ourselves to the analysis of whistler-mode waves we present in Fig. 5.4 the plots of N^2 versus θ for $P = 10^{-4}$; $Y^{-1} = 0.4$; $\beta_e = 0, 10^{-4}, 10^{-2}$; $A_e = 1, 2$. In Fig. 5.5 we present the same plots as in Fig. 5.4 but for $P = -10^{-4}$. The plots for $\beta_e = 0$ were computed using the Appleton–Hartree equation (see equation (2.6)).

As follows from Figs. 5.4 and 5.5 the forms of the curves appear to be extremely sensitive to the values of β_e so that the cold plasma approximation cannot be applied even when $\beta_e = 10^{-4}$ (a realistic value for magnetospheric and astrophysical conditions). The shapes of the curves corresponding to a

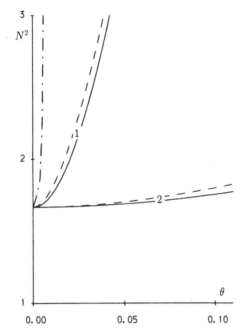

Fig. 5.5. The same as Fig. 5.4 but for $P = -10^{-4}$.

cold plasma appear to be quite different for $P = 10^{-4}$ and $P = -10^{-4}$, while the curves corresponding to a plasma with finite β_e are similar for these P. The general features of the curves corresponding to $Y^{-1} = 0.6$ are similar to those presented in Figs. 5.4 and 5.5.

The analysis similar to this section but for finite θ appears to be too complicated as in this case the Appleton–Hartree equation (2.6) cannot in general be simplified further. This simplification is possible when $\nu \gg 1$, but in this case relativistic corrections can be neglected as well. Hence, when generalizing the analysis of this section for finite θ, we restrict ourselves to relatively dense plasma with $\nu \gg 1$ when the whole problem can be considered within a non-relativistic approximation.

5.2 Propagation at a finite θ (low-temperature limit)

As was pointed out in Section 2.1, in the limit $\nu \gg 1$, but keeping the first-order terms with respect to ν^{-1}, we can write the expression for N_0^2 in the form (2.21), which is valid for θ not close to θ_{R0}. Assuming condition (5.1) valid and writing the expressions for ϵ_{ij} in the form (1.78), the relativistic term ϵ_{ij}^r being neglected (which is justified for $\nu \gg 1$) we obtain from (1.42) the following generalization of expression (2.21) (Sazhin, 1988b, 1990a):

$$N^2 = N_h^2 \equiv N_{0d}^2 \left(1 + \tilde{a}_c \nu^{-1} + \tilde{a}_r r + \tilde{a}_\beta \beta_e \right), \qquad (5.16)$$

where

$$\tilde{a}_\beta = \frac{Y^2 \sum_{i=0}^{4} \alpha_i \cos^i \theta}{2(Y \cos \theta - 1)^2 (Y^2 - 1)}, \qquad (5.17)$$

$$\alpha_0 = 1 + A_k,$$

$$\alpha_1 = -2Y + 4Y A_k,$$

$$\alpha_2 = -2Y^2 - 1 - (4Y^2 - 1)A_k,$$

$$\alpha_3 = 4Y(1 - A_k),$$

$$\alpha_4 = 4Y^2(1 - A_k),$$

$$A_k = A_e(Y^2 - 1)/(4Y^2 - 1),$$

β_e and A_e are the same as in Section 5.1, and other notations are the same as in equation (2.21).

The terms proportional to ν^{-1} and r have already been analysed in Section 2.1 (see Fig. 2.2). Here we generalize the analysis of that section by considering the term proportional to β_e. Plots of $\tilde{a}_\beta$ versus θ for $A_e = 1$ and 2 are shown in Fig. 5.6 for the same Y^{-1} as in Figs. 2.1 and 2.2. As follows from this figure, $\tilde{a}_\beta$ decreases with increasing θ for all A_e and Y under consideration. Also, $\tilde{a}_\beta$ decreases when A_e increases from 1 to 2. The values of $\tilde{a}_\beta$ for $Y^{-1} = 0.6$ and θ close to zero are outside the scale of the figure. When $\theta = 0$ and $Y^{-1} = 0.6$ then $\tilde{a}_\beta = 9.32$. Comparing Figs. 5.6 and 2.2 we can see that when $\tilde{a}_\beta > 0$ the effect of finite electron temperature increases the effect of finite electron density, and decreases it when $\tilde{a}_\beta < 0$.

In order to justify the usefulness and validity of equation (5.16) for the analysis of whistler-mode propagation in the magnetospheric plasma we compared the values of N_{0d} (see equation (2.11)), N_h (see equation (5.16)) and the values of N obtained from the exact numerical solution of the hot electromagnetic dispersion equation (N_n) (see Horne (1989) for details) for parameters relevant to the equatorial magnetosphere at $L = 4$, 6 and 6.6 (Sazhin & Horne, 1990). The results for all these L appeared to be roughly the same and so we will illustrate them only for $L = 6.6$ (geostationary orbit), where the electrons were assumed to consist of two isotropic Maxwellian components ($w = w_\perp = w_\parallel$ and $j = 0$ in (1.90)): cold with density $n_e = 1$ cm^{-3} and temperature $T_e = m_e w^2/2 = 1$ eV, and hot with density $n_e = 0.181$ cm^{-3} and temperature $T_e = 200$ eV. The corresponding curves of N_{0d}, N_h and N_n versus θ for $Y^{-1} = 0.6$ are shown in Fig. 5.7.

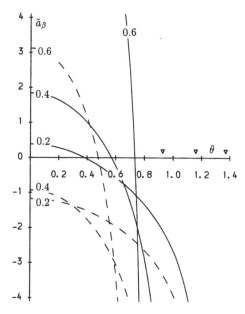

Fig. 5.6 Plots of $\tilde{a}_\beta$ (see equation (5.17)) versus θ for $Y^{-1} = 0.2$, 0.4 and 0.6 (curves indicated), $A_e = 1$ (solid) and $A_e = 2$ (dashed). $\triangledown$ near the θ axis indicate $\theta = \theta_{R0} = \arccos Y^{-1}$ for the Y^{-1} under consideration.

For clarity we only present results for a range of θ where the curves deviate significantly.

As follows from Fig. 5.7, the values of N_n and N_h almost coincide within the accuracy of plotting at $\theta < 49°$ ($\theta_R = 53.13°$), while the values of N_n and N_{0d} differ quite recognizably in the whole range of θ under consideration. Thus N_h seems to be almost the ideal approximation for N in this range of θ and it brings a considerable improvement when compared with the oversimplified approximation of N by N_{0d}. At $\theta > 49°$ the quasi-longitudinal approximation becomes invalid and the quasi-electrostatic approximation should be used (see Chapter 6).

The results for $Y^{-1} = 0.4$ appear to be essentially the same as for $Y^{-1} = 0.6$, while for $Y^{-1} = 0.8$ the approximation (5.16) brings no improvement over the use of N_{0d}. This is presumably due to the fact that at frequencies close to the electron gyrofrequency the decrement of damping of the waves is not negligibly small when compared with $\min(\omega_0, \Omega_0)$ and so equation (1.16) appears to be not strictly valid for whistler-mode waves. In this case there seems to be no reasonable analytical solution of the general equation (1.42) and so the whole problem of whistler-mode propagation needs to be approached by numerical methods.

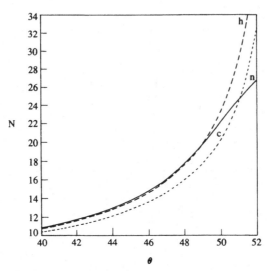

Fig. 5.7 Whistler-mode refractive index N as a function of θ measured in degrees for the electron distribution typical for the equatorial magnetosphere at $L = 6.6$ ($\nu = 10.577$ and $\beta_e = 6.5305 \times 10^{-4}$) and $Y^{-1} = 0.6$ ($\theta_R = 53.13°$). Numerical solutions of the hot electromagnetic dispersion equation (N_n) are shown by curve n, the analytical quasi-longitudinal solution (5.16) is shown by curve h, and the analytical solution corresponding to an infinitely dense cold plasma (see equation (2.11)) is shown by curve c (Sazhin & Horne, 1990).

In the low frequency limit ($Y^{-1} \ll 1$) the expressions for $\tilde{a}_c$, $\tilde{a}_r$ and $\tilde{a}_\beta$ reduce to:

$$\left. \begin{aligned} \tilde{a}_c &= \tilde{a}_{cl} \equiv \left(1 + \cos^2 \theta\right) / \left(2Y \cos \theta\right) \\ \tilde{a}_r &= \tilde{a}_{rl} \equiv - \left[Y \left(1 + \cos^2 \theta\right)\right] / \left(2 \cos \theta\right) \\ \tilde{a}_\beta &= \tilde{a}_{\beta l} \equiv \left[-2 - A_e + (4 - A_e) \cos^2 \theta\right] / 2 \end{aligned} \right\} . \qquad (5.18)$$

In the same limit $Y^{-1} \ll 1$ the expression for N_{0d}^2 reduces to $X/(Y \cos \theta)$. The expression for $\tilde{a}_{\beta l}$ at $|\theta| \ll 1$ reduces to $\tilde{a}_{\beta \| l} + \tilde{a}_{\theta l} \theta^2$, where $\tilde{a}_{\beta \| l}$ is equal to $1 - A_e$ (cf. equation (3.31) in the limit $Y^{-1} \ll 1$), and $\tilde{a}_{\theta l} = -1.5$. The latter result is consistent with equation (5.7).

The term $\tilde{a}_\beta \beta_e$ influences both the absolute value of N and the form of the curve $N(\theta)$. The latter influence seems to be particularly important as it can result in changes in the direction of wave group velocity. The general expression for the angle ψ between the whistler-mode group velocity $\mathbf{v}_g$ and the direction of an external magnetic field in a cold plasma, with the perturbations due to finite electron density and the contribution of ions taken into account, has already been derived in Section 2.2 (see equation (2.56)). Remembering (5.16), this expression can be generalized in a straightforward way so that the contributions of non-zero electron temperature and anisotropy are taken into account as well (Sazhin, 1990a):

$$\psi = \psi_0 + \tilde{\Delta}_c \nu^{-1} + \tilde{\Delta}_r r + \tilde{\Delta}_\beta \beta_e, \tag{5.19}$$

where

$$\tilde{\Delta}_\beta = \frac{Y^2 \sin\theta \sum_{i=0}^{4} \tilde{\gamma}_i \cos^i\theta}{(Y^2 - 1)(-Y^2 - 4 + 8Y\cos\theta - 3Y^2\cos^2\theta)(Y\cos\theta - 1)}, \tag{5.20}$$

$$\begin{aligned}
\tilde{\gamma}_0 &= 6Y A_k, \\
\tilde{\gamma}_1 &= -6Y^2 - 2 + 2(-2Y^2 + 1)A_k, \\
\tilde{\gamma}_2 &= 12Y(1 - A_k), \\
\tilde{\gamma}_3 &= 12Y^2(1 - A_k), \\
\tilde{\gamma}_4 &= -8Y^3(1 - A_k),
\end{aligned}$$

A_k is the same as in (5.17), and other notations are the same as in equation (2.56).

Plots of $\tilde{\Delta}_\beta$ versus θ for the same Y and A_e as in Fig. 5.6 are shown in Fig. 5.8. As follows from this figure, $\tilde{\Delta}_\beta > 0$ (except $\tilde{\Delta}_\beta = 0$ when $\theta = 0$) and increases with increasing θ. This means that the effect of non-zero electron temperature always tends to increase ψ except at $\theta = 0$. When $\theta = 0$ then ψ remains equal to 0, which means that the non-zero electron temperature does not influence the direction of whistler-mode group velocity at $\theta = 0$.

Equation (5.19) can be applied to different particular aspects of whistler-mode propagation in any realistic, and, in particular, magnetospheric plasma. In what follows, this equation will be applied to the analysis of thermal effects on whistler-mode propagation near the Storey angle, when ψ is maximal, and the Gendrin angle, when $\psi = 0$ at $\theta \neq 0$.

(a) Storey angle

As follows from the definition of the Storey angle given in Section 2.2, this angle is equal to ψ satisfying (2.51), i.e. it is equal to the maximal value of ψ within the quasi-longitudinal approximation. Having substituted (5.19) into (2.51) we can see that this equation is satisfied when:

$$\theta = \theta_s \equiv \theta_{s0} + \Delta\tilde{\theta}_{cs}\nu^{-1} + \Delta\tilde{\theta}_{rs}r + \Delta\tilde{\theta}_{\beta s}\beta_e, \tag{5.21}$$

where θ_{s0}, $\Delta\tilde{\theta}_{cs}$ and $\Delta\tilde{\theta}_{rs}$ are defined by (2.54), (2.62) and (2.63) respectively,

$$\Delta\tilde{\theta}_{\beta s} = -\frac{Y \sum_{i=0}^{6} \tilde{\kappa}_i \cos^i\theta_{s0}}{3(Y^2 - 1)(1 - Y\cos\theta_{s0})^3 \sin\theta_{s0}}, \tag{5.22}$$

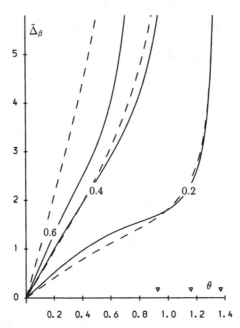

Fig. 5.8 Plots of $\tilde{\Delta}_\beta$ (see equation (5.20)) versus θ for $Y^{-1} = 0.2$, 0.4 and 0.6 (curves indicated), $A_e = 1$ (solid) and $A_e = 2$ (dashed). $\triangledown$ near the θ axis indicate $\theta = \theta_{R0} = \arccos Y^{-1}$ for the Y^{-1} under consideration.

$$\tilde{\kappa}_0 = 3Y^2 + 1 + (-4Y^2 - 1)A_k,$$
$$\tilde{\kappa}_1 = 3Y^3 - 11Y + (2Y^3 + 14Y)A_k,$$
$$\tilde{\kappa}_2 = -24Y^2 - 2 + (17Y^2 + 2)A_k,$$
$$\tilde{\kappa}_3 = (22Y^3 + 18Y)(1 - A_k),$$
$$\tilde{\kappa}_4 = (-8Y^4 + 18Y^2)(1 - A_k),$$
$$\tilde{\kappa}_5 = -32Y^3(1 - A_k),$$
$$\tilde{\kappa}_6 = 12Y^4(1 - A_k).$$

The expression for ψ_s in a hot plasma can be obtained from the equation:

$$\psi_s = \psi(\theta_{s0}) + \left.\frac{\partial\psi}{\partial\theta}\right|_{\theta=\theta_{s0}} \Delta\theta_s, \tag{5.23}$$

where

$$\Delta\theta_s = \Delta\tilde{\theta}_{cs}\nu^{-1} + \Delta\tilde{\theta}_{rs}r + \Delta\tilde{\theta}_{\beta s}\beta_e.$$

In view of the fact that:

$$\left.\frac{\partial\psi}{\partial\theta}\right|_{\theta=\theta_{s0}} = \left.\frac{\partial\tilde{\Delta}_c}{\partial\theta}\right|_{\theta=\theta_{s0}} \nu^{-1} + \left.\frac{\partial\tilde{\Delta}_r}{\partial\theta}\right|_{\theta=\theta_{s0}} r + \left.\frac{\partial\tilde{\Delta}_\beta}{\partial\theta}\right|_{\theta=\theta_{s0}} \beta_e, \tag{5.24}$$

and neglecting the contribution of second order terms with respect to ν^{-1}, r or β_e, we obtain from (5.23)

$$\psi_s = \psi_{s0} + \tilde{\Delta}_{cs}\nu^{-1} + \tilde{\Delta}_{rs}r + \tilde{\Delta}_{\beta s}\beta_e, \qquad (5.25)$$

where ψ_{s0}, $\tilde{\Delta}_{cs}$ and $\tilde{\Delta}_{rs}$ are defined by (2.55), (2.65) and (2.66) respectively,

$$\tilde{\Delta}_{\beta s} = \tilde{\Delta}_{\beta}(\theta = \theta_{s0}), \qquad (5.26)$$

$\tilde{\Delta}_{\beta}$ being defined by (5.20).

When deriving (5.21) and (5.25) we assumed that the corresponding corrections to θ_{s0} and ψ_{s0} were small. These equations are the straightforward generalizations of equations (2.59) and (2.60) for the case of whistler-mode propagation in a hot plasma.

Plots of $\Delta\tilde{\theta}_{cs}$, $\Delta\tilde{\theta}_{rs}$ and $\Delta\tilde{\theta}_{\beta s}$ versus Y^{-1} for $A_e = 1$ and 2 are shown in Fig. 5.9, while in Fig. 5.10 we present the corresponding plots of $\tilde{\Delta}_{cs}$, $\tilde{\Delta}_{rs}$ and $\tilde{\Delta}_{\beta s}$ for the same values of A_e. As follows from Fig. 5.9 the correction to θ_{s0} due to finite electron density tends to zero when $Y^{-1} \to 0$. At these Y^{-1} the correction due to ion effects is maximal. $\Delta\tilde{\theta}_{\beta s}$ monotonically increases with increasing Y^{-1} both for $A_e = 1$ and $A_e = 2$. Also, $\Delta\tilde{\theta}_{\beta s}$ increases when A_e increases.

In a similar way to θ_s, at $Y^{-1} > 0.1$ the thermal corrections to ψ_{s0} tend to compensate for the corrections to ψ_{s0} due to the effect of finite electron density as can be seen from Fig. 5.10. As Y^{-1} approaches 0.5 all corrections to ψ_{s0} tend to zero (note that when $Y^{-1} = 0$ then $\psi_{s0} = 0$). However, our theory is not valid in the immediate vicinity of $Y^{-1} = 0.5$ as in this case all the corrections to θ_{s0} become infinitely large.

The most straightforward applications of the results referring to the Storey angle to the Earth's magnetosphere might be related to refined interpretation of natural radio emissions having characteristic spectrograms of inverted-V shapes, which were observed by satellites in the topside ionosphere in the auroral region; see Gurnett & Frank (1972), also the discussion by Sazhin & Strangeways (1989).

(b) Gendrin angle

As follows from the definition of the Gendrin angle θ_G in Section 2.2, at $\theta = \theta_G$ the perpendicular component of whistler-mode group velocity and ψ are equal to zero. The expression for θ_G in a cold plasma, the effects of finite electron density and ions being taken into account, was obtained in Section 2.2 (see equation (2.67)). Remembering (5.19), expression (2.67) for

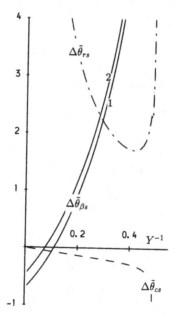

Fig. 5.9 Plots of $\Delta\tilde{\theta}_{cs}$ (see equation (2.62)) versus Y^{-1} (dashed), $\Delta\tilde{\theta}_{rs}$ (see equation (2.63)) versus Y^{-1} (dashed–dotted) and $\Delta\tilde{\theta}_{\beta s}$ (see equation (5.22)) versus Y^{-1} (solid) for $A_e = 1$ and $A_e = 2$ (curves indicated).

θ_G can be generalized so that the effects of non-zero electron temperature and anisotropy are taken into account as well, and written as:

$$\theta_G = \theta_{G0} + \Delta\tilde{\theta}_{cG}\nu^{-1} + \Delta\tilde{\theta}_{rG}r + \Delta\tilde{\theta}_{\beta G}\beta_e, \tag{5.27}$$

where θ_{G0}, $\Delta\tilde{\theta}_{cG}$ and $\Delta\tilde{\theta}_{rG}$ are the same as in (2.67) and (2.68), and

$$\Delta\tilde{\theta}_{\beta G} = \frac{2}{\sqrt{Y^2 - 4}}\left[6 + \frac{A_e(Y^2 + 6)}{4Y^2 - 1}\right]. \tag{5.28}$$

Plots of $\Delta\tilde{\theta}_{cG}$, $\Delta\tilde{\theta}_{rG}$ and $\Delta\tilde{\theta}_{\beta G}$ are shown in Fig. 5.11. As follows from this figure, the effects of non-zero electron temperature tend to compensate for the effects of finite electron density as was the case in Fig. 5.9 at $Y^{-1} > 0.1$ and Fig. 5.10. When Y^{-1} approaches 0.5 all the corrections tend to be infinitely large, while $\theta_{G0}(Y^{-1} = 0.5) = 0$. In this frequency range our equation (5.27) is invalid.

Energy focusing of whistler-mode waves at θ close to θ_{G0} is often associated with the corresponding local maximum in spectra of whistler-mode radio emissions observed in the ionosphere and the magnetosphere of the Earth (see e.g. Etcheto & Gendrin, 1970; Hayakawa, Parrot & Lefeuvre,

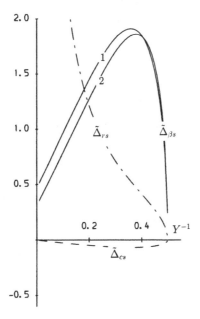

Fig. 5.10 The same as Fig. 5.9 but for $\tilde{\Delta}_{cs}$, $\tilde{\Delta}_{rs}$ and $\tilde{\Delta}_{\beta s}$ respectively (see equations (2.65), (2.66) and (5.26)).

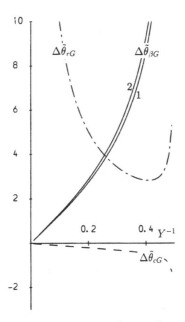

Fig. 5.11 The same as Fig. 5.9 but for $\Delta\tilde{\theta}_{cG}$, $\Delta\tilde{\theta}_{rG}$ and $\Delta\tilde{\theta}_{\beta G}$ respectively (see equations (2.68) and (5.28)).

1972). Observations of peak wave energy at θ close but not equal to θ_{G0} could be a good confirmation of the importance of our corrections to θ_{G0} in magnetospheric conditions.

5.3 Special cases of quasi-longitudinal propagation

As was shown in Section 1.7, the expressions for the elements of the non-relativistic plasma dielectric tensor $\hat{\epsilon}_{ij}$ can be simplified not only in a low-temperature limit (considered, in particular, in Sections 5.1 and 5.2) but also for some special values of ξ_n, namely when $\xi_1 = -0.924$ and $\xi_0 \gg 1$, or $\xi_0 = 0.924$ and $|\xi_1| \gg 1$, or $\xi_0 = -\xi_1 = 0.924$. In this section, using the results of Section 1.7, we obtain and analyse the corresponding expressions for the whistler-mode refractive index in the quasi-longitudinal approximation for these combinations of the parameters ξ_0 and ξ_1, following Sazhin & Sazhina (1988).

(a) $\xi_1 = -0.924;$ $\xi_0 \gg 1$

Having substituted the elements of $\hat{\epsilon}_{ij}$ obtained for these values of ξ_0 and ξ_1 (equations (1.115)–(1.120)) into the general dispersion equation (1.42) and imposing the condition (5.1) we obtain the following expression for N^2:

$$N^2 = N_{0d}^2 \left(1 + \tilde{a}_c \nu^{-1} + \tilde{a}_r r + \tilde{a}_{\beta 1} \beta_e \right), \tag{5.29}$$

where

$$\tilde{a}_{\beta 1} = \frac{\sum_{i=0}^{4} \tilde{\alpha}_i' \cos^i \theta}{4(Y \cos \theta - 1)^2}, \tag{5.30}$$

$$\tilde{\alpha}_0' = \frac{Y}{Y+1} + \frac{6Y^2 - 1}{4Y^2 - 1} A_e,$$

$$\tilde{\alpha}_1' = -\frac{2Y^2}{Y+1} + \frac{2Y}{4Y^2 - 1} A_e,$$

$$\tilde{\alpha}_2' = \frac{-5Y^3 - 4Y^2 - 2Y}{Y+1} + (-Y^2 - 2Y - 2) A_e,$$

$$\tilde{\alpha}_3' = \frac{-2Y^3 + 2Y^2}{Y+1} + \frac{8Y^4 - 2Y^2 - 2Y}{4Y^2 - 1} A_e,$$

$$\tilde{\alpha}_4' = \frac{7Y^3 + 4Y^2 + Y}{Y+1} + \frac{-4Y^4 + 8Y^3 + 3Y^2 - 2Y - 1}{4Y^2 - 1} A_e,$$

and other notations are the same as in (5.16).

The condition $\xi_1 = -0.924$ for the validity of (5.29) can be written as:

$$\frac{2\beta_e Y^2 \cos^2 \theta}{(Y \cos \theta - 1)(Y - 1)^2} \left(1 + \tilde{a}_c \nu^{-1} + \tilde{a}_r r + \tilde{a}_{\beta 1} \beta_e\right) = 0.924^{-2} \approx 1.171.$$

$$(5.31)$$

In the dense plasma limit $(\nu^{-1} = 0)$, with the contributions of ions neglected $(r = 0)$, and in the limit $\theta = 0$, condition (5.31) reduces to (4.9). The second condition $\xi_0 \gg 1$ is satisfied provided $|Y^{-1} - 1| \ll Y^{-1}$, i.e. when the wave frequency is sufficiently close to the electron gyrofrequency. As will be shown in Chapter 7, at these frequencies and for β_e in the range $(0.005, 0.015)$, when $A_e < 3$ then $|\gamma|$ at $\theta = 0$ becomes so large that we can no longer neglect the influence of damping on whistler-mode propagation. At $\theta \neq 0$ we can expect this damping to be even higher due to the contribution of Landau damping, and so equation (5.29) becomes invalid in the range of β_e under consideration and $A_e < 3$ both at $\theta = 0$ and $\theta \neq 0$. Thus our analysis of equations (5.29)–(5.31) will be restricted to A_e in the range $[3, 4]$.

The plots of $\tilde{a}_{\beta 1}$ versus θ for $Y^{-1} = 0.2$, 0.4 and 0.6 (the same as in Fig. 5.6) and $A_e = 3$ and 4 are shown in Fig. 5.12. As follows from this figure, $\tilde{a}_{\beta 1} < 0$ and $|\tilde{a}_{\beta 1}|$ increases with increasing θ and A_e. When $\theta = 0$ then $\tilde{a}_{\beta 1} = 0$ which agrees with the results of Section 4.1: for $\xi_1 = -0.924$ and $\theta = 0$ the whistler-mode refractive index in a hot plasma coincides with that in a cold plasma.

If we restrict our consideration to the dense plasma limit $(\nu^{-1} = 0)$, neglect the contribution of ions $(r = 0)$ and assume $|\theta| \ll 1$ then equation (5.29) can be simplified to:

$$N^2 = N_{0d\parallel}^2 \left(1 + \tilde{a}_{\Sigma 1} \theta^2\right), \qquad (5.32)$$

where

$$\tilde{a}_{\Sigma 1} = \frac{Y}{2(Y - 1)} + \beta_e \left[-\frac{3Y^2}{2(Y - 1)^2} - \frac{Y A_e}{(2Y - 1)(Y - 1)}\right], \qquad (5.33)$$

and $N_{0d\parallel} \equiv N_{0d}(\theta = 0)$ is the same as in equation (3.34).

The condition (5.33) in this case can be written as:

$$\frac{2\beta_e Y^2}{(Y - 1)^3} \left[1 + (\tilde{a}_{\Sigma 1} - 1)\theta^2\right] = 1.171. \qquad (5.34)$$

For $|\theta| \ll 1$ the plots of Y^{-1} versus β_e based on equation (5.34) are close to the corresponding plots of Y_x^{-1} versus β_e for $\theta = 0$ shown in Fig. 4.1 (see Sazhin & Sazhina, 1988, for details). The facts that $\tilde{a}_{\Sigma 1}$ decreases with increasing A_e is consistent with Fig. 5.12.

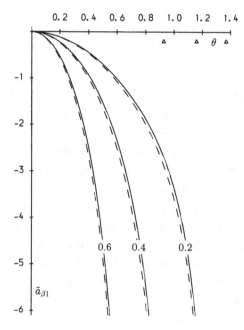

Fig. 5.12 The same as Fig. 5.6 but for $\tilde{a}_{\beta1}$ (see equation (5.30)), $A_e = 3$ (solid) and $A_e = 4$ (dashed).

(b) $\xi_0 = 0.924$; $|\xi_1| \gg 1$

This combination of ξ_0 and ξ_1 is possible only when $Y \gg 1$. For these Y the main contribution to wave damping or amplification comes from Landau damping or the Čerenkov instability. This damping or instability can be made arbitrarily small when $|\theta| \ll 1$ ($\theta \neq 0$). Imposing both these conditions $Y \gg 1$ and $|\theta| \ll 1$ along with inequality (5.1), neglecting the effects of finite electron density ($\nu^{-1} = 0$) and the contribution of ions ($r = 0$) (the latter two effects could be analysed in a similar way to that in Section 2.1: see equation (2.21)), and remembering the expressions for $\hat{\epsilon}_{ij}$ for these ξ_0 and ξ_1 (see equation (1.121)), we obtain from the general dispersion equation (1.42) the following expression for the squared whistler-mode refractive index:

$$N^2 = N^2_{0d\parallel} \left(1 + \tilde{a}_{\parallel0}\beta_e + \tilde{a}_{\Sigma0}\theta^2\right), \tag{5.35}$$

where:

$$\left.\begin{array}{l} \tilde{a}_{\parallel0} = 1 - A_e \\ \tilde{a}_{\Sigma0} = -\frac{1}{2}\nu Y - \frac{1}{2}\nu Y\beta_e(1 - A_e) \end{array}\right\} \tag{5.36}$$

when $A_e \neq 1$.

When $A_e = 1$ then higher order terms should be taken into account and expressions (5.36) should be replaced by:

$$\left.\begin{array}{l} \tilde{a}_{\|0} = Y^{-1} \\ \tilde{a}_{\Sigma 0} = -\frac{1}{2}\nu Y - \frac{1}{2}\nu\beta_e \end{array}\right\}. \tag{5.37}$$

In contrast to equations (5.29) and (5.32) expression (5.35) for N^2 does not reduce to N_{0d}^2 when $\beta_e = 0$ and $\theta \neq 0$.

The condition $\xi_0 = 0.924$ is satisfied when:

$$2\beta_e Y \left[1 + \tilde{a}_{\|0}\beta_e + (-1 - \tilde{a}_{\|0}\beta_e + \tilde{a}_{\Sigma 0})\theta^2\right] = 1.171. \tag{5.38}$$

For small β_e and θ^2 condition (5.38) can be simplified to:

$$\beta_e \approx (2Y)^{-1}. \tag{5.39}$$

(c) $\xi_0 = -\xi_1 = 0.924$

This combination of ξ_0 and ξ_1 is possible only if $Y = 2$. Similarly to subsection (b) we restrict our analysis to $|\theta| \ll 1$. In this case, having substituted expressions (1.124) for the elements of plasma dielectric tensor into the general dispersion equation (1.42), remembering (5.1), and neglecting the effects of finite electron density ($\nu^{-1} = 0$) and the contribution of ions ($r = 0$), we obtain the following expression for the squared whistler-mode refractive index within the quasi-longitudinal approximation

$$N^2 = N_{0d\|2}^2 \left(1 + \tilde{a}_{\Sigma 2}\theta^2\right), \tag{5.40}$$

where $N_{0d\|2}^2 = N_{0d\|}^2(Y = 2) = 4\nu$, and

$$\tilde{a}_{\Sigma 2} = -2\nu - \frac{8}{3}A_e\beta_e. \tag{5.41}$$

The condition $\xi_0 = -\xi_1 = 0.924$ is satisfied when

$$8\beta_e[1 - (1 + 2\nu + \frac{8}{3}A_e\beta_e)\theta^2] = 1.171. \tag{5.42}$$

When $\theta = 0$ then (5.42) can be simplified to

$$\beta_e \approx 0.146. \tag{5.43}$$

This value of β_e is consistent with the prediction of equation (4.9) at $Y = 2$.

Although the expressions for N^2 obtained in this section have a rather narrow area of direct application they might be useful for checking the results of numerical analysis of the whistler-mode dispersion equation or for other more sophisticated developments of whistler-mode theory.

5.4 Wave polarization

In this section we generalize expressions (2.75) and (2.76) for quasi-longitudinal whistler-mode polarization in a cold plasma to include the effects of
finite electron density and the contribution of ions along with the contributions of non-zero electron temperature and anisotropy of the electron distribution function. We start with the general expressions for the parameters α_1
and α_2 (see equations (1.152)) defined by expressions (1.157) and (1.158).
Then we substitute N^2 defined by (5.16) and the elements of $\hat{\epsilon}_{ij}$ defined by
(1.78) (with relativistic corrections neglected, but the contribution of ions
taken into account as in Section 2.3) into these expressions. As a result, after
very lengthy but straightforward algebra (see Sazhin, 1985, 1991a for some
details), we obtain the following final expressions for α_1 and α_2:

$$\alpha_1 = \alpha_{10}^0(1 + \alpha_{11}^{(c)}\nu^{-1} + \alpha_{11}^{(r)}r + \alpha_{11}^{(h)}\beta_e), \tag{5.44}$$

$$\alpha_2 = \alpha_{20}^0(1 + \alpha_{21}^{(c)}\nu^{-1} + \alpha_{21}^{(r)}r + \alpha_{21}^{(h)}\beta_e), \tag{5.45}$$

where

$$\alpha_{11}^{(h)} = \alpha_{11}^0 + \alpha_{11}^A A_e, \tag{5.46}$$

$$\alpha_{21}^{(h)} = \alpha_{21}^0 + \alpha_{21}^A A_e, \tag{5.47}$$

$$\alpha_{11}^0 = Y[(-Y^2 + 1) + (2Y^3 + 2Y)\cos\theta + (2Y^4 - 9Y^2 - 1)\cos^2\theta$$

$$+ (-2Y^3 - 2Y)\cos^3\theta$$

$$+ (-2Y^4 + 10Y^2)\cos^4\theta][2(Y\cos\theta - 1)^2(Y - \cos\theta)(Y^2 - 1)]^{-1},$$

$$\alpha_{11}^A = Y[(-5Y^2 - 1) + (8Y^3 - 2Y)\cos\theta + (-4Y^4 + 15Y^2 + 1)\cos^2\theta$$

$$+ (-8Y^3 + 2Y)\cos^3\theta$$

$$+ (4Y^4 - 10Y^2)\cos^4\theta][2(Y\cos\theta - 1)^2(Y - \cos\theta)(4Y^2 - 1)]^{-1},$$

$$\alpha_{21}^0 =$$

$$\frac{Y^2[-1 + 2Y\cos\theta + (2Y^2 - 3)\cos^2\theta + (-6Y^3 + 6Y)\cos^3\theta + 4Y^2\cos^4\theta]}{2\cos\theta(Y\cos\theta - 1)(Y - \cos\theta)(Y^2 - 1)},$$

$$\alpha_{21}^A =$$

$$\frac{Y^2[1 + 4Y\cos\theta + (-12Y^2 + 3)\cos^2\theta + (8Y^3 - 6Y)\cos^3\theta - 4Y^2\cos^4\theta]}{2\cos\theta(Y\cos\theta - 1)(Y - \cos\theta)(4Y^2 - 1)},$$

and other notations are the same as in (2.75) and (2.76).

The expressions for α_{11}^0 and α_{11}^A are written in the same form as in Sazhin (1985, 1991a). The expressions for α_{21}^0 and α_{21}^A are written in the form suggested by Sazhin (1991a). They are equivalent to the corresponding expressions derived by Sazhin (1985) but their forms are considerably simpler.

Plots of $\alpha_{11}^{(h)}$ versus θ and $\alpha_{21}^{(h)}$ versus θ are shown in Figs. 5.13 and 5.14 for the same values of Y^{-1} and A_e as in Figs. 2.9–2.12. As follows from Fig. 5.13, $\alpha_{11}^{(h)} > 0$ unless $\theta = 0$, and increases with increasing θ in a similar way to $\alpha_{11}^{(r)}$. When $\theta = 0$ then $\alpha_{11}^{(h)} = 0$, which means that non-zero electron temperature does not influence the circular polarization of whistler-mode waves propagating parallel to the magnetic field (cf. Figs. 2.9 and 2.10). In contrast to $\alpha_{11}^{(h)}$, $\alpha_{21}^{(h)}$ is always negative and its dependence on θ appears to be more complicated than in the case of $\alpha_{11}^{(h)}$. An increase of A_e from 1 to 2 decreases both $\alpha_{11}^{(h)}$ and $\alpha_{21}^{(h)}$ at $Y^{-1} = 0.2$ and increases them at $Y^{-1} = 0.4$ and 0.6. The absolute values of $\alpha_{21}^{(h)}$ are in general about an order of magnitude larger than those of $\alpha_{11}^{(h)}$. In particular the thermal corrections to α_2 for $\beta_e = 10^{-2}$ (typical value for magnetospheric conditions) can exceed 10%, which means that the influence of electron thermal motion on the value of α_2 can often be important and must be taken into consideration. In principle it also seems to be possible to use this parameter to estimate the electron temperature.

Expressions (5.44) and (5.45) for α_1 and α_2 can also be used for analysis of the wave magnetic field polarization, the ratio B_y/B_x, using expression (1.163). As follows from (1.164), the ratio $B_z/B_x = -\tan\theta$ depends neither on electron temperature nor on the anisotropy of the electron distribution function.

5.5 Propagation in the presence of electron beams

Our analysis in this chapter has so far been based on the assumption that the electron distribution function has the form (1.76) (weakly relativistic limit) or (1.90) (non-relativistic limit). In this section we will generalize some aspects of the analysis of Section 5.2 to the case when the electron distribution function has the form (1.130), i.e. when the contribution of electron beams is taken into account. As mentioned in Section 1.8 this distribution function is particularly important for the analysis of whistler-mode propagation in the magnetopause region of the Earth's magnetosphere. Using expressions (1.139) for the elements of the plasma dielectric tensor derived in Section 1.8 for this electron distribution function, we can generalize expression (5.16)

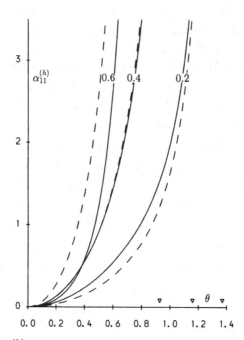

Fig. 5.13 Plots of $\alpha_{11}^{(h)}$ (see equation (5.46)) versus θ for $Y^{-1} = 0.2$, 0.4 and 0.6 (curves indicated), $A_e = 1$ (solid) and $A_e = 2$ (dashed). $\triangledown$ near the θ axis indicate $\theta = \theta_{R0} = \arccos Y^{-1}$ for these Y^{-1}.

for the quasi-longitudinal whistler-mode refractive index to (Sazhin, Walker & Woolliscroft, 1990b):

$$N^2 =$$

$$N_{0d}^2 \left(1 + \tilde{a}_c \nu^{-1} + \tilde{a}_r r + \sum_i \tilde{a}_{\beta_i} \beta_{e_i} + \tilde{a}_{b1} \nu^{1/2} \sum_i \tilde{v}_{0_i} \kappa_i + \tilde{a}_{b2} \nu \sum_i \tilde{v}_{0_i}^2 \kappa_i \right),$$

$$(5.48)$$

where $\tilde{a}_c$, $\tilde{a}_r$ and r are the same as in (5.16) with $\nu = \Pi_\Sigma^2/\Omega^2$, Π_Σ is the total electron plasma frequency, a_{β_i} and β_{e_i} are the same as in (5.16) and (5.17) but refer to each electron component separately,

$$\tilde{a}_{b1} = -\frac{Y^2 \cos^2 \theta}{(Y \cos \theta - 1)^{3/2}}, \qquad (5.49)$$

$$\tilde{a}_{b2} = \frac{Y^2 \left[1 - 2Y \cos \theta + (-2Y^2 - 1) \cos^2 \theta + 4Y \cos^3 \theta + 4Y^2 \cos^4 \theta \right]}{2 (Y \cos \theta - 1)^2 (Y^2 - 1)},$$

$$(5.50)$$

$\tilde{v}_{0_i} = v_{0_i}/c$, and summation is assumed over all i components. The contribution of the terms proportional to $\tilde{a}_c$, $\tilde{a}_r$ and $\tilde{a}_{\beta_i}$ was considered in

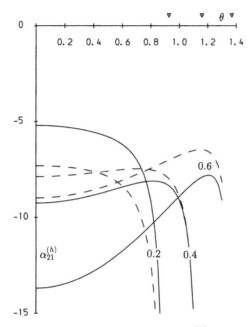

Fig. 5.14. The same as Fig. 5.13 but for $\alpha_{21}^{(h)}$ (see equation (5.47)).

Section 5.2. Here we concentrate our attention on the terms proportional to $\tilde{a}_{b1}$ and $\tilde{a}_{b2}$. Note that $\tilde{a}_{b2} = \tilde{a}_{\beta_i}$ when $A_{e_i} = 0$. This appears to be due to the fact that $w_{\parallel}^2$ is equal to $2\langle v_{\parallel}^2 \rangle$, where $\langle v_{\parallel}^2 \rangle$ is the mean square velocity of the ith electron component in the direction parallel to the magnetic field. In the case of a beam, $\langle v_{\parallel}^2 \rangle$ can be identified with v_0^2. Note also the factor 0.5 in the definition of β_e.

The plots of $\tilde{a}_{b1}$ versus Y^{-1} and $\tilde{a}_{b2}$ versus Y^{-1} for three values of θ are shown in Figs. 5.15 and 5.16 respectively. (This presentation is complementary to that in Fig. 5.6 where we have shown the plots of $\tilde{a}_{\beta}$ versus θ for different Y^{-1}). The term proportional to $\tilde{a}_{b1}$ is the dominant one if $\sum_i \tilde{v}_{0i} \kappa_i$ is not close to zero, i.e. when there is a net electron current in the plasma. As follows from Fig. 5.15, $\tilde{a}_{b1}$ is always negative, and its modulus increases for small Y^{-1} and for Y^{-1} approaching 1. The correction to N_{0d} due to the electron current depends on the direction of this current. In particular, when the net current flows in the positive direction, i.e. in the direction of wave propagation, it tends to decrease the value of N^2, and when it flows in the negative direction tends to increase it. The rate of this decrease or increase depends on values of $\tilde{a}_{b1}$, ν and the net current itself.

When the currents produced by different electron beams compensate each

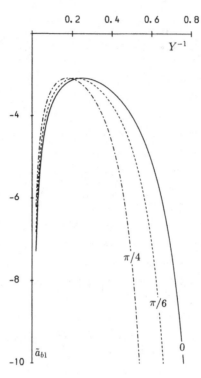

Fig. 5.15 Plots of $\tilde{a}_{b1}$ (see equation (5.49)) versus Y^{-1} for different θ (curves indicated).

other then the contribution of the term proportional to $\tilde{a}_{b1}$ can be neglected when compared with that proportional to $\tilde{a}_{b2}$. As follows from Fig. 5.16, $\tilde{a}_{b2}$ is positive and increases with increasing Y^{-1} for all values of θ (note that $\tilde{a}_{b2} \to 0$ when $\theta = \pi/4$ and $Y^{-1} \to 0$). This means that the electron beams producing no net current tend to increase the value of N^2. We can expect that for some values of θ, Y, ν and β_{e_i}, the contribution of the terms proportional to $\tilde{a}_c$, $\tilde{a}_{\beta_i}$, $\tilde{a}_{b1}$, and $\tilde{a}_{b2}$ can compensate for each other (cf. Figs. 1 and 2 of Sazhin, 1988b).

Equation (5.48) will be applied to the analysis of whistler-mode growth or damping in Chapter 7. Meanwhile, in the next chapter we analyse another limiting case of whistler-mode propagation based on the quasi-electrostatic approximation.

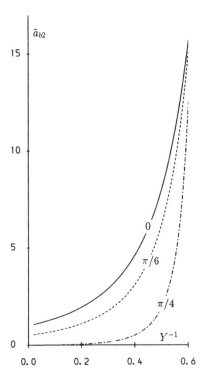

Fig. 5.16. The same as Fig. 5.15 but for $\tilde{a}_{b2}$ (see equation (5.50)).

Problems

Problem 5.1 When analysing thermal effects on whistler-mode propagation in a hot anisotropic plasma in the CGL (Chew, Goldberger & Low) approximation (valid at $\omega \ll \Omega$) Namikawa, Hamabata & Tanabe (1981) obtained the following dispersion equation (see their equation (26)):

$$\omega^2 \left(1 + \frac{c^2 k^2}{\Pi^2}\right)^2 = \frac{\Omega^2 c^4}{\Pi^4} k^4 \cos^2\theta \left\{\left[1 + \frac{1}{2}\beta_\perp + \beta_\parallel - \frac{1}{4}\beta_\perp^2 - \frac{3}{4}\beta_\parallel^2 + \beta_\perp \beta_\parallel \right.\right.$$

$$\left. + \left(\frac{1}{2}\beta_\perp - 2\beta_\parallel + \frac{1}{2}\beta_\perp^2 + \beta_\parallel^2 - \frac{3}{2}\beta_\perp \beta_\parallel\right)\cos^2\theta\right]$$

$$+ \frac{c^2 k^2}{\Pi^2}\left[\frac{3}{2}\beta_\perp + \frac{1}{4}\beta_\perp^2 - \frac{1}{4}\beta_\perp \beta_\parallel + \left(-\frac{3}{2}\beta_\perp - \frac{1}{2}\beta_\perp^2 - \beta_\parallel^2 + \frac{3}{2}\beta_\perp \beta_\parallel\right)\cos^2\theta\right.$$

$$\left.\left. + \left(\frac{1}{4}\beta_\perp^2 + \beta_\parallel^2 - \frac{5}{4}\beta_\perp \beta_\parallel\right)\cos^4\theta\right]\right\},$$

where:

$$\beta_{\|(\perp)} = \frac{n_e K T_{\|(\perp)}}{B_0^2/8\pi},$$

K is the Boltzmann constant, $T_{\|(\perp)}$ is the parallel (perpendicular) temperature (in K), B_0 is the induction of the external magnetic field, k is the magnitude of the wave vector, and other notations are the same as used so far. Show that this equation reduces to our equation (5.16) when taken in a low frequency ($Y \gg 1$) and dense plasma ($\nu \gg 1$) limit.

Problem 5.2   As can be seen in Fig. 5.3, our theory based on the hot plasma approximation predicts wave focusing for $\nu = 0.25$ and $Y^{-1} = 0.5$ in a cold plasma when $\beta_e \to 0$. At the same time this focusing is not predicted by equation (5.10) based on the cold plasma approximation ($\beta_e = 0$). Explain why.

6

Quasi-electrostatic approximation

6.1 The dispersion equation in a low-temperature limit

As was shown in Chapter 5, the quasi-electrostatic approximation becomes invalid when θ approaches the resonance cone angle θ_R defined by (2.23) or (2.24) (or θ_{R0} defined by (2.13) in the case of a dense plasma, the contribution of ions being neglected). However, at θ equal to or close to θ_R we can use another approximation based on the following assumptions:

(1) The wave refractive index N is assumed to be so large that only the contribution of the terms proportional to the highest powers of N in the dispersion equation is to be taken into account.

(2) The plasma temperature is assumed to be so low that inequalities (1.77) are valid and the elements of ϵ_{ij} can be written in the form (1.78), ϵ_{ij}^t and ϵ_{ij}^r being the perturbations of ϵ_{ij}^0.

The first assumption seems to be a straightforward one in a sufficiently low-temperature plasma, as in a cold plasma limit $N_0^2 \to \infty$ when $\theta \to \theta_R$. However, the second assumption does not seem to be an obvious one since $\epsilon_{ij}^t \sim N^2$ and $\epsilon_{ij}^t \to \infty$ as $N^2 \to \infty$. Thus N should be large enough to satisfy the first assumption but small enough to satisfy the second one. The validity of these assumptions can be checked by the actual value of N obtained from the solution.

The approximation based on these assumptions is usually called 'quasi-electrostatic' in order to link it with the electrostatic approximation based on the assumption that $N^2 \to \infty$ and the general dispersion equation (1.42) is satisfied when $A = 0$. In contrast to the electrostatic approximation, we partly take into account the contribution of the term BN^2 in (1.42) as well. As will be seen later, the contribution of this term sometimes significantly modifies the quasi-electrostatic solution when compared with the electrostatic one.

Having substituted (1.78) into expressions (1.43)–(1.45) for the coefficients A, B and C, and keeping only linear terms with respect to ϵ^t_{ij} and ϵ^r_{ij}, we obtain

$$A = A_0 + a^t_1 \beta_e N^2 + a^r_1 \beta_e, \qquad (6.1)$$

$$B = B_0 + b^t_1 \beta_e N^2 + b^r_1 \beta_e, \qquad (6.2)$$

$$C = C_0 + c^t_1 \beta_e N^2 + c^r_1 \beta_e, \qquad (6.3)$$

where A_0, B_0 and C_0 are defined by (2.1)–(2.3), the terms $a^t_1 \beta_e N^2$, $b^t_1 \beta_e N^2$ and $c^t_1 \beta_e N^2$ take into account the contribution of ϵ^t_{ij} (non-relativistic corrections to ϵ^0_{ij}) while the terms $a^r_1 \beta_e$, $b^r_1 \beta_e$ and $c^r_1 \beta_e$ take into account the contribution of ϵ^r_{ij} (relativistic corrections to ϵ^0_{ij}); the coefficients a^t_1, b^t_1, c^t_1, a^r_1, b^r_1, and c^r_1 depend neither on β_e nor on N.

In view of (6.1)–(6.3) we can rewrite equation (1.42) as:

$$a^t_1 \beta_e N^6 + (A_0 + a^r_1 \beta_e + b^t_1 \beta_e) N^4 + (B_0 + b^r_1 \beta_e + c^t_1 \beta_e) N^2 + (C_0 + c^r_1 \beta_e) = 0. \qquad (6.4)$$

Equation (6.4) can be used directly for the analysis of N in a similar way to that used for the analysis of almost parallel whistler-mode propagation at frequencies near the electron plasma frequency (see equation (5.13). In the non-relativistic limit (without the terms with the upper index 'r' in (6.4)) the direct numerical solution of (6.4) has been used by many authors (e.g. Aubry, Bitoun & Graff, 1970; Hashimoto, Kimura & Kumagai, 1977). In contrast to these authors we attempt to further simplify equation (6.4) so that its analytical solution may have a reasonably simple form. In order to do this we use our assumption $N^2 \gg 1$.

Firstly we consider the case when a^t_1 is not close to zero. Because of our assumption $N^2 \gg 1$ we can neglect the contribution of the terms proportional to a^r_1, b^t_1, b^r_1, c^t_1 and c^r_1 when compared with the term $a^t_1 \beta_e N^6$. After this we can neglect the term C_0 when compared with $B_0 N^2$ provided B_0 is not close to zero (which is true for whistler-mode waves unless X is close to unity). However, we cannot neglect $B_0 N^2$ when compared with $A_0 N^4$ since A_0 is close to zero at θ close to θ_R ($A_0 = 0$ when $\theta = \theta_R$). As a result equation (6.4) simplifies to:

$$a^t_1 \beta_e N^4 + A_0 N^2 + B_0 = 0. \qquad (6.5)$$

As one can see, equation (6.5) does not include relativistic terms and could be obtained within the non-relativistic approximation.

Let us now consider another limiting case when a^t_1 is so close to zero that we can neglect the term $a_1^t \beta_e N^6$ altogether. In this case the terms $a_1^r \beta_e N^4$ and $b_1^t \beta_e N^4$ can no longer be ignored and equation (6.4) reduces to:

$$(A_0 + a_1^r \beta_e + b_1^t \beta_e) N^2 + B_0 = 0. \tag{6.6}$$

In contrast to (6.5), equation (6.6) contains the relativistic term a_1^r.

In order to decide whether equation (6.5) or (6.6) is to be used for the analysis of quasi-electrostatic whistler-mode propagation in a hot plasma, we need to know the values of a_1^t as a function of Y, A_e and θ. The explicit expression for $a_1^t(Y, A_e, \theta)$ follows from (1.43) in a straightforward way (Sazhin, 1986b):

$$
\begin{aligned}
a_1^t &= \frac{Y^2}{(Y^2 - 1)^3 (4Y^2 - 1)} \\
&\times \left\{ \cos^4 \theta \left[10Y^2 - 49Y^4 + 39Y^6 - 12Y^8 + A_e(-10Y^2 + 14Y^4 - 4Y^6) \right] \right. \\
&+ \cos^2 \theta \left[-3 + 11Y^2 + 4Y^4 + A_e(3 + 4Y^2 - 11Y^4 + 4Y^6) \right] \\
&+ \left. A_e(-3 + 6Y^2 - 3Y^4) \right\}.
\end{aligned} \tag{6.7}
$$

Expression (6.7) can be simplified if we remember our initial assumption that θ is close to θ_R defined by (2.23) or (2.24). This allows us to write:

$$\theta = \theta_R + \theta', \tag{6.8}$$

where $|\theta'| \ll \min(\theta_R, 1)$, and all the angles are expressed in radians. Having substituted (6.8) into (6.7) we obtain:

$$a_1^t = a_0^t + a_\theta^t \theta', \tag{6.9}$$

where

$$a_0^t = \frac{\sum_{i=0}^{2} \alpha_i X^i}{X^2 (Y^2 - 1)^2 (4Y^2 - 1)}, \tag{6.10}$$

$$a_\theta^t = \frac{2[(-1 + X + Y^2)(1 - X)(1 - Y^2)]^{1/2} \sum_{i=0}^{1} \gamma_i X^i}{X^2 (Y^2 - 1)^3 (4Y^2 - 1)}, \tag{6.11}$$

$$
\begin{aligned}
\alpha_0 &= -12Y^8 + 51Y^6 - 88Y^4 + 59Y^2 - 10 \\
&\quad + A_e(-4Y^6 + 18Y^4 - 24Y^2 + 10), \\
\alpha_1 &= -24Y^6 + 82Y^4 - 87Y^2 + 17 + A_e(4Y^6 - 19Y^4 + 32Y^2 - 17), \\
\alpha_2 &= -12Y^4 + 31Y^2 - 7 + A_e(Y^4 - 8Y^2 + 7), \\
\gamma_0 &= 24Y^8 - 102Y^6 + 176Y^4 - 118Y^2 + 20 \\
&\quad + A_e(8Y^6 - 36Y^4 + 48Y^2 - 20), \\
\gamma_1 &= -\alpha_1.
\end{aligned}
$$

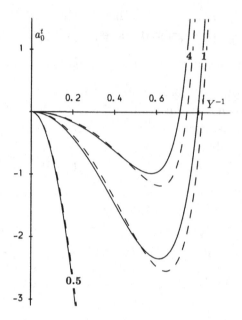

Fig. 6.1 Plots of a_0^t (see equation (6.10)) versus Y^{-1} for $A_e = 1$ (solid) and $A_e = 2$ (dashed); figures near the curves indicate the values of $\nu \equiv \Pi^2/\Omega^2$.

For whistler-mode propagation in a dense plasma ($X \gg 1$) the expressions for a_0^t and a_θ^t are simplified to:

$$a_0^t = \frac{7 - 3Y^2}{(Y^2 - 1)^2} + A_e \frac{Y^2 - 7}{(Y^2 - 1)(4Y^2 - 1)}, \tag{6.12}$$

$$a_\theta^t = 2(Y^2 - 1)^{-3/2} \left[\frac{6Y^4 - 19Y^2 + 17}{Y^2 - 1} + A_e \frac{-4Y^4 + 15Y^2 - 17}{4Y^2 - 1} \right]. \tag{6.13}$$

As follows from (6.9) the term proportional to a_θ^t gives a small correction to a_0^t provided (6.8) is valid. Hence, we can expect that the sign of a_1^t is determined by the sign of a_0^t provided a_0^t is not close to zero.

The plots of a_0^t determined by (6.10) versus Y^{-1} for different values of $\nu = X/Y^2$ are shown in Figs. 6.1 and 6.2. As follows from Fig. 6.1, the values of a_0^t for $\nu = 0.25$ are always negative. However, when $\nu \geq 1$ then a_0^t is negative for small Y^{-1}, but becomes positive at Y^{-1} close to unity. At a certain Y^{-1} in the interval $(0.69, \approx 0.8)$ $a_0^t = 0$. The value of Y^{-1} at which $a_0^t = 0$ decreases with increasing ν and decreasing A_e from 2 to 1.

The solutions of equations (6.5) and (6.6) will be considered in Sections 6.2 and 6.3 respectively.

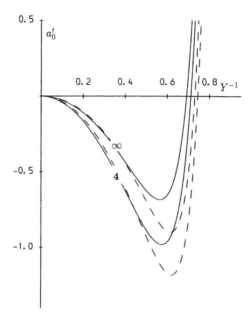

Fig. 6.2 The same as Fig. 6.1 but with a different scale of a_0^t axis and for different ν; the curves ∞ refer to $\nu \to \infty$.

6.2 The quasi-electrostatic solution for $a_1^t \neq 0$

As was pointed out in the previous section, for $a_1^t \neq 0$ the quasi-electrostatic whistler-mode propagation is described by equation (6.5). Two solutions of this equation are straightforward ones:

$$N_{1,2}^2 = \frac{-A_0 \pm \sqrt{A_0^2 - 4a_1^t \beta_e B_0}}{2a_1^t \beta_e}. \tag{6.14}$$

However, only the solutions corresponding to real and positive values of N^2 have physical meaning.

Writing θ in the form (6.8) and keeping only the terms proportional to θ' or β_e, but not to $\theta'\beta_e$, we can present the solution (6.14) in a simpler form:

$$N_{1,2}^2 = \frac{-A_{0R} \pm \sqrt{A_{0R}^2 - 4a_0^t \beta_e B_{0R}}}{2a_0^t \beta_e}, \tag{6.15}$$

where

$$A_{0R} = A_{00}\theta', \tag{6.16}$$

$$A_{00} = 2\sqrt{\frac{(X-1)(X+Y^2-1)}{Y^2-1}}, \tag{6.17}$$

$$B_{0R} = \frac{(X-1)(2X+Y^2-1)}{Y^2-1}, \tag{6.18}$$

and a_0^t was defined by equation (6.10).

In the case of a dense plasma ($X \gg 1$) the expressions for A_{00} and B_{0R} reduce to:

$$A_{00} = \frac{2X}{\sqrt{Y^2-1}}, \tag{6.19}$$

$$B_{0R} = A_{00}^2/2, \tag{6.20}$$

and the solution (6.15) simplifies to:

$$N_{1,2}^2 = -\frac{2Y^2\left(\theta' \pm \sqrt{\theta'^2 - 2a_0^t\beta_e}\right)}{a_0^t\tilde{w}_\parallel^2\sqrt{Y^2-1}}, \tag{6.21}$$

where $\tilde{w}_\parallel = w_\parallel/c$, and a_0^t is defined by (6.12).

In what follows we restrict ourselves to the analysis of solution (6.21), which appears to be the simplest from a mathematical point of view and is the most important for magnetospheric applications. The generalization of our analysis to equation (6.15) is straightforward.

As follows from (6.21), the structure of this solution appears to be different when $a_0^t < 0$ and $a_0^t > 0$ (when $a_0^t \approx 0$ then equation (6.21) becomes invalid). The solutions in these ranges of a_0^t will be considered separately in the next two subsections.

(a) The solution for $a_0^t < 0$

When $a_0^t < 0$ then $\theta'^2 - 2a_0^t\beta_e > 0$ and both solutions (6.21) are real. However, another requirement $N^2 > 0$ (when the solution (6.21) has a physical meaning) is satisfied only when we take '+' before the radical. Hence, in this case solutions (6.21) reduce to a single solution:

$$N^2 = -\frac{2Y^2\left(\theta' + \sqrt{\theta'^2 - 2a_0^t\beta_e}\right)}{a_0^t\tilde{w}_\parallel^2\sqrt{Y^2-1}}. \tag{6.22}$$

N^2 described by (6.22) monotonically increases with increasing θ'. When $\theta = \theta_R$ we put $\theta' = 0$ in (6.22) and reduce this equation to:

$$N^2 = N_{\lim}^2 \equiv \frac{2Y^2\sqrt{\nu}}{\sqrt{-a_0^t}\tilde{w}_\parallel\sqrt{Y^2-1}}. \tag{6.23}$$

In contrast to the case of wave propagation in a cold plasma, $N_{\lim}$ is finite when $\tilde{w}_\parallel \neq 0$.

Although equations (6.22) and (6.23) are obviously rather simple and convenient for practical applications, one should be cautious about the range of their applicability, i.e. the conditions under which they were derived should be checked. Alternatively we can directly compare the results predicted by equation (6.22) and those which follow from a rigorous numerical analysis of the wave dispersion equation (cf. Fig. 5.7 where the same comparison was made for quasi-longitudinal whistler-mode propagation).

Using the same Y and plasma parameters as in Fig. 5.7 but for θ in the vicinity of θ_R we present in Fig. 6.3 the results of the numerical solution of the dispersion equation and the quasi-electrostatic solution described by equation (6.21) (Horne & Sazhin, 1990). As follows from this figure, the lower branch of the numerical curve almost exactly coincides with the quasi-electrostatic solution at $\theta < \theta_R$. However, the upper branch of the numerical solution seems to have no analogue in the quasi-electrostatic solution (6.21). A comparison similar to that shown in Fig. 6.3 was made by Horne & Sazhin (1990) for other regions of the magnetosphere and other frequencies. In particular, in Fig. 6.4 we have shown the curves similar to those in Fig. 6.3 but for plasma parameters typical for the equatorial magnetosphere at $L = 4$ with $\Omega/2\pi = 12$ kHz, cold plasma density $n_c = 200$ cm^{-3} and temperature $T_c = 0.5$ eV, hot plasma parameters being the same as in Fig. 6.3. As follows from Fig. 6.4, the numerical solution in this case has only one branch which almost coincides with the quasi-electrostatic solution at $\theta < \theta_R$. In fact this coincidence between numerical and quasi-electrostatic solutions at $\theta < \theta_R$ appeared to be typical for other areas of the equatorial magnetosphere at $4 \leq L \leq 6.6$ considered by Horne & Sazhin (1990), while at $\theta > \theta_R$ these solutions often behaved in a quite different way. Thus our solution (6.22) seems to be reliable at $\theta < \theta_R$ but not at $\theta > \theta_R$. However, in the latter case a simplified electrostatic solution of the general dispersion equation (1.42), based on the assumption that $A = 0$, can be used for numerical analysis of wave propagation as was also shown by Horne & Sazhin (1990).

(b) The solution for $a_0^t > 0$

When $a_0^t > 0$ then $\theta'^2 - 2a_0^t\beta_e \geq 0$ only when $|\theta'| \geq \sqrt{2a_0^t\beta_e}$. This means that the solution (6.21) is complex at θ' close to zero, which contradicts our initial assumption that N is real. Hence the quasi-electrostatic solution describes no wave propagation when $|\theta'| < \sqrt{2a_0^t\beta_e}$. On the other hand, $N^2 < 0$ when $\theta' > \sqrt{2a_0^t\beta_e}$ which means that at these values of θ' the

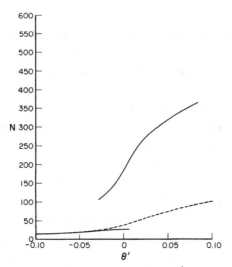

Fig. 6.3 Wave refractive index N as a function of $\theta' = \theta - \theta_R$, measured in radians for the same plasma parameters (typical for $L = 6.6$) and Y^{-1} as in Fig. 5.7. The numerical solution of the general electromagnetic dispersion equation is shown by the solid curve and the analytical solution of the quasi-electrostatic dispersion equation (6.21) is shown by the dashed curve. The resonance cone angle corresponds to $\theta' = 0$ (Horne & Sazhin, 1990).

waves, also, cannot propagate. However, at $\theta' \leq -\sqrt{2a_0^t\beta_e}$ (i.e. at $\theta \leq \theta_R - \sqrt{2a_0^t\beta_e}$) both solutions (6.21) correspond to $N^2 > 0$, i.e. describe actual quasi-electrostatic wave propagation. When $\theta' = -\sqrt{2a_0^t\beta_e}$, i.e. when $\theta = \theta_R - \sqrt{2a_0^t\beta_e}$, then N^2 is determined by equation (6.23).

As in subsection (a) we compare the results predicted by (6.22) with those which follow from a rigorous numerical analysis of the wave dispersion equation. First we take the same plasma parameters as were used for plotting the curves in Fig. 6.4 but for $Y^{-1} = 0.8$ when $a_0^t > 0$ (see Fig. 6.2). The corresponding plots are shown in Fig. 6.5. As one can see from this figure, the correspondence between the lower branch of the solution (6.21) and the numerical results is surprisingly good, while the upper branch seems to have no analogue in the numerical solution. This appears to be quite a persistent feature for other plasma parameters in the equatorial magnetosphere at $4 \leq L \leq 6.6$, although the upper branch in the numerical solution might appear for different plasma parameters (e.g. Tokar & Gary, 1985). Hence, we can conclude that when $a_0^t > 0$ then we can use only the lower branch of the solution (6.21) (corresponding to '+' before the radical) for the analysis

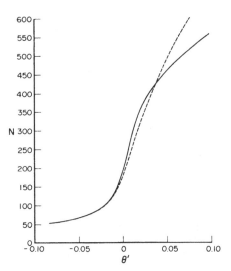

Fig. 6.4 The same as Fig. 6.3 but for plasma parameters typical for $L = 4$ ($Y^{-1} = 0.6$ and $\nu = 112.05$) (Horne & Sazhin, 1990).

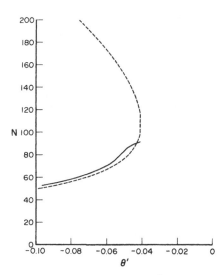

Fig. 6.5. The same as Fig. 6.4 but for $Y^{-1} = 0.8$ (Horne & Sazhin, 1990).

of whistler-mode propagation in the equatorial magnetosphere at $4 \leq L \leq 6.6$. The range of applicability of equation (6.21) in other regions of the magnetosphere or other plasma models needs a separate analysis which is beyond the scope of the present book.

In the next subsection we use expression (6.21) (or (6.22) when $a_0^t < 0$) for the analysis of the direction of whistler-mode group velocity.

(c) The direction of the group velocity

Having substituted (6.21) into equation (2.46) we obtain an expression for the angle $\psi = \theta + \theta_g$ in the form:

$$\psi = \theta_R + \theta' \pm \arctan[1/(2\sqrt{\theta'^2 - \tilde{\varphi}})], \qquad (6.24)$$

where

$$\tilde{\varphi} = 4a_0^t \beta_e B_{0R}/A_{00}^2, \qquad (6.25)$$

the upper (lower) sign in (6.24) corresponds to the upper (lower) sign in (6.21), as in the previous sections, and the angles are positive if measured clockwise.

Remembering our assumptions that $|\theta'| \ll 1$ and $\beta_e \ll 1$ (the latter implies that $|\tilde{\varphi}| \ll 1$) equation (6.24) can be simplified to:

$$\psi = \theta_R \pm \pi/2 + \theta' \mp 2\sqrt{\theta'^2 - \tilde{\varphi}}. \qquad (6.26)$$

When $a_0^t < 0$ then only the lower signs in (6.26) have physical meaning. When $a_0^t > 0$ then equation (6.26) is defined for $\theta' < -\sqrt{\tilde{\varphi}}$. In the case of a dense plasma ($X \gg 1$) the expression for $\tilde{\varphi}$ is simplified to $\tilde{\varphi} = 2a_0^t \beta_e$.

When $a_0^t < 0$ then $|\psi|$ attains its maximal value

$$|\psi_{cr}| = |\pi/2 - \theta_R - \sqrt{-3\tilde{\varphi}}| \qquad (6.27)$$

at $\theta' = -\sqrt{-\tilde{\varphi}/3}$. As follows from (6.27) the effect of non-zero electron temperature tends to decrease $|\psi_{cr}|$ so that the waves propagate closer to magnetic field lines.

When $a_0^t > 0$ then $|\psi|$ for the lower branch (lower sign in (6.26)) increases with increasing θ' (θ) until it reaches its maximal value

$$|\psi_{cr}| = |\pi/2 - \theta_R + \sqrt{\tilde{\varphi}}| \qquad (6.28)$$

at $\theta' = -\sqrt{\tilde{\varphi}}$. For the upper branch expression (6.26) predicts anomalous dispersion of the waves ($\psi \in (\pi/2, \pi)$ while $\theta \in (0, \pi/2)$). As follows from (6.28), for $a_0^t > 0$ the effects of finite electron temperature tend to increase ψ_{cr} in contrast to the case when $a_0^t < 0$.

Note that in contrast to the quasi-longitudinal approximation, thermal corrections to ψ in the quasi-electrostatic approximation are proportional to $\sqrt{\beta_e}$, not to β_e (cf. equations (5.19) and (6.26)–(6.28)). Hence, the influence of thermal effects on ψ appears to be more significant for quasi-electrostatic waves than for quasi-longitudinal ones. The effects of finite electron density and ions on ψ appear via the dependence of θ_R on ν^{-1} and r respectively (see equation (2.23)).

6.3 The quasi-electrostatic solution for $a_1^t = 0$

The solution of equation (6.6) describing quasi-electrostatic wave propagation when $a_1^t = 0$ is straightforward:

$$N^2 = -\frac{B_0}{A_0 + (a_1^r + b_1^t)\beta_e}. \tag{6.29}$$

This solution has a physical meaning when $N^2 > 0$.

As in Section 6.2 we write θ in the form (6.8) and keep only the terms proportional to θ' or β_e, but not $\theta'\beta_e$. As a result we present the solution (6.29) in a simpler form (Sazhin, 1987a):

$$N^2 = \frac{\tilde{\nu}}{\theta' + \theta_0'}, \tag{6.30}$$

where

$$\tilde{\nu} = -B_{0R}/A_{00}, \tag{6.31}$$

(B_{0R} and A_{00} are defined by (6.18) and (6.17) respectively),

$$\theta_0' = (a_1^r(\theta = \theta_R) + b_1^t(\theta = \theta_R))\beta_e/A_{00}(\theta = \theta_R), \tag{6.32}$$

$$a_1^r(\theta = \theta_R) \equiv a_0^r$$
$$= \frac{-3Y^4 + 3Y^2 + P(4Y^2 - 2) + A_e[-2Y^4 + 2Y^2 + P(6Y^2 + 2)]}{2(1 - P)(1 - Y^2)}, \tag{6.33}$$

$$b_1^t(\theta = \theta_R) \equiv b_0^r$$
$$= \left\{[-3Y^6 + 6Y^4 - 3Y^2 + P(3Y^4 + Y^2 + 8)](P - Y^2)(Y^2 - 1)^{-1}\right.$$
$$\left. + A_e P[-4Y^6 + 29Y^4 - 7Y^2 + P(-11Y^4 + Y^2 - 8)](4Y^2 - 1)^{-1}\right\}$$
$$\times [(1 - P)(1 - Y^2)^2]^{-1}. \tag{6.34}$$

The expression for a_0^r was obtained in a similar way to (6.10), non-relativistic perturbations of ϵ_{ij}^0 being replaced by relativistic ones. The expression for b_0^t was obtained in a similar way to (6.10) but for the temperature perturbation of the coefficient B in (1.42), $P = 1 - X = 1 - \nu Y^2$.

The range of validity of (6.30) is restricted to those Y, ν and A_e for which $a_0^t = 0$ (see Fig. 6.1). For $A_e \geq 1$ this condition can be satisfied only for $Y^{-1} \geq 0.69$. As already mentioned, the values of Y^{-1} when this condition is valid increase with increasing A_e and decreasing electron density.

As the parameters Y, P and A_e are bound by the condition $a_0^t = 0$, θ_0'/β_e appears to be a function of only two of these three parameters. For example,

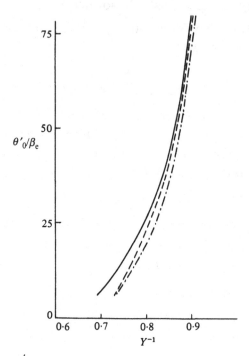

Fig. 6.6 Plots of θ_0'/β_e (see equation (6.32)) versus Y^{-1} for $A_e = 1$ (solid) and $A_e = 2$ (dashed); the dashed–dotted curve refers to the non-relativistic analogue of θ_0'/β_e (without the contribution of the term a_1^r) for $A_e = 2$. The divergence between relativistic and non-relativistic curves for $A_e = 1$ is about half that for $A_e = 2$ (Sazhin, 1987a).

we can fix the parameter A_e and consider P and θ_0'/β_e as functions of Y. Plots of θ_0'/β_e versus Y^{-1} for $A_e = 1$ and $A_e = 2$ are shown in Fig. 6.6. In the same figure we have also shown the plots of θ_0'/β_e versus Y^{-1} for $A_e = 2$ obtained within the non-relativistic approximation ($a_1^r = 0$). The divergence between relativistic and non-relativistic curves for $A_e = 1$ is about half that for $A_e = 2$.

As follows from Fig. 6.6, the values of θ_0'/β_e monotonically increase with increasing Y^{-1} both for $A_e = 1$ and $A_e = 2$, and are always positive. This means that these waves can propagate at $\theta < \theta_R - \theta_0'$. For $A_e = 1$ the value of θ_0'/β_e is slightly larger than for $A_e = 2$. Essentially the same results can be obtained from a non-relativistic treatment of the problem owing to the closeness of the relativistic and non-relativistic curves in Fig. 6.6. $N \to \infty$ when $\theta' \to \theta_0'$. Thus the general shapes of the curves N versus θ for a hot plasma when $a_0^t = 0$ are similar to that for a cold plasma, the resonance cone angle θ_R being replaced by $\theta_R - \theta_0'$. At θ' in the immediate vicinity of θ_0' the solution (6.30) becomes invalid as in this case we can no longer consider ϵ_{ij}^t as perturbations of ϵ_{ij}^0 (cf. equation (1.78)).

Similarly to Section 6.2 we can use equation (6.30) for the estimate of ψ. As a result we obtain:

$$\psi = \theta_R + \theta' + \arctan[1/(2(\theta' + \theta_0'))]. \tag{6.35}$$

Again as in Section 6.2 we simplify (6.35) to:

$$\psi = \theta_R - \pi/2 - \theta' - 2\theta_0'. \tag{6.36}$$

As follows from (6.36), the maximal value of $|\psi|$ is attained when $\theta' = -\theta_0'$ and is equal to:

$$|\psi_{cr}| = |\theta_R - \pi/2 - \theta_0'|. \tag{6.37}$$

The expression for $|\psi_{cr}|$ determined by (6.37) coincides with that for a cold plasma with the resonance cone angle equal to $\theta_R - \theta_0'$. In contrast to the case when $a_0^t \neq 0$, in the case considered in this section the thermal corrections to ψ are proportional to β_e and not to $\sqrt{\beta_e}$. Hence, the influence of thermal effects on ψ appears to be less significant when $a_0^t = 0$ than when $a_0^t \neq 0$.

6.4 An improved quasi-electrostatic approximation

In this section we attempt to improve the solutions discussed in the previous two sections by taking into account the contribution of more terms in the left-hand side of (6.4). Namely, looking for an improved approximation for the solution of equation (6.14) ($a_1^t \neq 0$) we add the contribution of the term C_0 and the terms proportional to $\beta_e N^4$ and reduce (6.4) to (Sazhin 1988e):

$$a_1^t \beta_e N^6 + (A_0 + a_1^r \beta_e + b_1^t \beta_e) N^4 + B_0 N^2 + C_0 = 0. \tag{6.38}$$

We suppose (and this will be consistent with the results) that the solution of (6.38) is close to (6.14) and can be presented as:

$$N^2 = N_q^2 (1 + \bar{\alpha}), \tag{6.39}$$

where N_q is the quasi-electrostatic solution (6.14), and $|\bar{\alpha}| \ll 1$. Remembering (6.14) we obtain after substituting (6.39) into equation (6.38):

$$\bar{\alpha} = -\frac{(a_1^r + b_1^t)\beta_e N_q^4 + C_0}{N_q^4 [2a_1^t \beta_e N_q^2 + A_{0R} + 2(a_1^r + b_1^t)\beta_e]}. \tag{6.40}$$

Considering wave propagation in a dense plasma where $X \gg 1$ and $\theta = \theta_R + \theta'$ ($|\theta'| \ll 1$) we obtain $A_{0R} = A_{00}\theta'$, where A_{00} is defined by (6.19),

$$C_0 = X^3/(Y^2 - 1), \qquad (6.41)$$

$a_1^t = a_1^t(\theta_R) \equiv a_0^t$ is defined by (6.12),

$$a_1^r = a_1^r(\theta_R) = \frac{2Y^2 - 1 + A_e(3Y^2 + 1)}{Y^2 - 1}, \qquad (6.42)$$

$$b_1^t = b_1^t(\theta_R) = \frac{X}{(Y^2 - 1)^2}\left[\frac{3Y^4 + 4Y^2 + 1}{Y^2 - 1} + A_e\frac{-11Y^4 - 1}{4Y^2 - 1}\right], \qquad (6.43)$$

and N_q are the quasi-electrostatic solutions (6.21) (negative and complex N_q^2 are disregarded). Note that expressions (6.42) and (6.43) for a_1^r and b_1^t are different from those defined by equations (6.33) and (6.34). The latter were obtained under the condition that $a_0^t = 0$ but without restrictions on the value of X.

Comparing (6.42) and (6.43) one can see that for $X \gg 1$ the contribution of a_1^r can be neglected when compared with b_1^t. As a result the expression for $\bar{\alpha}$ simplifies to:

$$\bar{\alpha} = \frac{\beta_e\sqrt{Y^2 - 1}(b_0^t + a_0^t\beta_e\xi_\theta^{-2})}{2(\xi_\theta - \theta' - b_0^t\beta_e\sqrt{Y^2 - 1})}, \qquad (6.44)$$

where

$$\xi_\theta = \theta' \pm \sqrt{\theta'^2 - 2a_0^t\beta_e}. \qquad (6.45)$$

Equation (6.44) has an especially simple form in two particular cases:
(1) $a_0^t < 0$; $\theta' = 0$:

$$\bar{\alpha} \approx \frac{(2b_0^t - a_0^t)\sqrt{Y^2 - 1}}{4\sqrt{2}\sqrt{-a_0^t}}\sqrt{\beta_e} \equiv \bar{\alpha}_0\sqrt{\beta_e}; \qquad (6.46)$$

(2) $a_0^t > 0$; $\theta' = -\sqrt{2a_0^t\beta_e}$:

$$\bar{\alpha} = -\frac{2b_0^t + a_0^t}{4b_0^t} \equiv \bar{\alpha}_1. \qquad (6.47)$$

The plots of $\bar{\alpha}_0$ versus Y^{-1} for different values of A_e are shown in Fig. 6.7. Remembering that in magnetospheric conditions β_e is in most cases below 0.02, and assuming that the condition $|\bar{\alpha}| \ll 1$ is valid when $|\bar{\alpha}| \lesssim 0.3$, we obtain that (6.39) is valid when $|\bar{\alpha}_0| \lesssim 2$, i.e. when $Y^{-1} \lesssim 0.59$ for

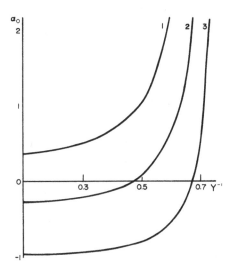

Fig. 6.7 Plots of $\bar{\alpha}_0$ (see equation (6.46)) versus Y^{-1} for different values of A_e (figures near the curves) (Sazhin, 1988e).

$A_e = 1$; $Y^{-1} \lesssim 0.67$ for $A_e = 2$; $Y^{-1} \lesssim 0.73$ for $A_e = 3$. For $A_e = 1$ this is roughly confirmed by the closeness between numerical and analytical results at $Y^{-1} \lesssim 0.6$ (see Figs. 6.3 and 6.4).

As follows from the numerical estimates, $\bar{\alpha}_1 < 0$ and $|\bar{\alpha}_1|$ is close to 0.5, being slightly above it. This means that our expression (6.39) is not strictly valid for $\theta' = -\sqrt{2a_0^t\beta_e}$. This also seems to be consistent with the divergence between numerical and analytical curves at $\theta' = -\sqrt{2a_0^t\beta_e}$ when $a_0^t > 0$ (see Fig. 6.5).

Now we improve the quasi-electrostatic solution (6.30) by keeping the term C_0 in equation (6.4). As a result, equation (6.4) reduces to equation (6.38) with $a_1^t = 0$. Again we look for the solution of equation (6.38) with N in the form (6.39). Substituting (6.39) into (6.38), remembering (6.30) and our assumption $a_0^t = 0$ we obtain

$$\bar{\alpha} = -C_0(\theta' + \theta_0')/(A_{00}\tilde{\nu}^2). \tag{6.48}$$

In the case of a dense plasma (6.48) reduces to

$$\bar{\alpha} = -\sqrt{Y^2 - 1}(\theta' + \theta_0')/2. \tag{6.49}$$

As follows from (6.49), the conditions $|\theta'| \ll 1$ and $\beta_e \ll 1$ imply $|\bar{\alpha}| \ll 1$ which confirms the validity of the presentation (6.39) with quasi-electrostatic whistler-mode refractive index N_q defined by (6.30). Both when a_0^t is not close to 0 and when $a_0^t = 0$, the smallness of $\bar{\alpha}$ can serve as an additional justification for the validity of the quasi-electrostatic approximation.

6.5 The quasi-electrostatic solution for $\xi_1 = -0.924$

The analysis of the previous four sections was restricted to quasi-electrostatic whistler-mode propagation in a low-temperature limit when the elements of the plasma dielectric tensor could be presented in the form (1.78). In this section we consider the solution similar to that in Section 6.2 but for a special case when $\xi_1 = (\omega - \Omega_0)/(k_\| w_\|) = -0.924$ but $\xi_0 = \omega/(k_\| w_\|) \gg 1$ (cf. the corresponding solution for quasi-longitudinal waves considered in Section 5.3). For this special case the elements of the plasma dielectric tensor reduce to ϵ_{ij}^{t1} defined by (1.115)–(1.120). Having substituted (1.115)–(1.120) into the general dispersion equation (1.42) and restricting our analysis to the dense plasma limit ($X \gg 1$) we can write the quasi-electrostatic solution of this equation in the form (6.21) with a_0^t replaced by (Sazhin & Sazhina, 1988):

$$
\begin{aligned}
a_{0\xi} &= \left[-24Y^6 - 48Y^5 - 14Y^4 + 8Y^3 + 5Y^2 + Y \right. \\
&+ \left. A_e(2Y^6 + 4Y^5 - 6Y^4 - 8Y^3 + 3Y^2 + 4Y + 1) \right] \\
&\times \left[2Y^4(Y+1)^2(4Y^2 - 1) \right]^{-1}.
\end{aligned}
\tag{6.50}
$$

The important restriction on our solution is that the condition $|\gamma| \ll \min (\omega_0, \Omega_0 - \omega_0)$ should be valid in order that we can neglect the influence of wave damping or amplification on their propagation. As in Section 5.3, this condition is satisfied only for sufficiently large $A_e \in [3, 4]$. Those Y^{-1} and β_e for which another restriction to our solution, namely $\xi_1 = -0.924$, is satisfied for $A_e = 3, 4$ for different θ' are shown in Fig. 6.8 for $\beta_e \in [0.005, 0.015]$ (typical values for magnetospheric conditions). Note that at $Y^{-1} < 0.65$ the third restriction on our model, namely $|\xi_0| \gg 1$, becomes invalid. $a_{0\xi} < 0$ for Y and A_e under consideration and so the general solution (6.21) for $a_{0\xi} \neq 0$ can be reduced to (6.22).

When $\xi_0 = 0.924$ and $|\xi_1| \gg 1$ then no quasi-electrostatic whistler-mode propagation can take place: the condition $A_0 = 0$ reduces to $\cot^2 \theta_R = -X/Y^2$, which is never satisfied.

Although the range of applicability of the quasi-electrostatic solution discussed in this section is much more restrictive than for the solutions discussed in Sections 6.2 and 6.3, it might be useful for checking the results

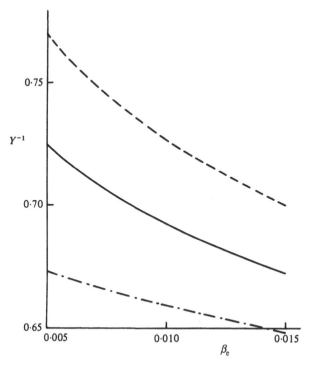

Fig. 6.8 Plots of Y^{-1} versus β_e satisfying the condition $\xi_1 = -0.924$ for $A_e = 3$ and $\theta' = \theta - \theta_R = -0.1$ rad (dashed), $\theta' = 0$ (solid) and $\theta' = 0.1$ rad (dashed–dotted). The corresponding plots for $A_e = 4$ almost coincide with those for $A_e = 3$, being slightly below them (Sazhin & Sazhina, 1988).

of numerical analysis as well as for obtaining a more complete picture of quasi-electrostatic whistler-mode propagation.

6.6 Wave polarization

As follows from equations (2.69) and (2.70), in the cold plasma limit $\alpha_1 = E_y/E_x \to 0$ and $\alpha_2 = E_z/E_x \to \cot\theta$ when $\theta \to \theta_R$. However, we might expect that the values for α_1 and α_2 would be different if we took into account the effects of finite electron temperature and anisotropy even when this temperature is quite low. We restrict ourselves to a limiting case $\theta = \theta_R$, $a_0^t < 0$ and $X \gg 0$, when the expression for N has a particular simple form (6.23). Having substituted (6.23) into the general expressions for the elements Λ_{ij} (see equation (1.156)) taken for $\theta = \theta_{R0} = \arccos(Y^{-1})$ we obtain (Sazhin, 1985):

$$\hat{\Lambda} = \hat{\Lambda}^{qe} \equiv X$$

$$
\times
\begin{pmatrix}
\frac{1}{Y^2-1} - \sqrt{\frac{2}{\beta_e|a_0|(Y^2-1)}}\frac{1}{Y^2} & \frac{iY}{Y^2-1} & \sqrt{\frac{2}{\beta_e|a_0|}}\frac{1}{Y^2} \\
-\frac{iY}{Y^2-1} & \frac{1}{Y^2-1} - \sqrt{\frac{2}{\beta_e|a_0|(Y^2-1)}} & 0 \\
\sqrt{\frac{2}{\beta_e|a_0|}}\frac{1}{Y^2} & 0 & -1 - \sqrt{\frac{2(Y^2-1)}{\beta_e|a_0|}}\frac{1}{Y^2}
\end{pmatrix}.
$$

$$(6.51)$$

After substituting (6.51) into the expressions (1.157) and (1.158) for α_1 and α_2 and neglecting higher-order terms with respect to β_e we obtain:

$$\alpha_1 = \alpha_1^{qe} \equiv -iY\sqrt{\frac{\beta_e|a_0^t|}{2(Y^2-1)}}, \tag{6.52}$$

$$\alpha_2 = \alpha_2^{qe} \equiv \frac{1}{\sqrt{Y^2-1}}\left[1 - Y^2\sqrt{\frac{\beta_e|a_0^t|}{2(Y^2-1)}}\right]. \tag{6.53}$$

As follows from (6.52) the effect of finite β_e results in the non-zero y component of the wave electric field being shifted in phase by $\pi/2$ with respect to its x component. The wave polarization in the (x, z) plane remains linear although α_2 is slightly reduced when $\beta_e \neq 0$.

It is convenient to consider wave polarization in a coordinate system x', y', z' such that $z' \parallel \mathbf{N}$ and y' coincides with y (see Fig. 6.9). In view of (6.52) and (6.53) we can write the components of the wave field in this system as (Sazhin, 1988f, 1989b):

$$E_{x'} = r_R E, \tag{6.54}$$

$$E_{y'} = -ir_R E, \tag{6.55}$$

where E is the modulus of the wave electric field in the direction of $\mathbf{N}$, and

$$r_R = \sqrt{0.5|a_0^t\beta_e|}. \tag{6.56}$$

In the case of plasma consisting of components with different β_e and a_0^t we can still use equations (6.54) and (6.55) with $a_0^t\beta_e$ replaced by $\sum_i a_{0_i}^t\beta_{e_i}$, where the summation is assumed over all the i components.

As follows from (6.54) and (6.55), quasi-electrostatic whistler-mode waves are circularly polarized with respect to $\mathbf{N}$ in a similar way as parallel propagating whistler-mode waves are circularly polarized with respect to the external magnetic field $\mathbf{B_0}$ (cf. Section 5.4).

It follows from (6.52)–(6.53) or (6.54)–(6.55) that the thermal effects on quasi-electrostatic whistler-mode polarization are proportional to $\sqrt{\beta_e}$,

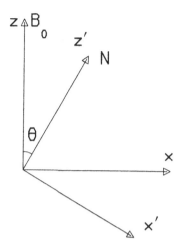

Fig. 6.9. The relative position of the axes (x, z) and (x', z') (Sazhin, 1989b).

i.e. they are more pronounced than the corresponding effects on quasi-longitudinal whistler-mode polarization (see equations (5.44) and (5.45)).

After substituting (6.52) and (6.53) into (1.163) and (1.164) we obtain the expressions for the wave magnetic field polarization at $\theta = \theta_R$ in the form:

$$B_y/B_x = -iY, \qquad (6.57)$$

$$B_z/B_x = -\sqrt{Y^2 - 1}. \qquad (6.58)$$

As follows from (6.57) and (6.58), the whistler-mode magnetic field polarization at $\theta = \theta_R$ does not depend on plasma temperature and anisotropy, at least within those approximations which have been used when deriving (6.52) and (6.53). Note that in a cold plasma at $\theta = \theta_R$ all the wave energy is concentrated in the wave electric field and the motion of particles, the energy of the wave magnetic field being zero. Hence, the ratios B_y/B_x and B_z/B_x for a cold plasma should be understood as the corresponding limits for $\theta \to \theta_R$.

Now we use expressions (6.54) and (6.55) for the analysis of a physical model of wave propagation, following Sazhin (1989b). Considering the components of the wave electric field $\mathbf{E}$ and the corresponding current $\mathbf{j}$ in the direction perpendicular to $\mathbf{N}$ ($\mathbf{E}_\perp$ and $\mathbf{j}_\perp$) we obtain from equations (2.84) and (2.85):

$$\mathbf{E}_\perp N^2 = \mathbf{E}_\perp - \frac{4\pi i}{\omega_0}\mathbf{j}_\perp. \qquad (6.59)$$

Remembering (6.54) and (6.55), after some rearrangements we obtain from equations (2.86) and (2.87):

$$j_{x'} = -\frac{1}{4\pi i} \frac{XE}{\sqrt{Y^2 - 1}}, \tag{6.60}$$

$$j_{y'} = \frac{1}{4\pi i} \frac{XE}{\sqrt{Y^2 - 1}}. \tag{6.61}$$

If we substitute equations (6.54)–(6.55) and (6.60)–(6.61) into (6.59) we can see that in the dense plasma limit this equation is satisfied when (6.23) is valid.

One can draw a parallel between this wave propagation and parallel whistler-mode propagation in a cold plasma. In both cases the waves are circularly polarized, although for quasi-electrostatic waves the axis of polarization coincides with $\mathbf{N}$, while for parallel propagating waves it coincides with $\mathbf{B}_0$. In both cases the wave energy is periodically transported from $\mathbf{j}_\perp$ to $\mathbf{E}_\perp$ and $\mathbf{B}_\perp$ and back to $\mathbf{j}_\perp$, the phase of $\mathbf{j}_\perp$ being shifted by $\pi/2$ with respect to the phases of $\mathbf{E}_\perp$ and $\mathbf{B}_\perp$. The equilibrium of the process is attained when N^2 satisfies equation (6.23).

6.7 Propagation in the presence of electron beams

The analysis of this chapter has so far been based on the assumption that the electron distribution function has the form (1.76) (weakly relativistic limit) or (1.90) (non-relativistic limit) (cf. Sections 5.1–5.4 where similar assumptions have been made for the analysis of quasi-longitudinal whistler-mode waves). However, in the actual conditions of the Earth's magnetosphere, in particular in the vicinity of the magnetopause, it is better to use a more general approximation (1.130) for the electron distribution function. The analysis of the quasi-longitudinal whistler-mode propagation in a plasma with this electron distribution function has been given in Section 5.5. Here we will generalize the analysis of Section 5.5 to the case of quasi-electrostatic whistler-mode propagation.

Using, as in Section 5.5, expressions (1.139) for the elements of the plasma dielectric tensor and assuming that the plasma is dense ($\Pi_\Sigma \gg \Omega_0$) we can generalize the solution of the quasi-electrostatic dispersion equation to:

$$N^2 = N_{qb}^2 \equiv \frac{Y^2 \nu \left(-\theta' \pm \sqrt{\theta'^2 - 2\sum_i \tilde{a}_{0i}\beta_{ei}}\right)}{\sqrt{Y^2 - 1}\sum_i \tilde{a}_{0i}\beta_{ei}}, \tag{6.62}$$

where

$$\tilde{a}_{0_i} = a_{0_i}^t + \Delta a_0 \hat{v}_{0_i}^2,$$ (6.63)

$a_{0_i}^t$ is the same as in (6.12) but refers to different electron components,

$$\Delta a_0 = \frac{2(-3Y^2 + 7)}{(Y^2 - 1)^2},$$ (6.64)

and $\hat{v}_{0_i} = v_{0_i}/w_{\parallel i}$; summation is assumed over all i components. When deriving (6.62) we assumed that $\sum_i \tilde{a}_{0_i}\beta_{e_i}$ was not close to zero, the case most important for practical applications. As in Section 6.2 the signs '+' and/or '−' in (6.62) are chosen so as to provide that N^2 is real and positive. Note that $\Delta a_0 = 2a_{0_i}^t$ when $A_{e_i} = 0$ (see the discussion after equation (5.50)).

In the limit $\hat{v}_{0_i} = 0$, equation (6.62) reduces to equation (1) of Sazhin (1988f) derived for quasi-electrostatic whistler-mode propagation in a hot anisotropic plasma without taking into account the contribution of electron beams.

The plots of Δa_0 versus Y^{-1} as well as a_0^t versus Y^{-1} for $A_e = 1$ and $A_e = 2$ are shown in Fig. 6.10. As follows from this figure, for $Y^{-1} \leq 0.65$ all the parameters Δa_0 and a_0^t are negative while for $Y^{-1} \geq 0.75$ they are positive; they change their signs for $0.65 < Y^{-1} < 0.75$. The parameter a_0^t describes the contribution of electron temperature and anisotropy, while the parameter Δa_0 describes the contribution of electron beams.

The sign of $\sum_i \tilde{a}_{0_i}\beta_{e_i}$ determines the character of the curves N versus θ' based on equation (6.62). When it is negative, N is a monotonically increasing function of θ' and is finite for $\theta' = 0$. When it is positive, then N is a two-valued function of θ' for $\theta' < \theta'_0 < 0$ and (6.62) has no solution for $\theta' > \theta'_0$ (cf. Section 6.2). For $\sum_i \tilde{a}_{0_i}\beta_{e_i}$ close to zero, equation (6.62) becomes invalid. In this case a different relativistic approach to the problem of quasi-electrostatic whistler-mode propagation is needed (cf. Section 6.3).

Now we generalize equation (6.62) to the case when the currents due to electron beams are not completely compensated. Remembering (1.140), (1.142), (1.144) and (1.43), this would result in the appearance of an additional term in the left-hand side of (6.5) equal to (see Sazhin, Walker & Woolliscroft, 1990b):

$$N^3 \frac{2Y^2 \nu \cos \theta_R \left[(2Y^2 - Y^4) \cos^2 \theta_R - 1 \right]}{(Y^2 - 1)^2} \sum_i \tilde{v}_{0_i} \kappa_i \equiv \hat{\phi} N^3,$$ (6.65)

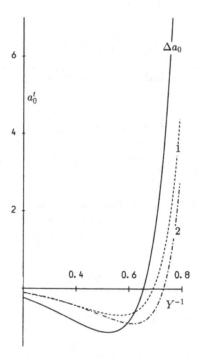

Fig. 6.10 Plots of a_0^t (see equation (6.12)) versus Y^{-1} for different A_e (dashed curves; values of A_e indicated) and the plot of Δa_0 (see equation (6.64)) versus Y^{-1} (solid curve Δa_0) (cf. Fig. 3 in Sazhin, Walker & Woolliscroft, 1990b).

where κ_i is the same as in (1.130) and

$$\hat{\phi} = -\frac{2\nu Y \sum_i \tilde{v}_{0_i}\kappa_i}{Y^2 - 1}. \tag{6.66}$$

Assuming $|\hat{\phi}| \ll 1$ and

$$N = N_{qb} + \Delta N, \tag{6.67}$$

where N_{qb} is determined by (6.62) and $|\Delta N| \ll N_q$, we can see that (6.5) is satisfied with the term $\hat{\phi}N^3$ taken into account when:

$$\Delta N = \frac{-\hat{\phi}N_{qb}^2}{2A_{0R} + 2N_{qb}^2 \sum_i \tilde{a}_{0_i}\beta_{e_i}}, \tag{6.68}$$

where A_{0R} and $\tilde{a}_{0_i}$ are determined by (6.19) and (6.63) respectively.

Equation (6.68) can be considerably simplified for $\theta = \theta_R$:

$$\Delta N (\theta = \theta_R) = \frac{\nu Y \sum_i \tilde{v}_{0_i}\kappa_i}{(Y^2 - 1) \sum_i \tilde{a}_{0_i}\beta_{e_i}}. \tag{6.69}$$

The sign of ΔN depends on the sign of $\sum_i a_{0_i}\beta_{e_i}$ and the direction of the net electron current. In particular, for $\sum_i a_{0_i}\beta_{e_i} < 0$ and $\sum_i \tilde{v}_{0_i}\kappa_i > 0$ the effect of this current tends to decrease the value of N_q determined by (6.62).

Problems

Problem 6.1 Comparing the numerical solution of equation (1.42) with the numerical solution of the electrostatic analogue of this equation ($A = 0$) at θ close to θ_R and $Y^{-1} < 1$, Horne & Sazhin (1990) found that these solutions almost coincided when $\theta > \theta_R$, except when θ was in the immediate vicinity of θ_R, while the electrostatic equation had no solution at $\theta < \theta_R$ for the parameters relevant to the Earth's magnetosphere ($\beta_e \ll 1$). Explain why.

Problem 6.2 Assuming an electron distribution function of the form (1.90) and imposing no restrictions on the electron thermal velocity (provided the non-relativistic approximation is valid), show that the electrostatic dispersion equation ($A = 0$ in equation (1.42)) can be written as:

$$1 + \frac{2X}{N^2\tilde{w}^2}\left[1 - \frac{2w_\parallel^2}{j!w_\perp^{2j+2}}\sum_{n=-\infty}^{n=+\infty}\mu_j^n Z(\xi_n)\right] = 0, \tag{6.70}$$

where the notation is the same as in equations (1.91)–(1.96). (For an isotropic Maxwellian plasma ($A_e = 1$) this equation reduces to that used in the numerical analysis by Horne & Sazhin (1990); see Problem 6.1).

7

Growth and damping of the waves

As was shown in Section 2.1, the general solution of the wave dispersion equation in a 'cold' plasma (see equation (2.4)) describes either wave propagation without damping or growth ($N_0^2 > 0$), or the absence of the waves ($N_0^2 \leq 0$; $N_0 = 0$ corresponds to plasma cutoffs). However, wave propagation in a plasma with non-zero temperature is accompanied, in general, by a change of amplitude of the waves, A_w, which is described by the increment of growth ($\gamma > 0$) or the decrement of damping ($\gamma < 0$) ($A_w \sim \exp(\gamma t)$). When $|\gamma| \ll \min(\omega_0, \Omega - \omega_0)$ (which is satisfied in most cases of whistler-mode propagation in the magnetosphere and will hereafter be assumed to be valid) then γ is described by equation (1.17). In this chapter, the latter equation will be applied to the analysis of whistler-mode growth ($\gamma > 0$) or damping ($\gamma < 0$) in different limiting cases of wave propagation considered in Chapters 3–6.

7.1 Parallel propagation (weakly relativistic approximation)

As was shown in Section 3.1, the general dispersion equation for parallel whistler-mode propagation in a weakly relativistic plasma, with the electron distribution function (1.76) (with $j = 0$), can be written in the form (3.10). This equation describes both wave propagation (see equation (1.16)) and the growth or damping (see equation (1.17)). The processes of parallel whistler-mode propagation in a weakly relativistic plasma in different limiting cases were analysed in Sections 3.2 and 3.3. Full analysis of whistler-mode growth or damping in a weakly relativistic plasma appears to be rather complicated even if we restrict ourselves to the case of parallel propagation, and we do not intend to present it in this book. Instead we restrict ourselves to the analysis of the marginal stability of the waves, i.e. to the analysis of the

144

conditions under which waves can propagate without change of amplitude, although a slight change of plasma parameters would result, in general, in their growth or damping. However, before doing this we need to consider some asymptotic properties of the generalized Shkarofsky function $\mathcal{F}_{q,p}$ (see equation (3.11)). This will be done in the next subsection following Sazhin & Temme (1990, 1991b).

(a) *Asymptotics of the generalized Shkarofsky function*

Putting $t = is$ we can rewrite the expression for $\mathcal{F}_{q,p}$ (see equation (3.11)) as:

$$\mathcal{F}_{q,p} = \int_0^{-i\infty} \exp(\varphi(s)) f(s) ds, \qquad (7.1)$$

where

$$\varphi(s) = -zs + \frac{as^2}{1+s},$$

$$f(s) = (1+s)^{-q}(1+bs)^{-p},$$

and both z and a are assumed to be real. We assume that the branch cuts of $f(s)$ are taken along the real interval $(-\infty, 0)$. We can rearrange the right-hand side of (7.1) so that $\mathcal{F}_{q,p}$ is written as:

$$\mathcal{F}_{q,p} = e^{z-2a} \int_0^{-i\infty} e^{a\psi(s)} f(s) ds, \qquad (7.2)$$

where

$$\psi(s) = (1-\mu)(s+1) + \frac{1}{s+1}, \qquad (7.3)$$

$$\mu = z/a.$$

To find an asymptotic expansion of $\mathcal{F}_{q,p}$ we use the saddle point method (Olver, 1974). Saddle points of ψ are determined by the condition:

$$\frac{d\psi(s)}{ds} = (1-\mu) - \frac{1}{(s+1)^2} = 0,$$

which is satisfied for

$$s = s_\pm \equiv -1 \pm \frac{1}{\sqrt{1-\mu}}. \qquad (7.4)$$

In view of the future application of our analysis to whistler-mode waves $(Y > 1)$ we consider $\mu \leq 0$. The analysis remains valid if $0 < \mu < 1$.

When $|\mu| \ll 1$ we have:

$$s_+ = \frac{\mu}{2} + O(\mu^2), \tag{7.5}$$

$$s_- = -2 - \frac{\mu}{2} + O(\mu^2). \tag{7.6}$$

$\psi(s)$ is real on the real axis and on the circle with centre $s = -1$ and radius $1/\sqrt{1-\mu}$. This circle passes through $s_\pm$. Hence, $\psi(s)$ is real and decreasing along the contour shown in Fig. 7.1. Now we deform the initial contour of integration from 0 to $-i\infty$ (see equation (7.2)) to this contour and write $\mathcal{F}_{q,p}$ as:

$$\mathcal{F}_{q,p} = \mathcal{F}_{q,p}^{(1)} + \mathcal{F}_{q,p}^{(2)} + \mathcal{F}_{q,p}^{(3)}, \tag{7.7}$$

where

$$\mathcal{F}_{q,p}^{(1)} = e^{z-2a} \int_0^{s_+} e^{a\psi(s)} f(s)\mathrm{d}s, \tag{7.8}$$

$$\mathcal{F}_{q,p}^{(2)} = e^{z-2a} \int_0^{-\pi} e^{a\psi(s)} f(s)\frac{\mathrm{d}s}{\mathrm{d}\tilde{\theta}}\mathrm{d}\tilde{\theta}, \tag{7.9}$$

$$\mathcal{F}_{q,p}^{(3)} = e^{z-2a} \int_{s_-}^{-\infty} e^{a\psi(s)} f(s)\mathrm{d}s. \tag{7.10}$$

In (7.9) we put $s = -1 + e^{i\tilde{\theta}}/\sqrt{1-\mu}$, $-\pi \le \tilde{\theta} \le 0$. This deformation of the contour of integration is possible only when the plasma is not very anisotropic ($\mu/2 < 1/A_e$), so that the contour does not pass through the singularity of $f(s)$. We expect that the condition $\mu/2 < 1/A_e$ is satisfied in most cases of whistler-mode propagation in the magnetosphere, and it will be assumed to be valid.

The term $\mathcal{F}_{q,p}^{(1)}$ gives the dominant contribution to the real part of $\mathcal{F}_{q,p}$. The term $\mathcal{F}_{q,p}^{(2)}$ contributes to the imaginary and real parts of $\mathcal{F}_{q,p}$, while the contribution of $\mathcal{F}_{q,p}^{(3)}$ can be neglected when compared with that of $\mathcal{F}_{q,p}^{(1)}$.

The analysis of the real part of $\mathcal{F}_{q,p}$ provides the basis for the analysis of whistler-mode propagation considered in Section 3.2 (see Sazhin & Temme (1990) for details). Here we concentrate our attention on the imaginary part of $\mathcal{F}_{q,p}$, i.e. on $\Im\mathcal{F}_{q,p}^{(2)}$ which will be eventually used for the analysis of marginally stable waves. For the semicircular part of the contour shown in Fig. 7.1 we have:

$$s + 1 = \frac{e^{i\tilde{\theta}}}{\sqrt{1-\mu}}, \quad -\pi \le \tilde{\theta} \le 0. \tag{7.11}$$

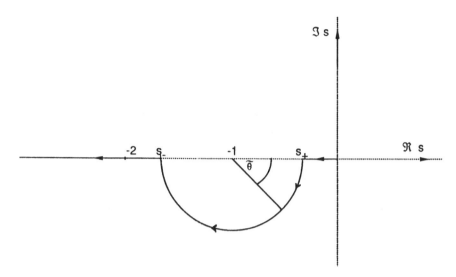

Fig. 7.1 The contour of integration in the complex s plane (see equations (7.7)–(7.10)) (Sazhin & Temme, 1990).

Equation (7.11) enables us to write:

$$\mathcal{F}_{q,p}^{(2)} = \mathrm{e}^{-ax_0^2} \int_0^{-\pi} \mathrm{e}^{-2a\sqrt{1-\mu}(1-\cos\tilde{\theta})} f(s) \frac{\mathrm{d}s}{\mathrm{d}\tilde{\theta}} \mathrm{d}\tilde{\theta}, \qquad (7.12)$$

where

$$x_0 = \sqrt{2 - \mu - 2\sqrt{1-\mu}}.$$

If we put $w = -\sin\left(\tilde{\theta}/2\right)$, we can rewrite (7.12) as:

$$\mathcal{F}_{q,p}^{(2)} = \mathrm{e}^{-ax_0^2} \int_0^1 \mathrm{e}^{-4a\sqrt{1-\mu}w^2} \tilde{g}(w) \mathrm{d}w, \qquad (7.13)$$

where

$$\tilde{g}(w) = f(s) \frac{\mathrm{d}s}{\mathrm{d}\tilde{\theta}} \frac{\mathrm{d}\tilde{\theta}}{\mathrm{d}w},$$

$$\frac{\mathrm{d}s}{\mathrm{d}\tilde{\theta}} = \mathrm{i}(s+1),$$

$$\frac{\mathrm{d}\tilde{\theta}}{\mathrm{d}w} = \frac{-2}{\sqrt{1-w^2}}.$$

The main contribution to the integral in (7.13) comes from the point $w = 0$. Expanding

$$\tilde{g}(w) = \sum_{k=0}^{\infty} c_k w^k, \tag{7.14}$$

we obtain:

$$\mathcal{F}_{q,p}^{(2)} \sim e^{-ax_0^2} \sum_{k=0}^{\infty} c_k \int_0^{\infty} e^{-4a\sqrt{1-\mu}w^2} w^k dw =$$

$$= \frac{1}{2} e^{-ax_0^2} \sum_{k=0}^{\infty} c_k \Gamma\left(\frac{k+1}{2}\right) \left[4a\sqrt{1-\mu}\right]^{-(k+1)/2} \tag{7.15}$$

for $a \gg 1$ and $\mu < 1$ (both these conditions are satisfied for whistler-mode waves in the magnetospheric plasma).

Using the definitions of s and w we can write

$$s = s_+ + \frac{1}{\sqrt{1-\mu}} \left[-2iw - 2w^2 + iw^3 + \frac{iw^5}{4} + \dots\right], \tag{7.16}$$

where s_+ is defined by (7.4). Furthermore we need:

$$\frac{ds}{d\theta} \frac{d\theta}{dw} = \frac{ds}{dw} = \frac{1}{\sqrt{1-\mu}} \left[-2i - 4w + 3iw^2 + \frac{5iw^4}{4} + \dots\right] \tag{7.17}$$

and

$$f(s) = f(s_+) \left[1 + f_1 w + f_2 w^2 + \dots\right], \tag{7.18}$$

where

$$f_1 = 2i \frac{\sqrt{1-\mu}(b-1)q - b(q+p)}{\sqrt{1-\mu}(b-1) - b},$$

$$f_2 = 2 \left\{ \frac{\sqrt{1-\mu}(b-1)b(2q^2 + 2qp - p)}{[\sqrt{1-\mu}(b-1) - b]^2} \right.$$

$$\left. + \frac{(1-\mu)[2q^2b - q^2] + q^2b^2(\mu - 2) - b^2[p^2 + 2pq]}{[\sqrt{1-\mu}(b-1) - b]^2} \right\}.$$

Having substituted (7.17) and (7.18) into (7.14) we obtain the following expressions for the first three coefficients in the expansion (7.14):

$$c_0 = \frac{-2i}{\sqrt{1-\mu}} f(s_+), \tag{7.19}$$

$$c_1 = \frac{-4[\sqrt{1-\mu}(b-1)(1-q) + b(q+p-1)]}{\sqrt{1-\mu}[\sqrt{1-\mu}(b-1) - b]} f(s_+), \tag{7.20}$$

$$c_2 = \frac{-i[\tilde{a} + \tilde{b}\mu + \tilde{c}\sqrt{1-\mu}]}{\sqrt{1-\mu}[\sqrt{1-\mu}(b-1)-b]^2} f(s_+), \qquad (7.21)$$

where

$$
\begin{aligned}
\tilde{a} &= b^2(-8q^2 - 4p^2 - 8qp + 16q + 8p - 6) + (2b-1)(4q^2 - 8q + 3), \\
\tilde{b} &= (b-1)^2(4q^2 - 8q + 3), \\
\tilde{c} &= 2b(b-1)(4q^2 + 4qp - 8q - 6p + 3).
\end{aligned}
$$

Having substituted (7.19) and (7.21) into (7.15) we obtain:

$$\Im F_{q,p} \sim -\frac{\sqrt{\pi} f(s_+)\exp(-ax_0^2)}{2\sqrt{a}(1-\mu)^{3/4}}\left[1 + \frac{\tilde{a} + \tilde{b}\mu + \tilde{c}\sqrt{1-\mu}}{16a(1-\mu)^{5/4}[\sqrt{1-\mu}(b-1)-b]^2}\right], \qquad (7.22)$$

where $\tilde{a}$, $\tilde{b}$ and $\tilde{c}$ are the same as in (7.21).

(b) Marginal stability of the waves

If we calculated $\Re F_{q,p}$ in the same asymptotic limit as we calculated $\Im F_{q,p}$ (Sazhin & Temme, 1990) then using equations (1.17) and (3.10) we could obtain the explicit expressions for the whistler-mode growth ($\gamma > 0$) or damping ($\gamma < 0$) (Sazhin, Sumner & Temme, 1992). However, these expressions appear to be too complicated for practical applications. In what follows in this subsection we shall restrict ourselves to the analysis of the conditions under which $\gamma = 0$ rather than to the calculation of the expressions for γ. These conditions will give us the frequency range for which the waves are unstable for a given anisotropy, or the range of anisotropies for which the waves are unstable at a given frequency. Both results are essential for the understanding of a wide range of wave phenomena in the Earth's magnetosphere, especially the origin of natural whistler-mode radio emissions (see Chapter 9).

As follows from (1.17), the condition for marginal stability ($\gamma = 0$) reduces to $\Im D = 0$, where D is equal to the difference between the left-hand side and the right-hand side of equation (3.10) for real values of the arguments in the case of parallel whistler-mode waves. As follows from (3.10), this condition can be written as:

$$\Im F_{1/2,2} = \frac{d\Im F_{3/2,2}}{dz}(A_e - 1)N^2. \qquad (7.23)$$

Having substituted (7.22) into (7.23) we obtain the equation of marginal stability in a more explicit form:

$$(b-1)N^2 f_{3/2}(s_+) \left[1 + \frac{\tilde{a}_{3/2} + \tilde{c}_{3/2}\sqrt{1-\mu}}{16a(1-\mu)^{5/4} \left[\sqrt{1-\mu}(b-1) - b \right]^2} \right]$$

$$= \left[1 + (b-1)N^2 \right] f_{1/2}(s_+) \left[1 + \frac{\tilde{a}_{1/2} + \tilde{c}_{1/2}\sqrt{1-\mu}}{16a(1-\mu)^{5/4} \left[\sqrt{1-\mu}(b-1) - b \right]^2} \right],$$

$$(7.24)$$

where $\tilde{a}_{1/2(3/2)} = \tilde{a}(q = 1/2(3/2))$, $\tilde{c}_{1/2(3/2)} = \tilde{c}(q = 1/2(3/2))$, $f_{1/2(3/2)} = f(q = 1/2(3/2))$.

Remembering our assumption that a is large we can expand A_e with respect to a^{-1} and write it as:

$$A_e = \bar{A}_e \sim \tilde{A}_{e0} + \frac{\tilde{A}_{e1}}{a} + \frac{\tilde{A}_{e2}}{a^2} + \dots . \tag{7.25}$$

Having substituted (7.25) into (7.24) we obtain the following expressions for $\tilde{A}_{e0}$ and $\tilde{A}_{e1}$:

$$\tilde{A}_{e0} = 1 + \frac{1}{N^2 \left(\sqrt{1-\mu} - 1 \right)}, \tag{7.26}$$

$$\tilde{A}_{e1} =$$

$$- \frac{N^2 \left(\sqrt{1-\mu} - 1 \right) + \mu N^2 - \sqrt{1-\mu}}{N^2 (1-\mu)^{5/4} \left[(N^2 - 1)(1-\mu)^{3/2} + N^2 (3\sqrt{1-\mu} + 3\mu - 4) + 4 - 3\mu \right]} . \tag{7.27}$$

The expression for A_{e2} is much more complicated and is only significant when we take into account one further term in (7.15).

Using the obvious physical result that the waves in an isotropic plasma ($A_e = 1$) are damped, we can see that whistler-mode waves are unstable when $A_e > \bar{A}_e$ at a given frequency or at frequencies ω less than the frequency at which $A_e = \bar{A}_e$.

At first sight expression (7.26) appears to contradict expression (19) of Sazhin & Temme (1991a) which in the limit $a \to \infty$ reduces to:

$$A_e = 1 + \frac{1}{(N^2 - 2) \left(\sqrt{1-\mu} - 1 \right)}. \tag{7.28}$$

However, expression (7.28) was derived under the assumption that $|A_e - 1| \ll 1$, which is satisfied only when $N^2 \gg 1$. In this case we can neglect 2 in the denominator of (7.28) when compared with N^2 and reduce (7.28) to (7.26) taken in the limit $|A_e - 1| \ll 1$.

In the limit $|\mu| \ll 1$ expressions (7.26) and (7.27) reduce to:

$$\tilde{A}_{e0} = -\frac{2}{\mu N^2} + 1 + \frac{1}{2N^2} + \frac{\mu}{8N^2} + ..., \tag{7.29}$$

$$\tilde{A}_{e1} = -\frac{1}{8N^2(N^2-1)\mu^3} + \frac{4}{(N^2-1)\mu^2} + \frac{4N^2-3}{4\mu N^2(N^2-1)} + ... \tag{7.30}$$

Taking the first three terms in the expansion (7.29) we can write it in a more explicit form:

$$\tilde{A}_{e0} = \frac{Y}{Y-1} + \frac{1}{2N^2}. \tag{7.31}$$

Expression (7.31) is equivalent to the corresponding expression derived by Sazhin & Temme (1990) in the limit $N^2 \gg z^2(Y-1)/4a \gg 1$ (see their expressions (5.3) and (5.6)).

In what follows we will return to expressions (7.26) and (7.27) valid for $\mu < 1$. The general expression for N^2 for whistler-mode waves with thermal and relativistic corrections taken into account can be written in the form (3.29). Restricting ourselves to considering frequencies at which the waves are marginally stable we can put in (3.30) $A_e = Y/(Y-1)$ and simplify expression (3.29) to:

$$N^2 = N_0^2 \left[1 + \frac{\nu(5Y-1)Y^2\tilde{r}}{4(Y-1)^3 N_0^2} \right]. \tag{7.32}$$

Having substituted (7.32) into (7.26) and (7.27) and neglecting higher-order terms with respect to $\tilde{r}$ we can rewrite the expression (7.25) as:

$$A_e = A_{e0} + A_{e1}\tilde{r}, \tag{7.33}$$

where

$$A_{e0} = 1 + \frac{1}{N_0^2 \left(\sqrt{1-\mu_0} - 1\right)}, \tag{7.34}$$

$$A_{e1} =$$
$$-\frac{N_0^2\left(\sqrt{1-\mu_0}-1\right) + \mu_0 N_0^2 - \sqrt{1-\mu_0}}{N_0^4(1-\mu_0)^{5/4}\left[(N_0^2-1)(1-\mu_0)^{3/2} + N_0^2(3\sqrt{1-\mu_0}+3\mu_0-4) + 4 - 3\mu_0\right]}$$

$$-\frac{\nu(5Y-1)Y^2}{4N_0^4(\sqrt{1-\mu_0}-1)(Y-1)^3}\left[1 + \frac{\mu_0}{2(\sqrt{1-\mu_0}-1)\sqrt{1-\mu_0}}\right], \tag{7.35}$$

$\mu_0 = 2(1-Y)/N_0^2$, and $\tilde{r}$ is the same as in (3.9).

Plots of A_{e0} versus Y^{-1} and A_{e1} versus Y^{-1} for $\nu = 0.5$, 1, 5 and 100 are shown in Figs. 7.2 and 7.3 respectively. We consider $Y^{-1} < 0.5$, when A_{e0}

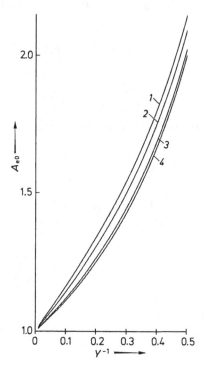

Fig. 7.2 Plots of A_{e0} (see equation (7.34)) versus Y^{-1} for $\nu = 0.5$ (curve 1), $\nu = 1$ (curve 2), $\nu = 5$ (curve 3) and $\nu = 100$ (curve 4) (Sazhin & Temme, 1991b).

is about or below 2 (typical values of this parameter in the Earth's magnetosphere, see e.g. Tsurutani *et al.* (1982), Bahnsen *et al.* (1985), Solomon *et al.* (1988)). Also, for the values of the parameters Y^{-1} and ν under consideration, the condition $\mu < 1$, on which our theory is based, is satisfied. As follows from Fig. 7.2 a decrease in electron density (parameter ν) tends to stabilize whistler-mode waves, in agreement with the earlier results of Sazhin & Temme (1991a), who reached the same conclusion for $Y^{-1} \ll 1$ and $|A_e - 1| \ll 1$.

At the same time, as follows from Fig. 7.3, the non-zero electron temperature (parameter $\tilde{r}$) tends to decrease A_e ($A_{e1} < 0$) thus destabilizing whistler-mode waves. However, in the actual conditions of the Earth's magnetosphere, when $\tilde{r}$ is almost always below 10^{-2} (electron energy below several keV), this reduction of the whistler-mode instability appears to be negligibly small.

Equation (7.25) can also be used for analysis of the whistler-mode marginal stability in a slightly different plasma model when we assume that the

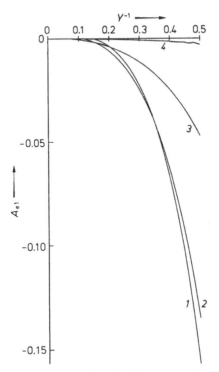

Fig. 7.3 The same as Fig. 7.2 but for A_{e1} (see equation (7.35)) (Sazhin & Temme, 1991b).

density of the 'hot' electrons responsible for whistler-mode growth or damping is well below the density of the 'cold' electrons responsible for whistler-mode propagation (cf. Kennel & Petschek, 1966; Etcheto *et al.*, 1973). In this case we can assume that $N = N_0$ and rewrite (7.33) as:

$$A_e = A_{e0} + \bar{A}_{e1}\tilde{r}, \qquad (7.36)$$

where A_{e0} is determined by (7.34),

$$\bar{A}_{e1} = \tilde{A}_{e1}(N = N_0)/N_0^2 \qquad (7.37)$$

and $\tilde{A}_{e1}$ is determined by (7.27). As follows from the plot of $\bar{A}_{e1}$ versus Y^{-1} shown in Fig. 7.4, in this case the effect of non-zero electron temperature (non-zero $\tilde{r}$) tends to stabilize whistler-mode waves. However, the efficiency of this process appears to be even smaller than in the case of destabilization due to $A_{e1} < 0$.

The results of this section are potentially most useful for analysis of the marginal stability of whistler-mode waves in those regions of the Earth's

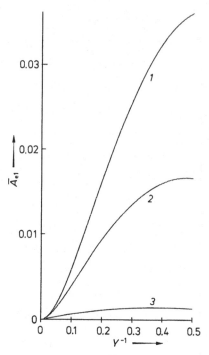

Fig. 7.4 The same as Fig. 7.2 but for $\bar{A}_{e1}$ (see equation (7.37)). The curve corresponding to $\nu = 100$ almost coincides with the Y^{-1} axis (Sazhin & Temme, 1991b).

magnetosphere where plasma is relatively rarefied ($\nu \lesssim 1$) (see e.g. Curtis, 1978). In the opposite case $\nu \gg 1$ we can use a simpler non-relativistic approach considered in the next section.

7.2 Parallel propagation (non-relativistic approximation)

As was pointed out in Section 3.1, in the non-relativistic limit, when the electron distribution function has the form (1.90), the dispersion equation (3.10) for parallel whistler-mode waves can be considerably simplified and reduced to (3.20). Remembering that $\Im Z(\xi) = \sqrt{\pi}\exp(-\xi^2)$, the condition for the non-relativistic marginal stability of whistler-mode waves can be reduced to:

$$A_e = \bar{A}_{e0} \equiv \frac{Y}{Y - 1}. \tag{7.38}$$

Waves are unstable when $A_e > \bar{A}_{e0}$ at a given frequency or when Y^{-1} is less than the Y^{-1} at which $A_e = \bar{A}_e$ (cf. the corresponding discussion in

Section 7.1). Comparing (7.38) with the corresponding weakly relativistic expressions (7.33) or (7.36) we can see that the non-relativistic expression (7.38) is valid only in a sufficiently dense plasma, which has already been pointed out in Section 7.1.

In contrast to Section 7.1 we will not restrict ourselves to considering the condition of the marginal stability of the waves (which is straightforward), but also analyse the explicit expression for γ. Remembering the definition of γ (expression (1.17)) and the non-relativistic dispersion equation for parallel whistler-mode waves (equation (3.20)) we obtain:

$$\tilde{\gamma} \equiv \frac{\gamma}{\omega_0} = \frac{[A_e + (1 - A_e)Y] \, X \sqrt{\pi} \exp(-\xi_1^2)}{N \tilde{w}_{\parallel} \omega_0 \, (\partial \Re D / \partial \omega_0)}, \tag{7.39}$$

where $\tilde{w}_{\parallel} = w_{\parallel}/c$, $\omega_0 = \Re \omega$ (subscript $_0$ for ω will be hereafter omitted),

$$\xi_1 = (1 - Y)/(N \tilde{w}_{\parallel}),$$

and

$$\Re D = X \left[(A_e - 1) + \frac{A_e + (1 - A_e)Y}{N \tilde{w}_{\parallel}} \Re Z(\xi_1) \right] - N^2. \tag{7.40}$$

When deriving (7.40) we took into account that $X = \nu Y^2 \gg 1$. Having substituted (7.40) into (7.39) we can write the latter equation in a more explicit form (Sazhin, 1990b):

$$\tilde{\gamma} = -\frac{[A_e + (1 - A_e)Y] \sqrt{\pi} \exp(-\xi_1^2)}{\tilde{\kappa}\xi_1 + \Re Z(\xi_1)(A_e + \tilde{\kappa}\xi_1^2)}, \tag{7.41}$$

where

$$\tilde{\kappa} = 2 \left[A_e + (1 - A_e)Y \right] / (Y - 1).$$

In what follows, expression (7.41) will be considered in more detail in different limiting cases similar to those considered in Section 4.1 (except for the metastable propagation when $\tilde{\gamma} = 0$). At the end of this section we shall briefly consider the problem of whistler-mode growth or damping in the case of wave propagation in a plasma with an arbitrary distribution function.

(a) Low-temperature limit (zero-order approximation)

The expression for $\tilde{\gamma}$ in a low-temperature limit can be obtained if we substitute (4.3) into (7.41) and keep only the lowest-order terms with respect to ξ_1^{-1} in the denominator of the right-hand side of (7.41). As a result we have:

$$\tilde{\gamma} = \frac{\xi_{00}(Y - 1)[A_e + (1 - A_e)Y]\sqrt{\pi} \exp(-\xi_{00}^2)}{Y} \equiv \tilde{\gamma}_L, \tag{7.42}$$

where

$$\xi_{00} = \frac{1 - Y}{\tilde{N}_{0d}\sqrt{2\beta_e}}, \tag{7.43}$$

$$\tilde{N}_{0d}^2 = \frac{Y^2}{Y - 1}. \tag{7.44}$$

When deriving (7.42) we neglected the contribution of thermal effects on the value of N. As follows from expression (7.42), $\tilde{\gamma}_L > 0$ (waves are unstable) when $Y^{-1} < (A_e - 1)/A_e$, and $\tilde{\gamma}_L < 0$ (waves are damped) when $Y^{-1} > (A_e - 1)/A_e$ (cf. the discussion about marginally stable waves in the beginning of this section).

$\tilde{\gamma}_L = 0$ when $Y^{-1} = (A_e - 1)/A_e$. Also, $\tilde{\gamma}_L \to 0$ when $Y^{-1} \to 0$, which corresponds to the increase of $|\xi_{00}|$ at $Y^{-1} \to 0$. (In fact, at sufficiently low frequencies we cannot neglect the contribution of ions which would result in wave damping at these frequencies: see Mann & Baumgärtel (1988) for details). At the intermediate frequency, $\tilde{\gamma}_L$ reaches its maximum as will be shown later in this section.

(b) *Low-temperature limit (first-order approximation)*

Taking the first-order temperature corrections in (7.41), we obtain

$$\tilde{\gamma} = \frac{\xi_{01}(Y - 1)[A_e + (1 - A_e)Y]\sqrt{\pi}\exp(-\xi_{01}^2)}{Y[1 + A_e(Y - 1)/(2Y\xi_{01}^2)]} \equiv \tilde{\gamma}_1, \tag{7.45}$$

where

$$\xi_{01} = \frac{1 - Y}{\tilde{N}_1\sqrt{2\beta_e}}, \tag{7.46}$$

and $\tilde{N}_1$ is determined by (4.4). This solution gives the exact value of $\tilde{\gamma}$ for Y^{-1} corresponding to $\xi_{01} = \xi_{11} \equiv -1.358$ (see Table 4.1) but can be used for approximate analysis of $\tilde{\gamma}$ in a wider range of frequencies, as will be shown later in this section.

(c) *Propagation at frequencies corresponding to $\xi_1 = -0.924$*

As was shown in Sections 1.7 and 4.1, when $\xi_1 = -0.924$ then $\Re Z(\xi_1)$ is determined by (4.7) and N is determined by (4.8). Having substituted (4.7) and (4.8) into (7.41) we obtain:

$$\tilde{\gamma} = \frac{\xi_{00}[A_e + (1 - A_e)Y]\sqrt{\pi}\exp(-\xi_{00}^2)}{A_e} \equiv \tilde{\gamma}_0, \tag{7.47}$$

where ξ_{00} is defined by (7.43). This expression is relevant to whistler-mode damping or growth at Y^{-1} corresponding to $\xi_{00} = \xi_{10} \equiv -0.924$ (see Table

4.1), but as in the case of expression (7.45) it can be used in a wider range of frequencies.

(d) Low-temperature limit (second-order approximation)

Taking into account the second-order temperature corrections in (7.41), i.e. having substituted (4.11) and (4.12) into (7.41) and keeping the first two terms with respect to ξ_1^{-1} in the denominator of the right-hand side of (7.41), we obtain:

$$\tilde{\gamma} = \frac{\xi_{02}(Y-1)[A_e + (1-A_e)Y]\sqrt{\pi}\exp(-\xi_{02}^2)}{Y\left\{1 + [3 - 2(Y-1)A_e/Y]/(2\xi_{02}^2)\right\}} \equiv \tilde{\gamma}_2, \tag{7.48}$$

where

$$\xi_{02} = \frac{1-Y}{\tilde{N}_2\sqrt{2\beta_e}}, \tag{7.49}$$

and $\tilde{N}_2$ is determined by (4.12). We expect this expression for $\tilde{\gamma}$ to be close to, but not exactly coincident with, that which follows from (7.41) for Y^{-1} corresponding to $\xi_{02} = \xi_{12} \equiv -1.685$ (see Table 4.1). The application of expression (7.48) in a wider range of frequencies can hardly be justified because of its complexity.

(e) Propagation at frequencies corresponding to $\xi_1 = -1.502$

As was shown in Section 4.1, at $\xi_1 = \xi_{1A} \equiv -1.502$ (see Table 4.1), $\Re Z(\xi_1)$ is determined by (4.13) and N is determined by (4.14). Having substituted (4.13) and (4.14) into (7.41) we obtain:

$$\tilde{\gamma} = \frac{(2\xi_{0A}^2 - 1)[A_e + (1-A_e)Y]\sqrt{\pi}\exp(-\xi_{0A}^2)}{\xi_{0A}(\tilde{\kappa} + 2A_e)} \equiv \tilde{\gamma}_A, \tag{7.50}$$

where

$$\xi_{0A} = \frac{1-Y}{\tilde{N}_A\sqrt{2\beta_e}}, \tag{7.51}$$

$\tilde{N}_A$ is determined by (4.14), and $\tilde{\kappa}$ is the same as in (7.41). Expression (7.50) gives an exact value of $\tilde{\gamma}$ at the frequency corresponding to $\xi_{0A} = \xi_{1A} \equiv -1.502$ (see Table 4.1), but can be used for the approximate analysis of $\tilde{\gamma}$ in a wider range of frequencies, as will be shown later from the comparison between different expressions for $\tilde{\gamma}$.

(f) Approximation of $\Re Z$ by $Z_B = -2\xi_1/(1+2\xi_1^2)$

As was shown in Section 4.1, Z_B determined by (4.15) provides a reasonably good approximation for $\Re Z$ if the modulus of the imaginary part of its

argument is well below the modulus of its real part. Assuming this is true
and $X \gg 1$ we can write N in the form (4.16). Having substituted (4.15)
and (4.16) into (7.41) we obtain (Sazhin, 1989d):

$$
\tilde{\gamma} = \frac{\sqrt{\pi}\left[(A_e - 1)Y - A_e\right]\left[(Y-1)^2 + \tilde{N}_B\beta_e\right]^2 \exp\left[-(Y-1)^2/(2\tilde{N}_B^2\beta_e)\right]}{\tilde{N}_B\sqrt{2\beta_e}\left[(Y-1)^2Y + (2A_eY - 2A_e - Y)\tilde{N}_B^2\beta_e\right]}
$$

$$
\equiv \tilde{\gamma}_B, \tag{7.52}
$$

where $\tilde{N}_B$ is determined by (4.16). As in the case of $\tilde{N}_B$, equation (7.52)
provides the worst approximation for a given frequency but is valid in the
widest range of frequencies when compared with the expressions for $\tilde{\gamma}$ consid-
ered in subsections (a)–(e). Note that when $|\tilde{\gamma}|$ is too large then the system
of equations (1.16) and (1.17), on which our analysis was based, becomes
invalid.

(g) Comparison of the results

The plots of $\tilde{\gamma}_A$, $\tilde{\gamma}_0$, $\tilde{\gamma}_L$, $\tilde{\gamma}_1$, $\tilde{\gamma}_2$ and $\tilde{\gamma}_B$ versus Y^{-1} are shown in Fig. 7.5 for
the same plasma parameters as in Fig. 4.5 $(A_e = 2)$ and $Y^{-1} \geq 0.5$. As in
Fig. 4.5 we indicate the values of Y^{-1} corresponding to ξ_{12}, ξ_{1A}, ξ_{11} and
ξ_{10} as well as the precise values of $\tilde{\gamma}$ for these Y^{-1}, calculated from (7.41).
These precise values coincide with $\tilde{\gamma}_0$ for $Y^{-1} = Y_0^{-1} \equiv 0.764$, with $\tilde{\gamma}_1$ for
$Y^{-1} = Y_1^{-1} \equiv 0.692$, and with $\tilde{\gamma}_A$ for $Y^{-1} = Y_A^{-1} \equiv 0.673$. The precise
value of $\tilde{\gamma}$ for $Y^{-1} = Y_2^{-1} \equiv 0.653$ is very close to that for the curves $\tilde{\gamma}_A$,
$\tilde{\gamma}_1$, $\tilde{\gamma}_L$ and $\tilde{\gamma}_2$ versus Y^{-1}. The curve $\tilde{\gamma}_B$ versus Y^{-1} almost coincides with
the curves $\tilde{\gamma}_1$ versus Y^{-1}, $\tilde{\gamma}_2$ versus Y^{-1}, and $\tilde{\gamma}_L$ versus Y^{-1} at $Y^{-1} \lesssim 0.65$
and is close to the curve $\tilde{\gamma}_0$ versus Y^{-1} for $Y^{-1} \gtrsim 0.65$.

Note that $|\tilde{\gamma}_0|$ at $Y^{-1} = Y_0^{-1}$ is so large that the influence of damping on
whistler-mode propagation cannot be neglected ((1.16) and (1.17) become
invalid). This point is shown only conventionally for the completeness of our
analysis. The value of $|\tilde{\gamma}_0|$ at $Y^{-1} = Y_0^{-1}$ is about a factor of 2 larger than
$|\tilde{\gamma}_L|$ at $Y^{-1} = Y_0^{-1}$. Hence, the use of a simple expression (7.42) could lead
to a considerable overestimate of the range of validity of (1.17).

Plots similar to those in Fig. 7.5 but for $0.4 \leq Y^{-1} \leq 0.5$ are shown in
Fig. 7.6. As can be seen in this figure, all the curves except those for $\tilde{\gamma}_B$
versus Y^{-1}, are close to each other and lie between the curves $\tilde{\gamma}_A$ versus
Y^{-1} and $\tilde{\gamma}_L$ versus Y^{-1}. The curve $\tilde{\gamma}_B$ versus Y^{-1} lies below other curves,
but its general shape is similar to that of the curves $\tilde{\gamma}_A$, $\tilde{\gamma}_0$, $\tilde{\gamma}_L$, $\tilde{\gamma}_1$ and $\tilde{\gamma}_2$
versus Y^{-1} .

Comparing Figs. 7.5 and 7.6 we can see that the expressions (7.45), (7.47),
(7.48), and (7.50) for $\tilde{\gamma}$ do not in fact bring any additional accuracy or

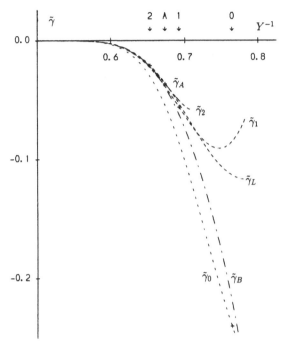

Fig. 7.5 The plots of $\tilde{\gamma}_L$ (see equation (7.42)), $\tilde{\gamma}_1$ (see equation (7.45)), $\tilde{\gamma}_0$ (see equation (7.47)), $\tilde{\gamma}_2$ (see equation (7.48)), $\tilde{\gamma}_A$ (see equation (7.50)), and $\tilde{\gamma}_B$ (see equation (7.52)) versus Y^{-1} for $Y^{-1} \geq 0.5$, $\beta_e = 0.01$ and $A_e = 2$. The meaning of the arrows is the same as in Figs. 4.4 and 4.5. The curve $\tilde{\gamma}_A$ versus Y^{-1} almost coincides with the curve $\tilde{\gamma}_2$ versus Y^{-1}.

convenience when compared with the traditional expression (7.42) for $Y^{-1} \leq 0.7$, although the values of $\tilde{\gamma}$ predicted by all these expressions appear to be surprisingly close. The estimate of $\tilde{\gamma}$ for greater Y^{-1} should be based on expression (7.47) rather than on any other approximation.

For $A_e = 1$ all $\tilde{\gamma}$ become negative (waves can only be damped) and $|\tilde{\gamma}|$ for $Y^{-1} > 0.5$ increases when compared with the values shown in Fig. 7.5. For $\beta_e = 0.001$ the threshold of instability for $A_e = 2$ would remain the same at $Y^{-1} = 0.5$, but all the curves at $Y^{-1} > 0.5$ would be shifted towards larger Y^{-1}.

(h) Arbitrary distribution function

Equation (7.39) for whistler-mode growth or damping can be generalized for the case of wave propagation in a plasma with an arbitrary distribution function f_0 provided $|\gamma| \ll \min(\omega, \Omega - \omega)$, using equation (3.7) written in a

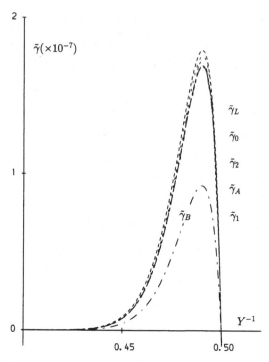

Fig. 7.6 The same as Fig. 7.5 but for $Y^{-1} < 0.5$. Curves at their peaks from bottom to top refer to the curves for $\tilde{\gamma}_B$, $\tilde{\gamma}_1$, $\tilde{\gamma}_A$, $\tilde{\gamma}_2$, $\tilde{\gamma}_0$ and $\tilde{\gamma}_L$ versus Y^{-1} respectively. Note the different scales of the axes.

non-relativistic limit and equation (1.17) (Rowlands, Shapiro & Shevchenko, 1966; Sazhin 1989e):

$$\tilde{\gamma} = \frac{\pi^2 \Pi^2 Y}{k \partial(\omega^2 \Re D)/\partial\omega} \int_0^\infty v_\perp^2 \mathrm{d}v_\perp \left[\frac{\partial f_0}{\partial v_\perp} + \frac{k v_\perp}{\Omega} \frac{\partial f_0}{\partial v_\parallel} \right] \Bigg|_{v_\parallel = v_R}, \qquad (7.53)$$

where

$$\Re D = \pi X \omega \int_0^\infty v_\perp^2 \mathrm{d}v_\perp P \int_{-\infty}^{+\infty} \frac{\left(1 - \frac{k v_\parallel}{\omega}\right) \frac{\partial f_0}{\partial v_\perp} + \frac{k v_\perp}{\omega} \frac{\partial f_0}{\partial v_\parallel}}{\omega - k_\parallel v_\parallel - \Omega} \mathrm{d}v_\parallel - N^2 + 1. \qquad (7.54)$$

In the case of low-temperature (with the temperature defined as an average kinetic energy of the electrons) and moderately dense plasma ($\nu > 1$), but with the contribution of ions taken into account, expression (7.54) can be rewritten as:

$$\Re D = X \left\{ \frac{1}{Y-1} + \frac{1}{Y^2 \nu} - r + \frac{Y^2}{(Y-1)^3} \left[\frac{Y}{Y-1} - A_e \right] \beta_e \right\} - N^2. \qquad (7.55)$$

Equation (7.55) reduces to (7.40) taken in the limit $|\xi_1| \gg 1$ if we neglect the contribution of ions and the effects of finite electron density.

If we assume that thermal, finite electron density and ion effects do not influence wave propagation so that the whistler-mode refractive index is defined by equation (2.11), then expression (7.53) can be simplified to:

$$\tilde{\gamma} = \pi^2 (Y - 1)|v_R| \int_0^\infty v_\perp^2 dv_\perp \left[\frac{\partial f_0}{\partial v_\perp} + \frac{k v_\perp}{\Omega} \frac{\partial f_0}{\partial v_\parallel} \right]\Bigg|_{v_\parallel = v_R}. \tag{7.56}$$

If we substitute f_0 in the form (1.90) into (7.56) then we obtain (7.42).

For low-frequency whistler-mode waves ($Y \gg 1$) expression (7.56) simplifies to:

$$\tilde{\gamma} = \pi^2 Y \int_0^\infty v_\perp^2 dv_\perp \left[v_\perp \frac{\partial f_0}{\partial v_\parallel} - v_\parallel \frac{\partial f_0}{\partial v_\perp} \right]\Bigg|_{v_\parallel = v_R}. \tag{7.57}$$

If we introduce new variables $v = \sqrt{v_\parallel^2 + v_\perp^2}$ and $\alpha_e = \arctan(v_\perp / v_\parallel)$ (electron pitch-angle) so that $f_0(v_\perp, v_\parallel) \equiv F(\alpha_e, v)$, then expression (7.57) can be written in a particularly simple form:

$$\tilde{\gamma} = \pi^2 Y |v_R|^3 \int_0^{\pi/2} \frac{\tan^2 \alpha_e}{\cos^2 \alpha_e} \frac{\partial F(\alpha_e, v)}{\partial \alpha_e} d\alpha_e \Bigg|_{v \cos \alpha_e = v_R}. \tag{7.58}$$

The expressions for $\tilde{N}^2$ and $\tilde{\gamma}$ under the same approximations as in Section 4.1 and in this section could also be derived for arbitrary N^2 (not necessarily $\gg 1$) but these expressions appear to be too complicated for practical applications.

7.3 Quasi-longitudinal propagation

When considering oblique whistler-mode growth or damping we restrict our analysis to low-temperature and dense plasma ($|\xi_1| \gg 1$ and $\nu \gg 1$). This limiting case is the most important for applications and is the simplest for analytical treatment. In a similar way to whistler-mode propagation (see Chapters 5 and 6) we consider the problem of oblique whistler-mode growth or damping in two approximations: quasi-longitudinal and quasi-electrostatic. The first approximation will be considered below, the second in the next section.

In a similar way to Sections 1.2, 7.1 and 7.2 we assume that wave growth or damping is not strong, so that equations (1.16) and (1.17) are valid and the electron distribution function is taken in the form (1.90). An explicit expression for N as a function of frequency, wave normal angle and plasma parameters in this case and in the quasi-longitudinal approximation has already been obtained in Section 5.2 (see equation (5.16)). Having substituted this expression, as well as (1.103), into the general expression for D (see equation (1.42)), we obtain (Sazhin, 1988a):

$$\Im D = \frac{\sqrt{\pi} X^3 \left[d_0 \exp(-\xi_0^2) + d_1 \exp(-\xi_1^2)\right]}{2N\tilde{w}_\parallel (Y+1)\cos^2\theta(Y\cos\theta - 1)^2}, \qquad (7.59)$$

where

$$d_0 = \frac{4Y\sin^2\theta}{N^2\tilde{w}_\parallel^2(Y-1)} - \frac{2A_e\sin^2\theta\left[(2Y^2-1)\cos^2\theta - 2Y\cos\theta + 1\right]}{Y(Y-1)}, \qquad (7.60)$$

$$d_1 = (Y-1)(\cos\theta+1)^2\left[(Y+1)\cos\theta - 1\right]^2 p_1$$

$$+(1+\cos\theta)[(Y+1)\cos\theta - 1][(-3Y^2+1)\cos^2\theta + (-Y^2+3Y)\cos\theta + (Y-1)]q_1 x, \qquad (7.61)$$

p_1, q_1 and x are the same as in (1.97)–(1.102), $\tilde{w}_\parallel = w_\parallel/c$, N is determined by (5.16), and $\xi_n = (1-nY)/(N\tilde{w}_\parallel)$. When deriving (7.59) we neglected the terms proportional to the multiples $\epsilon_{ij}^l \epsilon_{ij}^t$ and retained only the two lowest-order terms in d_0 and d_1.

Keeping only one lowest-order term in (7.60) and (7.61) we can simplify the expressions for d_0 and d_1 to:

$$d_0 = \frac{4Y\sin^2\theta}{N^2\tilde{w}_\parallel^2(Y-1)}, \qquad (7.62)$$

$$d_1 = (Y-1)(\cos\theta+1)^2\left[(Y+1)\cos\theta - 1\right]^2 p_1. \qquad (7.63)$$

If we furthermore neglect the influence of thermal effects on the value of $\partial \Re D/\partial \omega$ we obtain:

$$\frac{\partial \Re D}{\partial \omega} = \frac{2X^3 Y^2 \cos^2\theta}{\omega(Y^2-1)(Y\cos\theta - 1)^2}. \qquad (7.64)$$

Substituting expressions (7.59) (with d_0 and d_1 defined by (7.62) and (7.63) respectively) and (7.64) into (1.17) we obtain the following expression for $\tilde{\gamma} = \gamma/\omega$:

$$\tilde{\gamma} = \tilde{\gamma}_1 + \tilde{\gamma}_0, \qquad (7.65)$$

where

$$\tilde{\gamma}_1 = -\frac{\sqrt{\pi}(Y-1)^2\left[Y-(Y-1)A_e\right](1+\cos\theta)^2\left[(Y+1)\cos\theta-1\right]^2\exp(-\xi_1^2)}{4Y^3 N\tilde{w}_\parallel\cos^4\theta},$$

$$(7.66)$$

$$\tilde{\gamma}_0 = -\frac{\sqrt{\pi}\sin^2\theta\exp(-\xi_0^2)}{Y N^3\tilde{w}_\parallel^3\cos^4\theta}.$$

$$(7.67)$$

In the limit $\theta = 0$ and assuming $N = N_{0d}$, the expression for $\tilde{\gamma}$ defined by (7.65) reduces to that defined by (7.42). The relative contribution of the term $\tilde{\gamma}_0$ with respect to $\tilde{\gamma}_1$ increases when θ increases. Note that $\tilde{\gamma}_0$ is always negative, i.e. this term contributes to wave damping, while $\tilde{\gamma}_1$ is positive when A_e is larger than $\bar{A}_{e0}$ defined by (7.38) and negative when $A_e < \bar{A}_{e0}$. Hence, we can expect that the most potentially unstable whistler-mode waves are those propagating parallel to the magnetic field, provided the electron distribution function has the form (1.90).

As was mentioned in Section 7.1, in many practically important problems we need to know not the value of γ but rather the conditions for marginal stability of the waves, i.e. the conditions under which $\gamma = 0$. In the case of obliquely propagating whistler-mode waves the latter condition is not as trivial as in the case of parallel waves. As follows from (1.17), this condition reduces to $\Im D = 0$. Hence, in view of (7.59) and (5.16), the condition for marginal stability of oblique whistler-mode waves can be written as (Sazhin & Walker, 1989):

$$\tilde{d}_0 = -\tilde{d}_1\exp\left[\frac{(2-Y)(Y\cos\theta-1)\left[1-\tilde{a}_c\nu^{-1}-(\tilde{a}_{\beta 0}+A_e\tilde{a}_{\beta A})\beta_e\right]}{2Y\cos^2\theta\beta_e}\right],$$

where

$$(7.68)$$

$$\tilde{d}_0 = \frac{2\sin^2\theta(Y\cos\theta-1)}{Y(Y-1)}\left[\frac{1}{\beta_e}-\frac{\tilde{a}_c\nu^{-1}}{\beta_e}-\tilde{a}_{\beta 0}\right]-2A_e\sin^2\theta$$

$$\times\frac{(Y\cos\theta-1)}{Y(Y-1)}\left[\tilde{a}_{\beta A}+\frac{(2Y^2-1)\cos^2\theta-2Y\cos\theta+1}{Y\cos\theta-1}\right],\,(7.69)$$

$$\tilde{d}_1 = (Y-1)(\cos\theta+1)^2\left[(Y+1)\cos\theta-1\right]^2$$
$$- A_e\left\{\frac{(Y-1)^2(\cos\theta+1)^2\left[(Y+1)\cos\theta-1\right]^2}{Y}-\beta_e\sin^2\theta(1+\cos\theta)\right.$$
$$\times\left.\frac{\left[(Y+1)\cos\theta-1\right]\left[(-3Y^2+1)\cos^2\theta+(-Y^2+3Y)\cos\theta+(Y-1)\right]}{(Y\cos\theta-1)}\right\}$$

$$
- \frac{A_e^2 \beta_e \sin^2 \theta \, (j+2) \, (Y-1)}{2Y \, (j+1) \, (Y \cos \theta - 1)} \, (1 + \cos \theta) \, [(Y+1) \cos \theta - 1]
$$

$$
\times \; \left[\left(-3Y^2 + 1 \right) \cos^2 \theta + \left(-Y^2 + 3Y \right) \cos \theta + (Y-1) \right], \tag{7.70}
$$

$$
\tilde{a}_{\beta 0} = \frac{Y^2 \left[1 - 2Y \cos \theta - (2Y^2 + 1) \cos^2 \theta + 4Y \cos^3 \theta + 4Y^2 \cos^4 \theta \right]}{2 \, (Y \cos \theta - 1)^2 \, (Y^2 - 1)},
$$

$$
\tag{7.71}
$$

$$
\tilde{a}_{\beta A} = \frac{Y^2 \left[1 + 4Y \cos \theta - (4Y^2 - 1) \cos^2 \theta - 4Y \cos^3 \theta - 4Y^2 \cos^4 \theta \right]}{2 \, (Y \cos \theta - 1)^2 \, (4Y^2 - 1)},
$$

$$
\tag{7.72}
$$

and $\tilde{a}_c$ is the same as in (5.16); note that $\tilde{a}_\beta$ introduced in (5.16) is equal to $\tilde{a}_{\beta 0} + A_e \tilde{a}_{\beta A}$.

When deriving (7.68) it was assumed that $\tilde{d}_0$ and $\tilde{d}_1$ are defined by the general equations (7.60) and (7.61) respectively. Also, we neglected the contribution of ions, which is justified for rather high frequencies ($Y^{-1} \geq 0.45$) when equation (7.68) will be applied. The coefficients $\tilde{d}_0$ and $\tilde{d}_1$ are presented in such a form that the terms proportional to the powers of A_e are grouped together.

The conditions for the validity of equation (7.68) are the same as those for the validity of equation (5.16). In particular, the following inequalities should be valid: $\tilde{a}_c \nu^{-1} \ll 1$ and $(\tilde{a}_{\beta 0} + A_e \tilde{a}_{\beta A}) \beta_e \ll 1$.

In the limit $\beta_e \to 0$ and $\nu^{-1} \to 0$, equation (7.68) can be explicitly resolved with respect to A_e and has the form:

$$
\begin{aligned}
A_e \;=\; \hat{A}_{e0} &\equiv \frac{Y}{Y-1} \\
&+ \frac{2 \sin^2 \theta \, (Y \cos \theta - 1) \exp \left[(Y \cos \theta - 1) \, (Y - 2) \, / (2Y \beta_e \cos^2 \theta) \right]}{\beta_e \, (Y-1)^3 \, (\cos \theta + 1)^2 \, [(Y+1) \cos \theta - 1]^2}.
\end{aligned}
$$

$$
\tag{7.73}
$$

In the limit $\theta = 0$, the latter equation reduces to a trivial condition (7.38) for marginal stability of parallel propagating whistler-mode waves.

Equation (7.68) was solved numerically for different values of plasma and wave parameters. In Figs. 7.7 and 7.8 we show plots of A_e versus Y^{-1} for $j = 1$; $\theta = \pi/6$ and $\pi/12$; $\nu^{-1} = 0.1$; $\beta_e = 0.01$ and 0.02 . In the same figures we also show plots of $\hat{A}_{e0}$ versus Y^{-1} and $\bar{A}_{e0}$ (see equation (7.38)) versus Y^{-1} for the same parameters. The values of A_e lying above these curves correspond to unstable waves, those below to stable ones. As

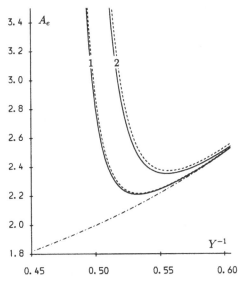

Fig. 7.7 The plots of A_e versus Y^{-1} for $j = 1$, $\nu = 10$, $\beta_e = 0.01$, $\theta = \pi/12$ (curve 1) and $\theta = \pi/6$ (curve 2) for the marginally stable quasi-longitudinal whistler-mode waves calculated from equation (7.68) (solid). The plots of $\hat{A}_{e0}$ versus Y^{-1} based on equation (7.73) (dashed) and the plot of $\bar{A}_{e0}$ versus Y^{-1} based on equation (7.38) (dashed–dotted) for the same values of the parameters.

follows from these figures, for Y^{-1} approaching 0.6 all the curves except those corresponding to $\theta = \pi/6$ and $\beta_e = 0.02$ appear to be very close to each other, which could justify application of the condition (7.38) for a rough estimate of the condition for marginal stability of whistler-mode waves. For Y^{-1} close to 0.5, the actual values of A_e are close to those predicted by equation (7.73) but deviate considerably from those predicted by equation (7.38). For Y^{-1} close to 0.45, equations (7.68) and (7.73) are not satisfied for any chosen values of the parameters, in contrast to the prediction of equation (7.38). At these values of Y^{-1}, whistler-mode waves can be unstable either for larger values of β_e and/or A_e or smaller values of θ. Decrease in ν^{-1} from 0.1 to 0.01 would slightly decrease the value of A_e, which cannot be shown within the scale of the figures.

As follows from Figs. 7.7 and 7.8, the difference between the curves of A_e versus Y^{-1} predicted by (7.68) and those predicted by (7.73) increases with increasing Y^{-1} and β_e. This can illustrate the general conclusion of Sazhin (1988a,b) that the thermal effects on whistler-mode propagation should be taken into account when considering wave growth or damping. Increase in θ results in an increase in A_e for which instability can develop.

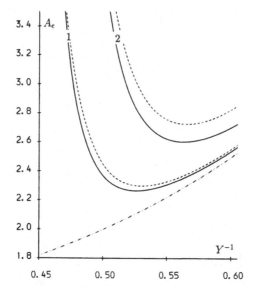

Fig. 7.8. The same as Fig. 7.7 but for $\beta_e = 0.02$.

Expressions (7.59) and (7.65) can be generalized in a straightforward way to the case when the electron distribution function has the form (1.130), i.e. when the contribution of beams and the electron populations with different temperatures are to be taken into account (Sazhin, Walker & Woolliscroft, 1990b). In this case Y should be replaced by

$$\tilde{Y} = \frac{\Omega}{\left(\omega - k_\parallel v_{0_i}\right)} \tag{7.74}$$

and ξ_n by

$$\tilde{\xi}_{n_i} = \frac{\omega - k_\parallel v_{0_i} - n\Omega}{k_\parallel w_{\parallel i}} = \xi_{n_i} - \hat{v}_{0_i}, \tag{7.75}$$

where $\hat{v}_{0_i} = v_{0_i}/w_{\parallel i}$, and the summation over all electron populations should be carried out.

As follows from the analysis of the final expression for $\tilde{\gamma}$ for the plasma consisting of two populations : hot (200 eV) and cold (2 eV) with equal densities ($\Pi_c^2/\Omega^2 = \Pi_h^2/\Omega^2 = 7.5$), equal but oppositely directed drift velocities ($|v_0| = 1.88 \times 10^8$ cm s^{-1}) and equal anisotropies ($A_e = 3$ and 4), the value of $\tilde{\gamma}$ depends on the direction of the beams, being larger when the hot electron population drifts in a negative direction with respect to the direction of wave propagation (see Sazhin, Walker & Woolliscroft (1990b) for details).

Now we consider another limiting case of quasi-longitudinal whistler-mode propagation. Namely, we assume that the waves propagate through plasma with the arbitrary distribution function f_0, but we neglect the influence of thermal effects on wave propagation (including $\Re D$) and consider the case when $\omega \ll \Omega$ and $\omega \ll \Pi$ (following Sazhin, 1991b). This allows us to write the solution of (1.16) for whistler-mode waves as (cf. equation (2.11)):

$$N^2 = X \left[\frac{1}{Y \cos \theta} + \frac{1}{Y^2 \cos^2 \theta} \right], \tag{7.76}$$

and the expression for $\partial \Re D / \partial \omega$ as (cf. equation (7.64)):

$$\frac{\partial \Re D}{\partial \omega} = \frac{2X^3}{\omega Y^2} \left(1 + \frac{2}{Y \cos \theta} \right). \tag{7.77}$$

In order to calculate $\Im D$ we present ϵ_{ij} as:

$$\epsilon_{ij} = \epsilon_{ij}^0 + \epsilon_{ij}^I, \tag{7.78}$$

where ϵ_{ij}^0 defined by expressions (1.79) results from the principal part of the integral with respect to $p_\|$ in (1.73), while ϵ_{ij}^I results from the residues of the same integral in the non-relativistic limit, thermal effects being neglected. If we assume that $Y \gg 1$ and $X \gg 1$ then the expressions for ϵ_{ij}^0 can be simplified to:

$$\epsilon_{ij}^0 = \left\| \begin{array}{ccc} \frac{X}{Y^2} & iX \left(\frac{1}{Y} + \frac{1}{Y^3} \right) & 0 \\ -iX \left(\frac{1}{Y} + \frac{1}{Y^3} \right) & \frac{X}{Y^2} & 0 \\ 0 & 0 & -X \end{array} \right\|. \tag{7.79}$$

The expressions for ϵ_{ij}^I follow from (1.73) in a straightforward way:

$$\epsilon_{ij}^I = -\frac{2i\pi^2 X\omega}{k_\|} \int_0^{+\infty} dv_\perp \int_{-\infty}^{+\infty} G_1 \sum_{n=-\infty}^{+\infty} \Pi_{ij}^{(n)} \delta \left(v_\| - \left(\frac{\omega}{k_\|} - \frac{n\Omega}{k_\|} \right) \right) dv_\|, \tag{7.80}$$

where

$$G_1 = \frac{\partial f_0}{\partial v_\perp} - \frac{k_\|}{\omega} \left(v_\| \frac{\partial f_0}{\partial v_\perp} - v_\perp \frac{\partial f_0}{\partial v_\|} \right) \equiv \tilde{R} f_0. \tag{7.81}$$

In what follows we assume that $|\epsilon_{ij}^I| \ll |\epsilon_{ij}^0|$ unless $\epsilon_{ij}^0 = 0$. Having substituted (7.78) and (7.76) into (1.42) and keeping only the first-order terms with respect to ϵ_{ij}^I we obtain the value of $\Im D$. Having substituted this value of $\Im D$ and $\partial \Re D / \partial \omega$ defined by (7.77) into (1.17) we obtain, after lengthy but straightforward calculations, (Sazhin, 1991b):

$$
\tilde{\gamma} \equiv \frac{\gamma}{\omega} = \frac{\pi^2 \omega Y^2}{k \cos \theta} \int_0^{+\infty} dv_\perp \int_{-\infty}^{+\infty} dv_\parallel \sum_{n=-\infty}^{+\infty} \delta \left(v_\parallel - \left(\frac{\omega}{k_\parallel} - \frac{n\Omega}{k_\parallel} \right) \right) G_1 \sum_{i=1}^{6} \kappa_i,
$$

(7.82)

where

$$
\kappa_1 = \frac{n^2 v_\perp^2}{\lambda_e^2} J_n^2 \frac{1}{Y \cos \theta},
$$

(7.83)

$$
\kappa_2 = \frac{2 v_\perp^2}{\lambda_e} n J_n J_n' \frac{1}{Y},
$$

(7.84)

$$
\kappa_3 = \frac{2 n v_\parallel v_\perp}{\lambda_e} J_n^2 \frac{\tan \theta}{Y^2},
$$

(7.85)

$$
\kappa_4 = \frac{v_\perp^2 J_n'^2 \cos \theta}{Y},
$$

(7.86)

$$
\kappa_5 = \frac{2 v_\parallel v_\perp J_n J_n' \sin \theta}{Y^2},
$$

(7.87)

$$
\kappa_6 = \frac{v_\parallel^2 J_n^2 \sin^2 \theta}{Y^3 \cos \theta},
$$

(7.88)

$k = |\mathbf{k}|$, and the argument of the Bessel functions is $\lambda_e = k_\perp v_\perp / \Omega$.

When deriving (7.82) we kept only the lowest-order terms with respect to Y^{-1} in each group of terms proportional to the elements of $\Pi_{ij}^{(n)}$, defined by (1.74), and assumed that θ was not close to the resonance cone angle θ_{R0} (see equation (2.13)). Note that if we neglected the term proportional to Y^{-2} in (7.76), then we would obtain a different expression for κ_6 ($\kappa_6 = (-1 - \cos^2 \theta) v_\parallel^2 J_n^2 / (Y^3 \cos \theta)$). The terms proportional to Y^{-3} in (7.77) and (7.79) do not influence the expressions (7.83)–(7.88).

If we keep only the terms κ_1, κ_2 and κ_4, which are of the lowest order with respect to Y^{-1} among all the terms κ_i, then we obtain an expression for γ similar to that obtained by Kennel (1966):

$$
\tilde{\gamma} = \tilde{\gamma}_{\text{Ken}} \equiv \frac{\pi^2 \Omega}{k} \int_0^{+\infty} v_\perp^2 dv_\perp \int_{-\infty}^{+\infty} dv_\parallel \sum_{n=-\infty}^{+\infty} \delta \left(v_\parallel - \left(\frac{\omega}{k_\parallel} - \frac{n\Omega}{k_\parallel} \right) \right) \vartheta_n G_1,
$$

(7.89)

where

$$
\vartheta_n = \left[\frac{(1 - \cos \theta) J_{n+1} + (1 + \cos \theta) J_{n-1}}{2 \cos \theta} \right]^2.
$$

(7.90)

Expressions (7.89) and (7.90) reduce to those given by Kennel (1966) (see his expressions (4.5) and (4.6)) if we formally replace n by $-n$ and remember that $J_{-n} = J_n(-1)^n$. For parallel propagation and $n = 1$ equation (7.89) reduces to (7.57).

However, the terms κ_3, κ_5 and κ_6 cannot be automatically neglected when compared with the terms κ_1, κ_2 and κ_4, as these terms contain another parameter λ_e which can be smaller and even (in the actual magnetospheric conditions) much smaller than Y^{-1}. Hence, in general, when considering growth or damping of oblique whistler-mode waves we should use the general equation (7.82) rather than its simplified version (7.89). The only case when this simplification of equation (7.79) is possible is when we can neglect the contribution of Landau resonance described by the term $n = 0$ in (7.82), as in this case the terms κ_1, κ_2 and κ_4 are indeed the dominant ones with respect to both Y^{-1} and λ_e. However, for the term describing the contribution of Landau resonance in (7.82) the situation appears to be very different, as in this case the terms κ_1, κ_2 and κ_3 are equal to zero, and the term κ_6 is the dominant one with respect to λ_e. This means that for the low-temperature plasma, when $\lambda_e \ll Y^{-1} \ll 1$ for most of the electrons, the term describing the contribution of Landau resonance in (7.82) can be simplified to:

$$\tilde{\gamma} = \tilde{\gamma}_0 \equiv \frac{\pi^2 \omega^2 \sin^2 \theta}{Y k^2 \cos^3 \theta} \int_0^{+\infty} dv_\perp v_\perp \left. \frac{\partial f_0}{\partial v_\parallel} \right|_{v_\parallel = \omega/k_\parallel}. \tag{7.91}$$

Comparing (7.89) with (7.91) we can see that for sufficiently low-temperature plasma, the term $\tilde{\gamma}_0$ dominates over the term $\tilde{\gamma}_{\text{Ken}}(n = 0)$. Hence, in the case when

$$\int_0^{+\infty} dv_\perp v_\perp \left. \frac{\partial f_0}{\partial v_\parallel} \right|_{v_\parallel = \omega/k_\parallel} < 0,$$

Landau damping of oblique whistler-mode waves appears to be more efficient than that predicted by equation (7.89). In general, a reasonably accurate approximation for $\tilde{\gamma}$ can be obtained if we write:

$$\tilde{\gamma} = \tilde{\gamma}_{\text{Ken}} + \tilde{\gamma}_0. \tag{7.92}$$

Having substituted (1.90) into (7.92), and neglecting the contribution of the higher-order resonances, we obtain expression (7.65) taken in the limit $Y \gg 1$.

Expressions (7.65), (7.82) and the following simplified versions of the latter expression can also describe whistler-mode growth or damping when these waves propagate through predominantly cold plasma, with density n_c, permeated by a dilute hot electron component with distribution function

f_h and density n_h, by multiplying the right-hand side of these expressions by $n_h/n_c \ll 1$. This model of the electron distribution function is widely used for the analysis of wave processes in the Earth's magnetosphere (e.g. Etcheto *et al.*, 1973).

7.4 Quasi-electrostatic propagation

When considering the problem of growth or damping of whistler-mode waves using the quasi-electrostatic approximation, we assume that the refractive index N is much greater than unity so that the imaginary part of the function D, as defined by equation (1.42), can be written as:

$$\Im D = N^4 \Im A, \tag{7.93}$$

where N^2 is defined by equation (6.14), (6.21) or (6.29) (depending on the values of a_0, θ and ν). The real part of the function D for values of θ close to the resonance cone angle θ_R (see equations (2.23) and (2.24)) can be written as (cf. equations (6.4) and (6.5)):

$$\Re D = (A_0 + A_1)\, N^4 + B_0 N^2, \tag{7.94}$$

where the coefficients A_0 and B_0 are given by equations (2.1) and (2.2). These coefficients correspond to the values of A and B in a cold electron plasma. A_1 is the thermal correction to A_0 (see equation (6.1)). In what follows we analyse quasi-electrostatic growth and damping in the limit of a dense plasma, which allows us to write the equations in a particularly simple form.

As in Sections 7.2 and 7.3 we restrict our analysis to the non-relativistic approximation and we first assume that the electron distribution function has the form (1.90). Restricting ourselves to the limiting case $\theta = \theta_R$ and remembering the definition of A and the expressions for ϵ_{xx}, ϵ_{xz} and ϵ_{zz} given by equations (1.91), (1.94) and (1.96), we can write the expression for $\Im D$ at $\theta = \theta_R$ as (Sazhin & Walker, 1989):

$$\Im D = \sqrt{\pi} X \left\{ \left[1 - \frac{N^2 \sin^2 \theta_R A_e \tilde{w}_\parallel^2}{2Y^2} \right] \frac{2N}{\tilde{w}_\parallel^3 \cos \theta_R} \exp\left(-\frac{1}{N^2 \tilde{w}_\parallel^2 \cos^2 \theta_R} \right) \right.$$

$$\left. + \left[1 + A_e \frac{1-Y}{Y} - \frac{N^2 \sin^2 \theta_R \tilde{w}_\parallel^2 A_e q_1}{2Y^2 (j+1)} \right] \frac{N^3 \sin^2 \theta_R}{2\tilde{w}_\parallel Y \cos \theta_R} \exp\left(-\frac{(Y-1)^2}{N^2 \tilde{w}_\parallel^2 \cos^2 \theta_R} \right) \right\},$$

$$\tag{7.95}$$

where q_1 is the same as in (1.97)–(1.102) and (7.61). In a moderately dense plasma θ_R is defined by equation (2.23).

The term $\partial \Re D / \partial \omega$ can be written as:

$$\frac{\partial}{\partial \omega} \Re D = -\frac{Y}{\omega} \frac{\partial}{\partial Y} \Re D, \tag{7.96}$$

where

$$\frac{\partial}{\partial Y} \Re D = \left(a_0^{t'} + 6 a_0^t Y^{-1} \right) \beta_e N^6 + \left(B_0' + 2 B_0 Y^{-1} \right) N^2 \tag{7.97}$$

(the prime superscript indicates the derivative with respect to Y), and B_0 is defined by equation (2.2). In a dense plasma limit we have:

$$a_0^{t'} = \frac{2Y \left(3Y^2 - 11 \right)}{(Y^2 - 1)^3} + \frac{4 A_e Y \left(-2Y^4 + 28 Y^2 - 17 \right)}{(Y^2 - 1)^2 (4Y^2 - 1)^2}, \tag{7.98}$$

$$B_0' = \frac{4 \nu^2 Y^3 (Y^2 - 2)}{(Y^2 - 1)^2}, \tag{7.99}$$

(note a small printing error in equation (77) of Sazhin, Walker & Woolliscroft (1990b)). When deriving (7.95) we took into account our assumption that $\theta = \theta_R$ so that we could set $A_0 = A_0' = 0$.

Having substituted (7.95) and (7.96) into (1.17) we obtain an explicit expression for $\tilde{\gamma} = \gamma / \omega$ which has a rather complicated form. In a similar way to Section 7.3, we restrict ourselves to the analysis of the marginal stability of the waves rather than the values of $\tilde{\gamma}$.

Remembering expression (6.22) for N^2 (we assume $a_0 < 0$, for otherwise the waves could not propagate at $\theta = \theta_R$) and (7.95), we can reduce the condition for the marginal stability of the waves ($\Im D = 0$) to:

$$1 - \frac{A_e \sin^2 \theta_R \sqrt{2 \beta_e}}{\sqrt{Y^2 - 1} \sqrt{-a_0^t}} + \frac{\sin^2 \theta_R Y \sqrt{2 \beta_e}}{2 \sqrt{Y^2 - 1} \sqrt{-a_0^t}} \left\{ 1 + A_e \frac{(1 - Y)}{Y} - \frac{A_e \sin^2 \theta_R \sqrt{2 \beta_e}}{\sqrt{Y^2 - 1} \sqrt{-a_0^t}} \right.$$

$$\left. \times \left[1 - \frac{(j + 2)(Y - 1) A_e}{2 (j + 1) Y} \right] \right\} \exp \left[\frac{(2 - Y) \sqrt{Y^2 - 1} \sqrt{-a_0^t}}{2Y \sqrt{2 \beta_e} \cos^2 \theta_R} \right] = 0 \quad (7.100)$$

where θ_R and a_0^t are defined by equations (2.23) and (6.12) respectively.

In Fig. 7.9, we show the results of numerical analysis of equation (7.100) for $\nu^{-1} = 0.1$ and different β_e and j. As in Figs. 7.7 and 7.8, the values of Y^{-1} and A_e lying above the curves correspond to unstable waves, and those below them to damped ones. The curves for $\nu^{-1} = 0.01$ coincide with those for $\nu^{-1} = 0.1$ within the accuracy of plotting, being slightly below them.

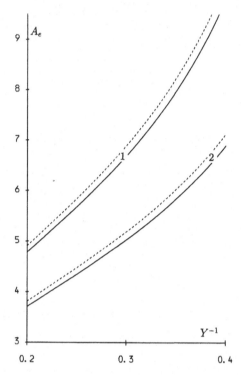

Fig. 7.9 The plots of A_e versus Y^{-1} for $j = 2$ (dashed) and $j = 4$ (solid), $\nu = 10$, $\beta_e = 0.01$ (curve 1) and $\beta_e = 0.02$ (curve 2) for the marginally stable quasi-electrostatic whistler-mode waves at $\theta = \theta_R$ (see equation (7.100)).

The chosen values of j, β_e and Y^{-1} guarantee the validity of our equations ($|\tilde{\gamma}|$ is not too large).

As follows from Fig. 7.9, the values of A_e corresponding to marginal stability increase with decreasing β_e and j, being unrealistically high ($A_e > 3$) for magnetospheric conditions. This is roughly consistent with the results of numerical analysis by Ohmi & Hayakawa (1986) for $\omega > \Omega/2$. Hence, we can expect that quasi-electrostatic whistler-mode waves at $\theta = \theta_R$ are stable in magnetospheric conditions, unless an additional mechanism for their excitation is provided.

In a similar way to Section 7.3, expressions (7.95) and (7.96) can be generalized in a straightforward way when the contribution of beams and electron populations with different temperatures are taken into account, i.e. the electron distribution function has the form (1.130) (Sazhin, Walker & Woolliscroft, 1990b). In this case the expression (7.95) for $\Im D$ is generalized to:

$$\Im D = \sum_i \sqrt{\pi} X_i \left\{ \frac{2N}{\tilde{w}_{\|i}^3 \cos \theta_R} \left[1 - \frac{N^2 \sin^2 \theta_R A_{e_i} \tilde{w}_{\|i}^2}{2Y^2} \right. \right.$$

$$\times \left. \left(1 - \frac{3 \left(j_i + 2 \right) N^2 \sin^2 \theta_R A_{e_i} \tilde{w}_{\|i}^2}{16 \left(j_i + 1 \right) Y^2} \right) \right]$$

$$\times \left(1 - N \cos \theta_R \tilde{v}_{0_i} \right) \exp \left[- \left(\xi_{0_i} - \hat{v}_{0_i} \right)^2 \right]$$

$$+ \frac{N^3 \sin^2 \theta_R}{2 \tilde{w}_{\|i} Y \cos \theta_R} \left[1 + A_{e_i} \frac{\left(1 - \tilde{Y}_i \right)}{\tilde{Y}_i} - \frac{N^2 \sin^2 \theta_R \tilde{w}_{\|i}^2 A_{e_i} q_{1_i}}{2 \left(j_i + 1 \right) Y^2} \right]$$

$$\times \left. \exp \left[- \left(\xi_{1_i} - \hat{v}_{0_i} \right)^2 \right] \right\}, \tag{7.101}$$

where

$$\xi_{0_i} = \frac{1}{N \tilde{w}_{\|i} \cos \theta_R}, \qquad \xi_{1_i} = \frac{1 - Y}{N \tilde{w}_{\|i} \cos \theta_R},$$

$$q_{1_i} = \left(j_i + 1 \right) - \frac{A_{e_i} \left(Y - 1 \right) \left(j_i + 2 \right)}{2Y}, \quad \hat{v}_{0_i} = v_{0_i}/w_{\|i}, \quad \tilde{w}_{\|i} = w_{\|i}/c,$$

and N and $\tilde{a}_{0_i}$ are defined by equations (6.62) and (6.63) respectively.

The expressions for $\partial \Re D / \partial \omega$ can be written in a form similar to (7.96) but with a_0^t replaced by $\tilde{a}_{0_i}$ (see equation (6.63)), N defined by (6.62) and with the summation performed over all electron populations.

As follows from the numerical analysis for the same plasma model as in Section 7.3, electron beams tend to destabilize plasma when a hot electron component drifts in the direction opposite to that of wave propagation. However, even in this case instability occurs only when the plasma is extremely anisotropic ($A_{e_i} \gtrsim 4$, $j_i \gtrsim 2$) (see Sazhin, Walker & Woolliscroft (1990b) for details).

Expression (7.82) can be generalized for the case of quasi-electrostatic whistler-mode propagation in a plasma with an arbitrary distribution function f_0. To do this we substitute the general expressions for ϵ_{xx}, ϵ_{xz}, and ϵ_{zz} given by (1.73) into (1.43) and assume that $\Im D$ is defined by equation (7.93). As a result, after some straightforward algebra, we obtain:

$$\Im D = -\frac{2\pi^2 N^4 X \omega}{k \cos \theta} \sum_{n=-\infty}^{+\infty} \int_0^\infty J_n^2 dv_\perp \int_{-\infty}^{+\infty} dv_\parallel \left(\frac{n\Omega}{k} + v_\parallel \cos \theta \right)^2$$

$$\times \delta \left(v_\parallel - \left(\frac{\omega}{k_\parallel} - \frac{n\Omega}{k_\parallel} \right) \right) G_1, \tag{7.102}$$

where G_1 is defined by (7.81). Expressions (7.95) and (7.101) can be obtained from (7.102) in a straightforward way if we take into account only the terms corresponding to $n = 0$ and $n = 1$ in (7.102), and assume that the argument of the Bessel function, λ_e, is well below unity and that f_0 is determined by (1.90) and (1.130) respectively.

7.5 A physical model

The emphasis in this chapter has so far been on the formal mathematical analysis of the whistler-mode dispersion equation. No attempt to discuss a physical background of wave growth or damping has been made. In some cases, including whistler-mode growth or damping in a weakly relativistic plasma and quasi-electrostatic whistler-mode growth or damping, we do not fully understand the physical background of the processes and are forced to rely on formal analysis only. However, in other cases alternative approaches to the problem of whistler-mode growth or damping, based on a physical analysis of the energy exchange between waves and particles, have been developed. In what follows we discuss some of these models referring to quasi-longitudinal whistler-mode growth or damping. Firstly, we shall consider a qualitative approach to the problem and attempt to clarify why the energy exchange between waves and electrons is most likely to take place when $\omega - k_\parallel v_\parallel - n\Omega = 0$ ($n = 0, \pm 1, \pm 2, \ldots$). Then we shall present direct calculations of the energy exchange between waves and electrons and obtain the same expression (7.92) for $\tilde{\gamma}$ which follows from a formal analysis. Finally we shall discuss a geometrical criterion for instability which will allow us to look at it from a different perspective.

(a) Qualitative analysis

In the absence of other mechanisms of energy source or sink, we can antici- pate that growth or damping of obliquely propagating whistler-mode waves can be related only to the energy exchange between waves and particles. For the analysis of this exchange we express the whistler-mode electric field $\mathbf{E}$ as a sum of two vectors $\mathbf{E}_\perp$ and $\mathbf{E}_\parallel$:

$$\mathbf{E} = \mathbf{E}_\perp + \mathbf{E}_\parallel, \tag{7.103}$$

where $\mathbf{E}_\perp$ is perpendicular to $\mathbf{B}_0$, while $\mathbf{E}_\parallel$ is parallel to $\mathbf{B}_0$.

The interaction of $\mathbf{E}_\parallel$ with the electrons is effective only for the case of the Čerenkov resonance. This interaction is in many respects similar to the resonance interaction between a Langmuir wave and electrons (e.g. Stix, 1962; Lacina, 1972; Subramaniam & Hughes, 1986) and its quantitative analysis will be given in the next subsection. Meanwhile, we consider some qualitative features of the interaction between the electrons and $\mathbf{E}_\perp$.

$\mathbf{E}_\perp$ can be considered as a superposition of the right-handed ($\mathbf{E}_R$) and the left-handed ($\mathbf{E}_L$) circular polarized waves, i.e.

$$\mathbf{E}_\perp = \mathbf{E}_R + \mathbf{E}_L. \tag{7.104}$$

Presenting $\mathbf{E}_{R(L)}$ in a complex form:

$$E_{R(L)} = E_x + (-)iE_y \tag{7.105}$$

we can write

$$E_{R(L)} = |\mathbf{E}_{R(L)}| \exp\left[+(-)\left(i\omega t - i\mathbf{k}\mathbf{r} + i\varsigma_{R(L)}\right)\right], \tag{7.106}$$

where $\varsigma_{R(L)}$ are the initial phases of the waves. In what follows in this subsection we restrict ourselves to the interaction of the electrons with E_R; their interaction with E_L might be considered in a similar way.

The unperturbed electron velocity can be written as:

$$\mathbf{v} = |v_\perp| \cos(\Omega t + \varsigma_e)\mathbf{i}_x + |v_\perp| \sin(\Omega t + \varsigma_e)\mathbf{i}_y + v_\parallel \mathbf{i}_z, \tag{7.107}$$

where $\mathbf{i}_{x,y,z}$ are unit vectors along the corresponding axes of the right-handed coordinate system x, y, z ($\mathbf{B}_0 \parallel \mathbf{i}_z$), and ς_e is the initial phase of the electrons (cf. equations (3.42)). The component of $\mathbf{v}$ in the direction perpendicular to $\mathbf{B}_0$ can also be presented in a complex form:

$$v_\perp = |v_\perp| \exp(i\Omega t + i\varsigma_e). \tag{7.108}$$

Integrating (7.107) with respect to t we obtain the expression for the unperturbed trajectory of the electron:

$$\mathbf{r} = \frac{|v_\perp|}{\Omega} \sin(\Omega t + \varsigma_e)\mathbf{i}_x - \frac{|v_\perp|}{\Omega} \cos(\Omega t + \varsigma_e)\mathbf{i}_y + v_\parallel t \mathbf{i}_z + \mathbf{r}_0, \tag{7.109}$$

where $\mathbf{r}_0$ is the position of the centre of the electron gyration. Without loss of generality we can assume that $\mathbf{k}$ lies in the (x, z) plane, $\mathbf{r}_0 = 0$, $\varsigma_e = 0$ and $\varsigma_R = \varsigma_L = \varsigma$.

From (7.106), (7.108) and (7.109) we obtain the average power absorbed ($\varepsilon > 0$) or emitted ($\varepsilon < 0$) by the electron:

$$\varepsilon = -|e||\mathbf{E}_\perp||v_\perp|I, \tag{7.110}$$

where

$$I = \lim_{T \to \infty} \frac{1}{T} \int_0^T \cos\left(\chi(t) + \varsigma\right) dt, \tag{7.111}$$

$$\chi(t) = \lambda_e \sin \Omega t + (-k_\| v_\| + \omega - \Omega)t, \tag{7.112}$$

and $\lambda_e \equiv k_\perp |v_\perp|/\Omega$.

The phase shift between $\mathbf{E}_\perp$ and $\mathbf{v}_\perp$ is accounted for by the parameter ς. In general the sign and the absolute value of ε depend on ς and $\chi(t)$. However, for the time being we restrict ourselves to the analysis of the conditions under which $\varepsilon \neq 0$, which, in its turn, can be satisfied only when $I \neq 0$. We begin the analysis of the latter condition with the analysis of the function $\chi(t)$ determined by (7.112). In this subsection we shall only consider the case $\varsigma = 0$. The contribution of different ς will be considered in the next subsection.

In Fig. 7.10a we show the plots of the function $\chi(t)$ for $b \equiv -k_\| v_\| + \omega - \Omega = 0$, $\lambda_e = 1$ (curve 1) and $\lambda_e = 2.4$ (curve 2). As can be seen from this figure, for curve 1 the condition $|\chi(t)| < \pi/2$ is valid for all t; thus $I \neq 0$. It is easy to see that the latter inequality is also valid for other values of λ_e in the range $0 < \lambda_e < \pi/2$. At the same time, for curve 2, the part of the period during which $\cos \chi > 0$ (solid part of the curve) is roughly equal to the part of the period during which $\cos \chi < 0$ (dashed part of the curve). Thus, taking into account the periodicity of $\chi(t)$, we can expect that I is close to zero. When λ_e increases further, I changes its sign and finally reaches another value for which $I = 0$ etc. All these behaviour patterns of $I(\lambda_e)$ are consistent with the properties of the integrand in (7.89) for $n = 1$. Note that the condition $b = 0$ reduces to the well known condition for the cyclotron resonance $\omega - k_\| v_\| - \Omega = 0$.

If $b \neq n\Omega$ (n being an integer) and $b \neq 0$ then the time intervals during which $\cos \chi > 0$ and $\cos \chi < 0$ will change from one period to another. Thus $I = 0$ and no regular energy exchange between whistler-mode waves and electrons can take place. If $b = n\Omega$ then the behaviour of $\cos \chi$ is similar for all electron gyroperiods and for both $b = 0$ and $b \neq 0$, and we can replace $T \to \infty$ in (7.111) by $T = 2\pi/\Omega$. Let us, for example, consider the case $b = \Omega$ corresponding to the second harmonic resonance ($\omega - k_\| v_\| - 2\Omega = 0$). The corresponding curves $\chi(t)$ are shown in Fig. 7.10b for $\lambda_e = 1$ (curve 1) and $\lambda_e = 3.8$ (curve 2). As follows from this figure, for curve 1 the interval of t during which $\cos \chi > 0$ (solid part of the curve) is smaller than the interval of t during which $\cos \chi < 0$ (dashed part of the curve), and thus we can expect that $I \neq 0$. When λ_e is reduced, the corresponding parts of this curve become more equal; for $\lambda_e = 0$ the curve $\chi(t)$ degenerates

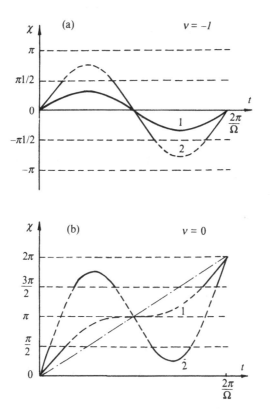

Fig. 7.10 (a) The curves $\chi(t)$ (see equation (7.112)) for the cyclotron resonance, $\omega - k_\parallel v_\parallel - \Omega = 0$, $\lambda_e \equiv k_\perp v_\perp/\Omega = 1$ (curve 1) and $\lambda_e = 2.4$ (curve 2). (b) The curves $\chi(t)$ for the second order cyclotron resonance, $\omega - k_\parallel v_\parallel - 2\Omega = 0$, $\lambda_e = 1$ (curve 1), $\lambda_e = 3.8$ (curve 2) and $\lambda_e = 0$ (dashed–dotted line). (The curves $\chi(t)$ for the Čerenkov resonance, $\omega - k_\parallel v_\parallel = 0$ could be described by the same curves as in (b) if we formally replaced χ by $-\chi$.) The solid (dashed) parts of the curves correspond to those parts of the electron gyroperiod when $\cos \chi > 0$ ($\cos \chi < 0$) (Sazhin, 1982b).

into a straight line (dashed–dotted line in Fig. 7.10b) for which these parts are equal, and thus $I = 0$. When λ_e increases and approaches 3.8, the corresponding parts of this curve are also equalized and $I = 0$ for $\lambda_e = 3.8$. When λ_e increases further then $|I|$ begins to increase again etc. As in the case $b = 0$ these results are consistent with the properties of the integrand in (7.89) for $n = 2$. The case of other resonances including the Čerenkov resonance can be similarly considered, although in the latter case the main source of the energy exchange between electrons and waves comes from the interaction between $v_\parallel$ and $E_\parallel$. This will be rigorously proved in the next

subsection where we shall give a quantitative analysis of the energy exchange
between whistler-mode waves and electrons.

(b) Quantitative analysis

In this subsection we give a quantitative analysis of the interaction between
oblique whistler-mode waves and electrons leading to wave growth or damping. As in the previous subsection we present the wave electric field in the
form (7.103). Firstly, we neglect the contribution of $\mathbf{E}_\parallel$ and consider the
energy exchange between electrons and $\mathbf{E}_\perp$. In this case the whistler-mode
electric field, in the system moving parallel to $\mathbf{B}_0$ with velocity $\omega/k_\parallel$, is
close to zero (see Appendix of Sazhin, 1982b) and the change of the electron
kinetic energy in the laboratory reference system (W_1) can be obtained from
the equation:

$$\frac{\mathrm{d}W_1}{\mathrm{d}t} = m_e \frac{\omega}{k_\parallel} \frac{\mathrm{d}v_\parallel}{\mathrm{d}t}. \tag{7.113}$$

Here $v_\parallel$ changes due to the Lorentz force:

$$\frac{\mathrm{d}v_\parallel}{\mathrm{d}t} = \frac{e}{m_e c} \mathbf{v}_\perp \times \mathbf{B}_\perp, \tag{7.114}$$

where $\mathbf{B}_\perp$ is the component of the wave magnetic field perpendicular to
$\mathbf{B}_0$; in a similar way to (7.103) we have expressed the electron velocity $\mathbf{v}$ as
$\mathbf{v}_\perp + \mathbf{v}_\parallel$, $v_\parallel = \mathrm{sign}(\mathbf{v}_\parallel \cdot \mathbf{B}_0)|\mathbf{v}_\parallel|$.

As was done with $\mathbf{E}_\perp$ (see equations (7.104)–(7.106)) we consider $\mathbf{B}_\perp$ as
a superposition of the right-handed $(\mathbf{B}_R)$ and the left-handed $(\mathbf{B}_L)$ circular
polarized waves, i.e. we write expressions (7.104)–(7.106) with $\mathbf{E}$ replaced by
$\mathbf{B}$. From the analysis of oblique whistler-mode polarization in a cold plasma
(see Section 2.3) it follows that:

$$|\mathbf{B}_{R(L)}| = \frac{1 + (-)\cos\theta}{2} |\mathbf{B}|, \tag{7.115}$$

where $|\mathbf{B}|$ is the amplitude of the total magnetic field of the wave. Then
having substituted (7.109) into (7.106) (with $\mathbf{E}$ replaced by $\mathbf{B}$) we obtain:

$$B_{R(L)} = |\mathbf{B}_{R(L)}| \exp\left[+(-)\left(\mathrm{i}\omega t - \mathrm{i}k_\parallel v_\parallel t - \frac{\mathrm{i}k_\perp |v_\perp|}{\Omega}\sin(\Omega t + \varsigma_e) + \mathrm{i}\varsigma\right)\right]. \tag{7.116}$$

Substituting (7.108) and (7.116) into (7.114) we obtain:

$$\frac{\mathrm{d}v_\parallel}{\mathrm{d}t} = \frac{e}{m_e c} \Re\left[-\mathrm{i}(B_R + B_L)v_\perp^*\right], \tag{7.117}$$

where * indicates the complex conjugate value of the parameter.

In what follows, we put $\varsigma_e = 0$ as it is essential for us to know only the relative phase between $B_\perp$ and $v_\perp$ which is described by ς. In view of (7.108) and (7.116), equation (7.117) can be rewritten in a more explicit form:

$$\frac{dv_\parallel}{dt} = \frac{e|v_\perp||\mathbf{B}|}{2m_e c} \sum_{n=-\infty}^{+\infty} [(1-\cos\theta)J_{n+1} + (1+\cos\theta)J_{n-1}]$$

$$\times \cos\left[\omega t - k_\parallel v_\parallel t - n\Omega t + \varsigma - \pi/2\right], \qquad (7.118)$$

where the argument of the Bessel functions is the same as in (7.82). Equation (7.118) formally reduces to equation (2) of Section 7.2 of Stix (1962) describing the electron dynamics in the field of a Langmuir wave if we replace

E	by	$|v_\perp||\mathbf{B}| \sum_{n=-\infty}^{+\infty} [(1-\cos\theta)J_{n+1} + (1+\cos\theta)J_{n-1}]/(2c)$,
ω	by	$\omega - n\Omega$,
kv_0	by	$k_\parallel v_\parallel$,
kz_0	by	$-\varsigma + \pi/2$.

Following very closely the analysis of Stix (1962) (see also Problem 7.1), that is, solving (7.118) by the method of successive approximations for the wave field $\mathbf{B}$ switched on at $t = 0$, keeping only the second-order terms, averaging over the phase ς for a given electron distribution function f_0, introducing the new variables

$$\left.\begin{array}{l} \hat{\alpha} = k_\parallel v_\parallel - \omega + n\Omega \\ g(v_\perp, \hat{\alpha}) = f_0\left(v_\perp, (\hat{\alpha} + \omega - n\Omega)/k_\parallel\right) \\ h_n = [(1-\cos\theta)J_{n+1} + (1+\cos\theta)J_{n-1}]/2 \end{array}\right\}, \qquad (7.119)$$

and neglecting the terms vanishing as $t \to \infty$, we have from (7.118):

$$\left\langle \frac{dv_\parallel}{dt} \right\rangle_{\varsigma,\mathbf{v}} = -\frac{\pi n_e e^2 |\mathbf{B}|^2}{m_e^2 c^2} \int_0^\infty dv_\perp v_\perp^3$$

$$\times \sum_{n=-\infty}^{+\infty} h_n^2 P \int_{-\infty}^{+\infty} d\hat{\alpha}\, g(v_\perp, \hat{\alpha}) \frac{\sin\hat{\alpha}t}{\hat{\alpha}^2}, \qquad (7.120)$$

where n_e is the electron density, and P denotes the principal value of the integral. Subscripts ς and $\mathbf{v}$ mean that the parameter is averaged with respect to ς and electron velocities $\mathbf{v}$.

As follows from (7.120), whistler-mode waves exchange their energy mainly with electrons having small $\hat{\alpha}$. Thus $g(v_\perp, \hat{\alpha})$ can be expanded:

$$g(v_\perp, \hat{\alpha}) = g(v_\perp, 0) + \hat{\alpha}\left.\frac{dg(v_\perp, \hat{\alpha})}{d\hat{\alpha}}\right|_{\hat{\alpha}=0} + \frac{\hat{\alpha}^2}{2}\left.\frac{d^2 g(v_\perp, \hat{\alpha})}{d\hat{\alpha}^2}\right|_{\hat{\alpha}=0} + \dots . \quad (7.121)$$

If we keep only the second-order terms in (7.121) then after substituting (7.121) into (7.120) and (7.120) into (7.113) we have:

$$\left\langle \frac{dW_1}{dt} \right\rangle_{\varsigma,\mathbf{v}} = -\frac{\pi^2 n_e e^2 |\mathbf{B}|^2 \omega^2}{m_e c^2 k_\parallel^3} \int_0^\infty dv_\perp v_\perp^2$$

$$\times \int_{-\infty}^{+\infty} dv_\parallel \sum_{n=-\infty}^{n=+\infty} h_n^2 \delta \left(v_\parallel - \left(\frac{\omega}{k_\parallel} - \frac{n\Omega}{k_\parallel} \right) \right) G_1, (7.122)$$

where G_1 and h_n were defined by (7.81) and (7.119) respectively (see also the analysis by Lutomirski (1970) of parallel whistler-mode waves).

Now we consider the change of electron energy due to the contribution of $\mathbf{E}_\parallel$ in (7.103). $\mathbf{E}_\parallel \equiv \mathbf{E}_{\parallel 0} \cos(\omega t - \mathbf{kr} + \varsigma_\parallel)$ changes only the component of the electron energy in the direction parallel to $\mathbf{B}_0$ (W_2), and if we assume that $\lambda_e \ll 1$ then this change can be obtained as was done by Stix (1962) (see his Section 7.2 and also our Problem 7.2, and cf. our analysis of equation (7.113)). As a result we have:

$$\left\langle \frac{dW_2}{dt} \right\rangle_{\varsigma_\parallel,\mathbf{v}} = -\frac{\pi^2 n_e e^2 |\mathbf{E}_{\parallel 0}|^2 \omega}{m_e k^2 \cos^2 \theta} \int_0^\infty dv_\perp v_\perp \frac{\partial f_0}{\partial v_\parallel}\bigg|_{v_\parallel = \omega/k_\parallel}. \tag{7.123}$$

When deriving (7.123) we assumed that $k_\parallel > 0$.

From analysis of the low-frequency whistler-mode polarization (see Section 2.3) it follows that:

$$|\mathbf{E}_{\parallel 0}| = \frac{\sin \theta |\mathbf{B}|}{Y \cos \theta N}. \tag{7.124}$$

Hence we can rewrite (7.123) as:

$$\left\langle \frac{dW_2}{dt} \right\rangle_{\varsigma_\parallel,\mathbf{v}} = -\frac{\pi^2 n_e e^2 \omega |\mathbf{B}|^2 \sin^2 \theta}{m_e k^2 N^2 Y^2 \cos^4 \theta} \int_0^\infty dv_\perp v_\perp \frac{\partial f_0}{\partial v_\parallel}\bigg|_{v_\parallel = \omega/k_\parallel}. \tag{7.125}$$

The whistler-mode energy, determined by the equation:

$$U = |\mathbf{B}|^2 / 8\pi \tag{7.126}$$

is connected with $\langle dW_1/dt \rangle_{\varsigma,\mathbf{v}}$ and $\langle dW_2/dt \rangle_{\varsigma_\parallel,\mathbf{v}}$ by the equation:

$$\left\langle \frac{dW_1}{dt} \right\rangle_{\varsigma,\mathbf{v}} + \left\langle \frac{dW_2}{dt} \right\rangle_{\varsigma_\parallel,\mathbf{v}} = -2\gamma U. \tag{7.127}$$

From (7.122) and (7.125)–(7.127) we obtain expression (7.92) for $\tilde{\gamma}$ in which the term $\tilde{\gamma}_{\mathrm{Ken}}$ describes the contribution of $\langle dW_1/dt \rangle_{\varsigma,\mathbf{v}}$, while the term $\tilde{\gamma}_0$ describes the contribution of $\langle dW_2/dt \rangle_{\varsigma_\parallel,\mathbf{v}}$. Hence, the physical meaning of the terms $\tilde{\gamma}_{\mathrm{Ken}}$ and $\tilde{\gamma}_0$ is as follows: the first results from the energy exchange between the electrons and $\mathbf{E}_\perp$, while the second results from a similar exchange between the electrons and $\mathbf{E}_\parallel$. In the case of Landau damping (or the corresponding instability) the second process is the dominant one.

(c) Geometrical analysis

As follows from (7.89) and (7.92), except when $n = 0$, the value and the sign of $\tilde{\gamma} \equiv \tilde{\gamma}_{\text{Ken}}$ are controlled by ϑ_n defined by (7.90) and G_1 defined by (7.81) when the plasma density, wave frequency and the harmonic n are fixed. ϑ_n is always positive or zero. Thus, ϑ_n can control only the absolute value of $\tilde{\gamma}$, while G_1 determines both the value and the sign of $\tilde{\gamma}$. This allows us to work out a simple criterion for instability, generalizing the results earlier obtained by Gendrin (1981) for the case of parallel whistler-mode propagation. Almost repeating Gendrin's (1981) analysis it can be shown that for a 'regular' electron distribution function ($\nabla_v f_0 < 0$, where $\nabla_v f_0$ is the projection of the gradient of the unperturbed electron distribution function f_0 in the velocity space on $v_\parallel$) whistler-mode waves are unstable only in the case when the diffusion curve for the electrons in the $(v_\perp, v_\parallel)$ space (D curve), determined by the equation

$$v_\perp^2 + (v_\parallel - (\omega/k_\parallel))^2 = \text{const} \qquad (7.128)$$

(the signs of $k_\parallel$ and $v_\parallel$ depend on the resonance condition for $v_\parallel$; see Chapter 8 for more detailed discussion on electron diffusion), can be placed between the curves of constant density (F curve) determined by the equation

$$f_0(v_\parallel, v_\perp) = \text{const}, \qquad (7.129)$$

and those of constant energy (E curve) determined by the equation

$$v_\perp^2 + v_\parallel^2 = \text{const} \qquad (7.130)$$

in the same velocity space, when all these three curves intersect at the same point. For a 'non-regular' electron distribution function ($\nabla_v f_0 > 0$) instability occurs when the D curve is not between the E and F curves. $\tilde{\gamma} = 0$ when $\vartheta_n = 0$ or when the D curve coincides with the F or E curve. When deriving these criteria, it was assumed that $\nabla_v f_0$ does not change its sign in the vicinity of the point in the velocity space considered.

Strictly speaking, the above mentioned results were obtained for a fixed value of $v_\perp$. However, for realistic distribution functions, and, in particular, for the bi-Maxwellian distribution, the character of the mutual positions of the F, E and D curves does not depend on $v_\perp$ (Gendrin, 1981) and the sign of γ obtained for an arbitrary $v_\perp$ is conserved after integration over $v_\perp$.

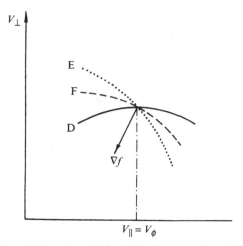

Fig. 7.11 The D, F and E curves in the velocity space for the Čerenkov interaction between whistlers and electrons (see equations (7.128)–(7.130)). $v_\phi \equiv v_{ph} = \omega/k_\parallel$ (Sazhin, 1982b).

Let us now illustrate how the criteria described above can be applied to the analysis of the Čerenkov whistler-mode growth or damping. Taking into account that $v_\parallel = \omega/k_\parallel$ we can present the D (solid) (see equation (7.128)) and E (dotted) (see equation (7.130)) curves for fixed $v_\perp$ as shown in Fig. 7.11. In the same figure we also show the realistic F curve (dashed) (see equation (7.129)). The actual shape of the F curve can change from being tangential to being perpendicular to the D curve. The curves coincide in the vicinity of the intersection point of the F and E curves only in the first degenerate case, and so the D curve is not located between the F and E curves. Thus we may expect that whistler-mode waves can be unstable on Čerenkov resonance only when $\nabla_v f_0 > 0$. As we assumed that $\nabla_v f_0$ does not change its sign, the criterion $\nabla_v f_0 > 0$ reduces to $\partial f_0/\partial v_\parallel > 0$. In the opposite case, when $\partial f_0/\partial v_\parallel < 0$, the waves are damped, which agrees with the results of the previous analysis.

The relative position of the E, D and F curves for other resonances ($n \neq 0$ in equation (7.82)) can be deduced from the corresponding derivatives for these curves ($dv_\perp/dv_\parallel$) calculated from equations (7.128)–(7.130). Assuming that f_0 is determined by (1.90) with $j = 0$ (bi-Maxwellian distribution) we can see that the D curve is between the E and F curves and thus $\gamma > 0$ when

$$v_\parallel < v_\parallel - (\omega/k_\parallel) < v_\parallel w_\perp^2/w_\parallel^2 \qquad (7.131)$$

or

$$v_{\parallel} > v_{\parallel} - (\omega/k_{\parallel}) > v_{\parallel} w_{\perp}^2 / w_{\parallel}^2. \qquad (7.132)$$

The first system of inequalities can be satisfied only for $\omega/k_{\parallel} < 0$, i.e. for $n > 0$ in equations (7.82) or (7.89). The second system can be satisfied only for $\omega/k_{\parallel} > 0$, i.e. for $n < 0$ in the same equation. In other words, the condition for whistler-mode growth at the corresponding resonances can be written as:

$$A_e > \frac{nY}{nY - 1} \qquad (7.133)$$

for $n > 0$, and

$$A_e < \frac{nY}{nY - 1} \qquad (7.134)$$

for $n < 0$. These criteria for whistler-mode growth are in agreement with those which follow from (7.82) or (7.89). In particular, from (7.133) and (7.134) it follows that at low frequencies ($Y \gg 1$) the whistler-mode instability corresponding to $n > 0$ can occur only when $A_e > 1$, whilst the instabilities corresponding to $n < 0$ occur for $A_e < 1$. Note that the actual contribution of the resonances corresponding to $n \neq 1, 0$ is negligibly small when compared with the resonances corresponding to $n = 1, 0$. Hence our geometrical analysis is mainly useful for understanding some of the physical processes that occur with oblique whistler-mode growth or damping, which have not so far been discussed, rather than for practical diagnosis of unstable waves.

When considering whistler-mode growth in this chapter, it was implicitly assumed that the propagating waves are monochromatic and no interference takes place between the modes corresponding to different frequencies and wave numbers. If, however, this kind of interference is taken into account, then there can appear two different ways for the further development of the instability. Firstly, the wave amplitude can infinitely increase when $t \to \infty$ at a given point of space. Secondly, this amplitude can increase infinitely in the reference system connected with the propagating wave, remaining finite at any specific point in space. The instabilities in these cases are known as absolute and convective respectively (Sturrock, 1958; Briggs, 1964; Derfler, 1967, 1969; Akhiezer et al., 1975). As was pointed out by Lee & Crawford (1970), most realistic whistler-mode instabilities in the Earth's magnetosphere are of the convective type.

Finally we should note that the instabilities analysed in this chapter give us only the rate of the initial wave growth and say nothing about the final amplitude of the wave. The latter problem is considered in the next chapter where some of the non-linear effects connected with whistler-mode waves are discussed.

Problems

Problem 7.1 Derive equation (7.26) under the assumption that $\tilde{A}_{e0} - 1 \ll 1$ and using the following presentation of the Shkarofsky function (Robinson, 1986):

$$\Im \mathcal{F}_q = -\pi e^{z-2a} \left[(a-z)/a\right]^{(q-1)/2} I_{q-1}\left(2\sqrt{a}\sqrt{a-z}\right). \qquad (7.135)$$

Problem 7.2 Show details of the derivation of equation (7.123).

8

Non-linear effects

The linear theory of whistler-mode propagation, growth and damping considered so far has been based on the assumption that waves with different frequencies and wave numbers do not interact with each other (superposition principle) and that the waves do not cause any systematic change in the background particle (electron) distribution function f_0. A self-consistent analysis of both these processes creates, in general, a very complicated problem even for modern computers (see e.g. Nunn, 1990). However, in many practically important cases we can develop an approximate analytical theory which takes into account some of these processes and neglects others. This theory, known as the non-linear theory, has been developed during the last 30 years and its results are summarized in numerous monographs and review papers, such as those by Kadomtsev (1965), Vedenov (1968), Sagdeev & Galeev (1969), Tsytovich (1972), Karpman (1974), Akhiezer *et al.* (1975), Hasegawa (1975), Vedenov & Ryutov (1975), Galeev & Sagdeev (1979), Bespalov & Trakhtengertz (1986), Zaslavsky & Sagdeev (1988), Petviashvili & Pokhotelov (1991) and many others. I have no intention of amending this long list of references by one more contribution. Instead I will restrict myself to illustrating the methods of non-linear theory by two particularly simple examples: the quasi-linear theory of whistler-mode waves (Section 8.1) and the non-linear theory of monochromatic whistler-mode waves (Section 8.2).

8.1 Quasi-linear theory

(a) Basic equations

A theory which assumes that the wave amplitude is so small that the superposition principle remains valid, but large enough to provide a non-negligible change of the background electron distribution function f_0 under the influence of the waves, is known as the quasi-linear theory (Akhiezer *et al.*, 1975).

This theory was first developed in the pioneering paper by Romanov & Filippov (1961), and it has been widely used up to the present time (see e.g. Villalón *et al.*, 1989; Gail *et al.*, 1990; Jancel & Wilhelmsson, 1991). At the same time a number of authors have pointed out that mode-coupling terms may yield non-negligible corrections to the quasi-linear terms (see e.g. Laval & Pesme (1983) and the references therein). This means that although quasi-linear theory can be used for a qualitative analysis of waves and particle dynamics in plasma, caution is needed when applying it quantitatively.

In order to obtain the equation describing this change of f_0 we substitute (1.46) into (1.25) and keep not only linear terms, as was done when deriving equation (1.49), but also non-linear terms proportional to $|E_1|f_1$. The resultant equation still appears to be too complicated for analysis, but it can be further simplified if we assume that: (i) $|\Delta k_\parallel| \ll k_0$ (the width of the wave packet, $|\Delta k_\parallel|$, is well below the wave number referring to the main harmonic k_0); (ii) the time interval under consideration, Δt, is well above the time of phase mixing of the packet $t_{ph} = 1/(k_0 \Delta v_R)$; (iii) $\Delta t \ll 1/|\gamma(\mathbf{k})|$, where $\gamma(\mathbf{k})$ is the increment of growth ($\gamma(\mathbf{k}) > 0$) or the decrement of damping ($\gamma(\mathbf{k}) < 0$) of the kth harmonic. Remembering that

$$\Delta v_R = \left(\frac{\partial \omega(\mathbf{k})}{\partial k_\parallel} - v_\parallel \right) \frac{\Delta k_\parallel}{k_\parallel} \bigg|_{k_\parallel = k_0} ,$$

we can see that all these assumptions are non-contradictory when:

$$1 \gg \frac{|\Delta k_\parallel|}{k_0} \gg \frac{|\gamma(\mathbf{k})|}{k_\parallel |v_\parallel| - \frac{\partial \omega(\mathbf{k})}{\partial k_\parallel}} \bigg|_{k_\parallel = k_0} . \tag{8.1}$$

Δt can be identified with the quasi-linear diffusion time t_D. If the wave is presented as a superposition of a finite number n of harmonics then we need to impose one more assumption that the regions of trapping corresponding to different harmonics in the packet overlap. This assumption does not contradict the previous assumptions when n is sufficiently large (Lifshitz & Pitaevsky, 1979; Galeev *et al.*, 1980).

Condition (8.1) is not satisfied for waves with linear or almost linear dispersion (see Akhiezer *et al.* (1975) for an alternative approach), but this is not the case for whistler-mode waves.

In view of conditions (8.1) the equation for f_0 can be written as (Akhiezer *et al.*, 1975):

$$\frac{\partial f_0}{\partial t} = \frac{\pi e^2}{m_e^2} \sum_{n=-\infty}^{+\infty} \int d^3 k \frac{1}{v_\perp} \tilde{R} \left\{ v_\perp |E_{kx} \frac{n}{\lambda_e} J_n(\lambda_e) + iE_{ky} J_n'(\lambda_e) \right.$$

$$\left. + E_{kz} \frac{v_\parallel}{v_\perp} J_n(\lambda_e)|^2 \times (\tilde{R} f_0) \right\} \delta(\omega(\mathbf{k}) - n\Omega - k_\parallel v_\parallel) + S_e + P_e, \quad (8.2)$$

where $\tilde{R}$ was defined by (7.81), $\lambda_e \equiv k_\perp v_\perp / \Omega$ is the same as in (1.60)–(1.62), $E_{kx(y,z)} \equiv \sqrt{\left\langle E_{x(y,z)}^2 \right\rangle_\mathbf{k}}$ are the root mean squared spectral densities (in $\mathbf{k}$ space) of the components of the wave electric field, integration in (8.2) is assumed over all wave vectors $\mathbf{k}$ (see Lifshitz & Pitaevsky (1979) for a more detailed discussion on the definition of $\langle E^2 \rangle$), $S_e > 0$ describes the contribution of the electron source, and $P_e < 0$ refers to the loss of electrons.

Equation (8.2) can be considerably simplified if we restrict ourselves to considering wave propagation parallel to the magnetic field. In this case the electrostatic mode can be separated from the electromagnetic whistler-mode, so that when describing the latter waves we can put $E_z = 0$. Remembering (2.69) and (5.44), we can write the equation $E_y = -iE_x$ for parallel whistler-mode waves and we can reduce equation (8.2) for these waves to:

$$\frac{\partial f_0}{\partial t} = \frac{\pi e^2}{m_e^2} \int dk \frac{1}{v_\perp} \tilde{R} \left\{ v_\perp E_k^2 (\tilde{R} f_0) \delta(\omega(k) - \Omega - k v_\parallel) \right\} + S_e + P_e. \quad (8.3)$$

As in Chapters 3 and 4, we assume that $k = k_\parallel$; the integral in (8.3) and the corresponding power spectral densities $E_k^2 \equiv |E_k|^2 = \langle E^2 \rangle_k$ are assumed to be one-dimensional (the mean square of the wave electric field is defined as $\overline{E^2} = \int E_k^2 dk$). The integral in (8.3) could be written in the form of the sum over different harmonics ($\int dk = \sum_k$).

Alternatively, equation (8.3) can be written as:

$$\frac{\partial f_0}{\partial t} = \frac{\pi e^2}{m_e^2} \int dk \left[\frac{\Omega}{v_\perp} \left(\frac{\partial}{\partial v_\perp} v_\perp \right) + k v_\perp \frac{\partial}{\partial v_\parallel} \right] \frac{B_k^2}{k^2 c^2}$$

$$\times \delta(\omega(k) - n\Omega - k v_\parallel) \left(\Omega \frac{\partial}{\partial v_\perp} + k v_\perp \frac{\partial}{\partial v_\parallel} \right) f_0 + S_e + P_e, \quad (8.4)$$

where $B_k^2 \equiv |B_k|^2 = \langle B^2 \rangle_k$ is the magnetic field power spectral density. Equation (8.4) follows from (8.3), if we remember that

$$B_k^2 = E_k^2 N^2 \quad (8.5)$$

(equation (8.5) follows from the Maxwell equation (1.27) as $\mathbf{k} \perp \mathbf{E}$ for parallel whistler-mode waves), and that only the contribution of $v_{\parallel}$ close to $(\omega - \Omega)/k_{\parallel}$ is essential in equation (8.3).

The wave energy spectral density with respect to wave numbers in equations (8.3) and (8.4) can be replaced by the corresponding wave energy spectral densities with respect to angular frequencies E_{ω}^2 and B_{ω}^2 or with respect to cyclic frequencies $E_f^2 = 2\pi E_{\omega}^2$ and $B_f^2 = 2\pi B_{\omega}^2$ if we remember that:

$$\left. \begin{array}{l} v_g E_{\omega}^2 = E_k^2 \\ v_g B_{\omega}^2 = B_k^2 \end{array} \right\}, \tag{8.6}$$

where v_g is the wave group velocity.

$w(k)$ in equation (8.4) can be expanded in the vicinity of $w(k) = \omega_R \equiv kv_{\parallel} + \Omega$:

$$w(k) = \omega_R + \Delta k \partial \omega / \partial k|_{\omega = \omega_R} + \dots . \tag{8.7}$$

Keeping only the first two terms in this expansion and substituting (8.7) into (8.4) enables us to perform an explicit integration with respect to k. As a result we obtain:

$$\frac{\partial f_0}{\partial t} = \frac{\pi e^2}{m_e^2} \frac{\Omega^2}{k^2 c^2} \left[\frac{1}{v_{\perp}} \frac{\partial}{\partial v_{\perp}} v_{\perp} + \frac{kv_{\perp}}{\Omega} \frac{\partial}{\partial v_{\parallel}} \right]$$

$$\times \frac{B_k^2}{|v_{\parallel} - v_g|} \left[\frac{\partial}{\partial v_{\perp}} + \frac{kv_{\perp}}{\Omega} \frac{\partial}{\partial v_{\parallel}} \right] f_0|_{v_{\parallel} = v_R = \frac{\omega - \Omega}{k_{\parallel}}} + S_e + P_e. \tag{8.8}$$

The expression for $k = N\omega/c$ can be derived from (3.34) without relativistic effects taken into account. v_g can be taken in the form (4.29). The corresponding expression for v_R follows from (3.34) and can be written as (see equation (8) of Sazhin, 1989e):

$$v_R = v_{R0} \left[1 - \frac{Y-1}{2\nu Y^2} + \frac{(Y-1)r}{2} - \frac{Y^2 \beta_e}{2(Y-1)^2} \left(\frac{Y}{Y-1} - A_e \right) \right], \tag{8.9}$$

where

$$v_{R0} = -\frac{c(Y-1)^{3/2}}{Y\sqrt{\nu}}. \tag{8.10}$$

Strictly speaking, the parameters A_e and β_e in equations (3.34), (4.29) and (8.9) are time-dependent in quasi-linear theory in which, in contrast to linear theory, f_0 is no longer assumed to be constant. However, as N, v_g and v_R are only weakly dependent on A_e within the reasonable range of this parameter in magnetospheric conditions, and β_e changes only slightly under the influence of whistler-mode waves, in what follows we shall assume that both A_e and β_e are constant in all equations referring to wave propagation.

Equation (8.8) describing the evolution of f_0 under the influence of the wave field is complementary to equation (7.53) describing the evolution of the wave field for a particular value of f_0. The solution of these equations depends on the particular problem under consideration. We might look for a temporal evolution of, or a steady state solution for, either f_0 or B_f^2 (or $B_{k(\omega)}^2$, or $E_{k(\omega \text{ or } f)}^2$) under different initial or boundary conditions. In view of the potential application to the interpretation of whistler-mode wave spectra in magnetospheric conditions, in what follows we shall concentrate our attention on the solutions for B_f^2. These solutions will be found under the following assumptions.

First we assume that the plasma is homogeneous and the contribution of boundary effects can be ignored. Then we write the following equation:

$$\frac{\partial B_{k(\omega \text{ or } f)}^2}{\partial t} = 2\gamma_{k(\omega \text{ or } f)} B_{k(\omega \text{ or } f)}^2, \tag{8.11}$$

where $\gamma_{k(\omega \text{ or } f)}$ is the temporary wave growth or damping at a particular k (ω or f). In this case the contribution of the source, S_e, and loss, P_e, terms can be neglected as well.

Secondly we assume that the waves are concentrated inside a magnetic tube so that their total amplification during the process of their propagation from one hemisphere to the other is compensated for by the loss at reflection from the ionospheres. This model is particularly useful for the conditions in the inner magnetosphere where this area might be related to a magneto-spheric field tube with the boundaries at the ionospheric level. In this case we cannot, in general, write a simple equation for $B_{k(\omega \text{ or } f)}^2$ as a function of t and γ, as the contribution of S_e and P_e cannot be neglected. However, we can write a separate equation for γ in the form:

$$2 \int_{-s_{\text{ion}}}^{+s_{\text{ion}}} \frac{\gamma(s) \mathrm{d}s}{v_g(s)} = \ln(1/R_w), \tag{8.12}$$

where R_w is the wave energy reflection coefficient, which is frequency-dependent in general; integration in (8.12) is assumed along the whole magnetic field line from one hemisphere to the other. In the latter model, with appropriate choice of S_e and P_e, we shall look for a stationary solution of equation (8.8), i.e. we shall assume from the very beginning that $\partial f_0/\partial t = 0$ in this equation.

These models will be described in some detail in subsections (b) and (c).

(b) A model for homogeneous plasma

As was mentioned in the previous subsection, in the case of parallel whistler-mode propagation in a homogeneous plasma we assume that $S_e = P_e = 0$ in (8.8) and link $B^2_{k(\omega \text{ or } f)}$ with $\gamma_{k(\omega \text{ or } f)}$ by equation (8.11). The solution of equations (7.53), (8.8) (or modifications of the latter equation for B^2_ω or B^2_f) and (8.11) depends on the particular problem under consideration. In what follows in this subsection we assume that we know the initial electron distribution function, $f^0_0 \equiv f_0(t = 0)$, and the initial wave spectral density. From these initial conditions and equations (7.53), (8.8) and (8.11) we then obtain the expression for the wave spectral density which results from the evolution of f_0 from its initial unstable state to the final stable state during the process of the instability development.

We begin with some simplifications of equations (7.53) and (8.8) which can be achieved by introducing the new variables (Andronov & Trakhtengertz, 1964; Akhiezer *et al.*, 1975):

$$\left. \begin{array}{c} \hat{w} = v^2_\perp + v^2_\| - 2\int_0^{v_\|} v_{\text{ph}}(v'_\|)\mathrm{d}v'_\| \\ \hat{v} = v_\| \end{array} \right\}, \tag{8.13}$$

where $v_{\text{ph}} = \omega/k$ is the phase velocity of the wave which is assumed to be a function of $v_\|$.

In view of (8.13), and neglecting the contribution of the terms S_e and P_e, we can rewrite equations (7.53) and (8.8) as:

$$\gamma \equiv \tilde{\gamma}\omega = \frac{\pi^2 \Pi^2}{2\partial \Re(\omega^2 D)/\partial \omega} \int_{\hat{w}_{\min}}^\infty v^2_\perp(\hat{w}, \hat{v}) \frac{\partial f_0}{\partial \hat{v}} \mathrm{d}\hat{w}, \tag{8.14}$$

$$\frac{\partial f_0}{\partial t} = \frac{\partial}{\partial \hat{v}} v^2_\perp(\hat{w}, \hat{v}) D_d \frac{\partial f_0}{\partial \hat{v}}, \tag{8.15}$$

where:

$$D_d = \frac{\Omega^2 v_g B^2_f}{2B^2_0 |\hat{v} - v_g|}, \tag{8.16}$$

(subscript $_d$ after D in (8.15) and (8.16) is added to indicate that D_d in these equations refers to the diffusion coefficient and not to the parameter D used in equations (1.16) and (1.17)), $\Re D$ is taken in the form (7.55), and $\hat{w}_{\min}$ is the minimal value of $\hat{w}$ for a given $\hat{v}$ which is attained for $v_\perp = 0$. When deriving (8.16) we considered D_d to be a function of B^2_f. However, remembering (8.5) and (8.6) we could derive in a straightforward way the expressions for D_d as functions of $E^2_{k(\omega \text{ or } f)}$ or $B^2_{k(\omega)}$.

Integrating (8.15) twice, once with respect to $\hat{v}$ from $\hat{v}_{\lim}$ to $\hat{v}$ keeping $\hat{w} = \text{const}$, and then with respect to $\hat{w}$ from $\hat{w}_{\min}$ to ∞ keeping $\hat{v} = \text{const} = v_R$, and remembering (8.14) and (8.11), we obtain:

$$\frac{\partial B_f^2}{\partial t} = \frac{2\pi^2 \Pi^2 B_0^2 |v_R - v_g|}{\Omega^2 v_g \partial \Re(\omega^2 D)/\partial \omega} \int_{\hat{w}_{\min}}^{\infty} d\hat{w} \int_{\hat{v}_{\lim}}^{v_R} d\hat{v} \frac{\partial f_0}{\partial t}, \qquad (8.17)$$

where $\hat{v}_{\lim}$ is defined as a maximal value of $|\hat{v}|$ attained for $\hat{w} = \text{const}$.

Integrating both sides of equation (8.17) with respect to t and neglecting the value of $B_f^2(t = 0)$ we obtain:

$$B_f^2(t) = \frac{2\pi^2 \Pi^2 B_0^2 |v_R - v_g|}{\Omega^2 v_g \partial \Re(\omega^2 D)/\partial \omega} \int_{\hat{w}_{\min}}^{\infty} d\hat{w} \int_{\hat{v}_{\lim}}^{v_R} \left[f_0(t, \hat{w}, \hat{v}) - f_0^0(\hat{w}, \hat{v}) \right] d\hat{v}, \qquad (8.18)$$

where $f_0^0(\hat{w}, \hat{v}) = f_0(t = 0, \hat{w}, \hat{v})$.

As $f_0(t, \hat{w}, \hat{v})$ in the right hand side of (8.18) is itself a function of $B_f^2(t)$, this equation tells us little about the temporal evolution of B_f^2. However, it can give us asymptotic solutions for $B_f^2(t \to \infty)$ if we make physically reasonable assumptions about $f_0(t \to \infty, \hat{w}, \hat{v})$. These solutions were probably first discussed by Rowlands, Shapiro & Shevchenko (1966) more than 25 years ago. In what follows I consider a different and much simpler approach to the analysis of (8.18) in the limit $t \to \infty$ based on my earlier paper (Sazhin, 1989e). The relative simplicity of my approach will enable us to consider a wide range of applications of the final equation to the interpretation of whistler-mode observations in the magnetosphere of the Earth (see Chapter 9).

Remembering the definitions of $\hat{v} = v_R = (\omega - \Omega)/k$ and $v_{\text{ph}} = \omega/k$, we can write:

$$v_{\text{ph}} = \frac{1}{1 - Y} \hat{v}. \qquad (8.19)$$

Substituting (8.19) into (8.13) we obtain

$$\hat{w} = v_\perp^2 + \frac{Y}{Y - 1} \hat{v}^2. \qquad (8.20)$$

Assuming $Y \gg 1$ we can simplify equation (8.20) to:

$$\hat{w} = v_\perp^2 + (1 + Y^{-1}) \hat{v}^2. \qquad (8.21)$$

Introducing the parameter

$$v_t = \sqrt{v_\perp^2 + \hat{v}^2} \qquad (8.22)$$

we obtain from (8.21):

$$v_t^2 = \hat{w} - Y^{-1}\hat{v}^2, \tag{8.23}$$

$$\hat{v}^2 = v_t^2 \cos^2 \alpha_e = \hat{w}\cos^2\alpha_e - Y^{-1}\hat{v}^2 \cos^2\alpha_e, \tag{8.24}$$

$$\hat{v} = \sqrt{\hat{w}}\cos\alpha_e \left(1 - \frac{Y^{-1}\cos^2\alpha_e}{2}\right), \tag{8.25}$$

$$d\hat{v} = \sqrt{\hat{w}}\left(-1 + \frac{3}{2}Y^{-1}\cos^2\alpha_e\right)\sin\alpha_e d\alpha_e, \tag{8.26}$$

$$\hat{w} = \frac{v_R^2}{\cos^2\alpha_e}\left(1 + Y^{-1}\cos^2\alpha_e\right), \tag{8.27}$$

$$d\hat{w} = \frac{2v_R^2\sin\alpha_e d\alpha_e}{\cos^3\alpha_e}, \tag{8.28}$$

where α_e is the pitch-angle of resonant electrons (cf. equation (7.58)); when deriving (8.26) we assumed that $\hat{w} = \text{const}$, while in (8.28) it is assumed that $\hat{v} = \text{const} = v_R$. Making the physically reasonable assumption that, in the limit $t \to \infty$, f_0 is an isotropic Maxwellian we can set:

$$f_0(t \to \infty) = \frac{1}{\pi^{3/2}w^3}\exp\left(-\frac{v_t^2}{w^2}\right), \tag{8.29}$$

where v_t is defined by (8.22). Also we assume that

$$f_0^0 = \frac{1}{\pi^{3/2}w^3}K_m \exp\left(-\frac{v_t^2}{w^2}\right)\sin^m\alpha_e, \tag{8.30}$$

where m is an integer, and K_m is the normalization coefficient: $K_1 = 4/\pi$; $K_2 = 3/2$; $K_3 = 16/3\pi$; $K_4 = 15/8$. In view of (8.22) and (8.25) we obtain for $Y^{-1} \ll 1$:

$$\exp\left(-\frac{v_t^2}{w^2}\right) = \exp\left(-\frac{\hat{w}}{w^2}\right)\left(1 + \frac{\hat{w}Y^{-1}\cos^2\alpha_e}{w^2}\right). \tag{8.31}$$

Having substituted (8.29) and (8.30) into (8.18) we obtain in view of (8.26) and (8.31) the following expression for the internal integral:

$$\int_{v_{\text{lim}}}^{v_R}dv\left(f_0 - f_0^0\right) = \left(\frac{1}{\pi^{3/2}w^3}\right)\sqrt{\hat{w}}\exp\left(-\frac{\hat{w}}{w^2}\right)\left[M_m + \left(\frac{\hat{w}}{w^2} - \frac{3}{2}\right)Y^{-1}\tilde{M}_m\right], \tag{8.32}$$

where

$$
\left.
\begin{aligned}
M_m &= \int_{-1}^{\cos \alpha_e} \left(1 - K_m(1 - t^2)^{m/2}\right) dt \\
\tilde{M}_m &= \int_{-1}^{\cos \alpha_e} t^2 \left(1 - K_m(1 - t^2)^{m/2}\right) dt
\end{aligned}
\right\}.
\tag{8.33}
$$

When deriving (8.32) we took into account that $v_R < 0$ and $\alpha_e \in (-\pi/2, -\pi]$, so that the value of $v_{\lim}$ corresponds to $\alpha_e = -\pi$. Note that both M_m and $\tilde{M}_m$ are positive.

Remembering (3.34), (4.29), (7.55) and (8.9), we obtain in the limit $Y \gg 1$:

$$
\frac{\partial \Re(\omega^2 D)}{\partial \omega} = \frac{\Pi^2}{\omega Y} \left[1 + 2Y^{-1} + (1 - A_e)\beta_e\right],
\tag{8.34}
$$

$$
v_g = \frac{2c}{\sqrt{Y}\nu} \left[1 - \frac{3Y^{-1}}{2} + \frac{(A_e - 1)\beta_e}{2} - \frac{Yr}{2}\right],
\tag{8.35}
$$

$$
v_R = -\frac{c\sqrt{Y}}{\sqrt{\nu}} \left[1 - \frac{3Y^{-1}}{2} + \frac{(A_e - 1)\beta_e}{2} + \frac{Yr}{2}\right],
\tag{8.36}
$$

$$
|v_R - v_g| = \frac{c\sqrt{Y}}{\sqrt{\nu}} \left[1 + \frac{Y^{-1}}{2} + \frac{(A_e - 1)\beta_e}{2} + \frac{Yr}{2}\right].
\tag{8.37}
$$

When deriving (8.34)–(8.37) we neglected the contribution of the terms proportional to r when compared with those proportional to Yr.

Having substituted (8.27), (8.28), (8.32) and (8.34)–(8.37) into (8.18), we obtain the final expression for $B_f^2(t \to \infty)$ in the form (Sazhin, 1989e):

$$
B_f^2 = \frac{2\sqrt{\pi}\Omega m_e^2 c^2}{e^2} q^{3/2} \sum_{i=0}^{3} \phi_i,
\tag{8.38}
$$

where:

$$
\phi_0 = Y^{5/2} \int_{-1}^{0} \frac{d\xi}{\xi^4} \exp\left(-\frac{qY}{\xi^2}\right) M_m,
\tag{8.39}
$$

$$
\phi_1 = Y^{3/2} \int_{-1}^{0} \frac{d\xi}{\xi^4} \exp\left(-\frac{qY}{\xi^2}\right) \left[M_m \left(\frac{\xi^2}{2} - qY\right) + \tilde{M}_m \left(\frac{qY}{\xi^2} - \frac{3}{2}\right)\right],
\tag{8.40}
$$

$$
\phi_2 = \eta \phi_0,
\tag{8.41}
$$

$$
\phi_3 = -qY^{7/2}\kappa \int_{-1}^{0} \frac{d\xi}{\xi^6} \exp\left(-\frac{qY}{\xi^2}\right) M_m,
\tag{8.42}
$$

$$q = \frac{v_A}{w} = \frac{c^2\Omega^2}{\Pi^2 w^2},$$

$$\eta = -4.5Y^{-1} + 2.5\beta_e(A_e - 1) + 2.5Yr,$$

$$\kappa = -3Y^{-1} + \beta_e(A_e - 1) + Yr,$$

and v_A is the Alfvén velocity.

Equation (8.38) in fact gives us the required solution to the problem under consideration. In other words, if we assume that the free energy of electrons with the initial distribution function (8.30) is transferred into whistler-mode waves propagating parallel to the magnetic field then the resultant whistler-mode wave spectrum is determined by equation (8.38). In fact equation (8.38) giving the final spectrum of the waves is complementary to the corresponding equations for $\tilde{\gamma}$ given in Chapter 7 (see e.g. equations (7.53) and (7.56)–(7.58)) describing the initial stages of this spectrum formation.

Equation (8.38) can be generalized in a straightforward way to the case when, besides the electrons with the initial distribution function (8.30) (energetic electrons), we take into account the contribution of low-energy (background) electrons which do not contribute to the process of instability development, but influence wave propagation. If the densities of energetic and background electrons are n_{1e} and n_{0e} respectively, then the expression (8.38) should be multiplied by (n_{1e}/n_{0e}) with the parameters referring to whistler-mode propagation determined for the mixture of background and injected electrons.

As already mentioned, when deriving (8.18) we neglected the time dependence of β_e and A_e. In (8.38) both these parameters are understood as their asymptotic values at $t \to \infty$. However, in this case we can put $A_e = 1$ which leads to an independence of the parameters η and κ from β_e in (8.39)–(8.42).

In the limit $Y \to \infty$ and $\beta_e \to 0$ the contribution of $\phi_{1,2,3}$ can be neglected when compared with ϕ_0, and equation (8.38) reduces to that given by Roux & Solomon (1971) (see their equation (2.2)). The terms ϕ_1, ϕ_2 and ϕ_3 describe different corrections to ϕ_0. In particular, the term ϕ_1 takes into account the fact that diffusion trajectories of the electrons deviate from those corresponding to $v_t = \text{const}$, i.e. from those which would take place for $Y \to \infty$. The term ϕ_2 describes the corrections due to finite Y and β_e for the term

$$\frac{|v_R - v_g||v_R|^3}{v_g \partial \Re(\omega^2 D)/\partial \omega}.$$

The term ϕ_3 describes the corrections due to finite Y and non-zero β_e of the exponential term.

The expressions for M_m and $\tilde{M}_m$ appear to be especially simple for $m = 2$:

$$M_2 = \frac{1}{2}(\xi^3 - \xi), \qquad (8.43)$$

$$\tilde{M}_2 = \frac{1}{30}(9\xi^5 - 5\xi^3 + 4). \qquad (8.44)$$

The plots of ϕ_0 versus Y^{-1} and $\phi_\Sigma = \sum_{i=0}^{3} \phi_i$ versus Y^{-1} for $q = 0.05$, $m = 2$, and $A_e = 1$ are shown in Fig. 8.1. As one can see from this figure, both plots are rather close to each other, so that for the qualitative analysis of B_f^2 it seems possible to replace ϕ_Σ by ϕ_0 in equation (8.38). In this case the latter equation coincides exactly with that obtained by Roux & Solomon (1971), who neglected the contribution of the terms of the order of Y^{-1} as well as temperature, finite electron density and ion corrections from the very beginning. It can be shown (see Sazhin, 1989e, for details) that the closeness between our results appears due to the fact that the contributions of the terms ϕ_2 and ϕ_3 effectively compensate each other, while the term ϕ_1 is negligibly small when compared with the terms ϕ_2 and ϕ_3.

The curves ϕ_0 versus Y^{-1} and ϕ_Σ versus Y^{-1} for $q = 0.1$ and $q = 0.02$ have essentially the same shape as the curves shown in Fig. 8.1, although $\phi_{0(\Sigma)\mathrm{max}}$ decreases with increasing q, while the Y^{-1} at which $\phi_{0(\Sigma)} = \phi_{0(\Sigma)\mathrm{max}}$ increases with increasing q (see Figs. 1 and 3 of Sazhin, 1989e).

As follows from Fig. 8.1, the peak of wave intensity is achieved at $Y^{-1} \approx 0.05$ which seems to justify our assumption $Y^{-1} \ll 1$. However, the intensity of the waves at $Y^{-1} > 0.15$ is not negligibly small, which means that the reliability of our model for $q = 0.05$ is a marginal one, although it is more reliable than the model of Roux & Solomon (1971). The contribution of the waves at $Y^{-1} > 0.15$ increases for larger q and decreases for smaller q. Hence, our model is more reliable for smaller q than for larger ones.

Note that the formation of the wave spectrum (8.38) is accompanied by the formation of the 'plateau' ($\partial f_0 / \partial \hat{v} = 0$ for any given $\hat{w}$) on the electron distribution function. This follows from equation (8.15) if we put $\partial f_0 / \partial t = 0$ therein.

Our main assumption that the plasma is homogeneous, so that the boundary conditions do not interfere with the process of instability development, makes it rather restrictive for magnetospheric applications. It is likely to be applicable to the interpretation of whistler-mode observations in the magnetosheath or the distant tail of the magnetosphere. However, the interpretation of similar observations in the inner magnetosphere, where the contribution of reflections from the ionospheres cannot be neglected in general, requires a different model which will be considered in the next subsection.

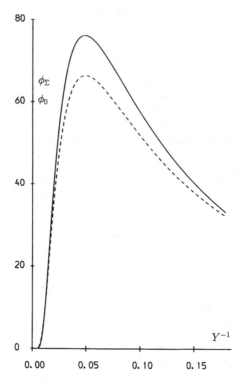

Fig. 8.1 Plots of $\phi_\Sigma = \sum_{i=0}^{3} \phi_i$ (see equations (8.39)–(8.42)) versus Y^{-1} (solid) and ϕ_0 (see equation (8.39)) versus Y^{-1} (dashed) for $q = 0.05$, $m = 2$ and $A_e = 1$.

(c) A model for a closed magnetospheric field tube

The model which is described below was first suggested by Etcheto *et al.* (1973) and further developed by Sazhin (1984, 1987c, 1989e). We restrict our analysis to the version of this model suggested by Sazhin (1989e).

We assume that the waves propagate strictly along the magnetospheric magnetic field line until they reach the ionosphere. Then they are reflected and continue their propagation along the same field line but in the opposite direction until they reach the opposite ionosphere and are reflected there, etc. The total amplification during wave propagation from one hemisphere to the other is assumed to be compensated by the loss at reflection from the ionospheres so that equation (8.12) remains valid.

Although the magnetic field tube is naturally inhomogeneous, the wavelength of whistler-mode waves is generally well below the characteristic scale of inhomogeneity. Thus we can assume that the equation for $\tilde{\gamma}$ (equation (7.53)) and the diffusion equation for f_0 (equation (8.8)) derived for a homogeneous plasma remain valid at any point along the magnetic field tube. B_k^2 in the latter equation depends on the position along the field line, in general. However, in our future analysis we neglect this dependence and assume that $B_k^2 = \langle B_k^2 \rangle$, where $\langle B_k^2 \rangle$ is the average value of B_k^2 along the field line, roughly equal to its value at the equatorial plane. Also we assume that $\partial f_0 / \partial t$ in equation (8.8) can be neglected so that the number of injected electrons (term S_e) is compensated by electrons precipitated into the loss cone (term P_e). We take these functions in the form:

$$S_e \equiv S_e(\alpha_e, v_t) = \frac{K_m}{\pi^{3/2} n_{0e} w^3} \left(\frac{\mathrm{d}n_{ie}}{\mathrm{d}t} \right) \exp \left(-\frac{v_t^2}{w^2} \right) \sin^m \alpha_e, \qquad (8.45)$$

$$P_e \equiv P_e(\alpha_e, v_t) = g(\alpha_e) f_0(\alpha_e, v_t)/t_B, \qquad (8.46)$$

where $\mathrm{d}n_{ie}/\mathrm{d}t$ is the rate of electron density change due to the penetration of electrons into the inner magnetosphere, $g(\alpha_e) = 0$ when $\alpha_e > \alpha_L$, $g(\alpha_e) = 1$ when $\alpha_e \leq \alpha_L$, α_L is the electron pitch-angle corresponding to the edge of the loss cone, and t_B is the time needed for the electrons to precipitate into the ionosphere.

Considering the time scale of the processes to be much greater than t_B we can assume that the density of penetrating electrons does not depend on the position along the field line unless $m \gg 1$ (cf. Sazhin, 1987d). Thus we can approximately consider S_e to be independent of s. If the diffusion processes take place mainly in the vicinity of the equatorial plane of the magnetosphere, we can also assume that α_L, t_B and P_e do not depend on the position along the field line, being equal to the corresponding values in the equatorial magnetosphere. Similarly to subsection (*b*), the self-consistency of our model is achieved when most of the wave energy is concentrated at $Y^{-1} \ll 1$.

As in subsection (*b*), we introduce the variables $\hat{w}$ and $\hat{v}$ defined by equations (8.13) and write equation (8.8) for $\alpha_e > \alpha_L$ as

$$\frac{\partial}{\partial \hat{v}} v_\perp^2(\hat{w}, \hat{v}) D_d \frac{\partial f_0}{\partial \hat{v}} = -S_e, \qquad (8.47)$$

where D_d is the same as in (8.15) and is defined by (8.16).

Remembering that $\alpha_L \ll 1$ in the magnetospheric conditions (except at the very low L shells, say $L \lesssim 1.5$) we can neglect the contribution of electrons at $\alpha_e < \alpha_L$ to the process of instability development, as was originally suggested by Etcheto *et al.* (1973). Integrating equation (8.47) twice, once with respect to $\hat{v}$ keeping $\hat{w} = $ const, the second time with respect to $\hat{w}$ keeping $\hat{v} = \text{const} = v_R$ (cf. the analysis of equation (8.15)), we obtain:

$$D_d \int_{\hat{w}_{\min}}^{\infty} v_\perp^2 \frac{\partial f_0}{\partial \hat{v}} d\hat{w} = \int_{\hat{w}_{\min}}^{\infty} d\hat{w} \int_{\hat{v}}^{\hat{v}_{\min}} S_e d\hat{v}, \tag{8.48}$$

where $\hat{v}_{\min}$ is the minimal value of $\hat{v}$ for a given $\hat{w}$, and $\hat{w}_{\min}$ is the same as in (8.14). Substituting (8.48) into (8.14) we obtain:

$$\gamma = \frac{\pi^2 \Pi^2}{2 D_d \partial \Re(\omega^2 D)/\partial \omega} \int_{\hat{w}_{\min}}^{\infty} d\hat{w} \int_{\hat{v}}^{\hat{v}_{\min}} S_e d\hat{v}. \tag{8.49}$$

Having substituted (8.49) into (8.12) and remembering the definition of D_d (see equation (8.16)) we obtain:

$$B_f^2 = \frac{2\pi^2 m_e^2 c^2}{\ln(1/R_w) e^2} \int_{-s_{\text{ion}}}^{+s_{\text{ion}}} ds \frac{\Pi^2 |v_R - v_g|}{v_g^2 \partial \Re(\omega^2 D)/\partial \omega} \int_{\hat{w}_{\min}}^{\infty} d\hat{w} \int_{\hat{v}}^{\hat{v}_{\min}} S_e d\hat{v}. \tag{8.50}$$

As in subsection (b) we restrict our analysis to the case when most of the wave energy is concentrated at $Y^{-1} \ll 1$, retaining the terms of the order of Y^{-1}. Having substituted (8.23), (8.26), (8.28) and (8.45) into (8.50) we obtain:

$$B_f^2 = \frac{4\sqrt{\pi} m_e^2 c^2}{\ln(1/R_w) e^2 n_{0e} w^3} \frac{dn_{ie}}{dt} \int_{-s_{\text{ion}}}^{+s_{\text{ion}}} ds \frac{\Pi^2 |v_R - v_g| |v_R|^3}{v_g^2 \partial \Re(\omega^2 D)/\partial \omega}$$

$$\times \int_{-1}^{0} \frac{d\xi}{\xi^4} \left\{ \nu_m + Y^{-1} \left[\left(\frac{\xi^2}{2} - \frac{v_R^2}{w^2} \right) \nu_m + \left(\frac{v_R^2}{\xi^2 w^2} - \frac{3}{2} \right) \tilde{\nu}_m \right] \right\} \exp\left(-\frac{v_R^2}{w^2 \xi^2} \right), \tag{8.51}$$

where

$$\nu_m = K_m \int_{\xi}^{0} (1 - t^2)^{m/2} dt, \tag{8.52}$$

$$\tilde{\nu}_m = K_m \int_{\xi}^{0} t^2 (1 - t^2)^{m/2} dt. \tag{8.53}$$

Now we assume a dipole model of the magnetic field and replace integration with respect to s by integration with respect to the latitude along the magnetospheric field line, λ, ($\lambda = 0$ at the magnetospheric equator), remembering that (Sazhin, 1984):

$$ds = R_E L \cos \lambda \sqrt{1 + 3 \sin^2 \lambda} d\lambda, \tag{8.54}$$

$$\Omega = \Omega_{\text{eq}} \sqrt{1 + 3 \sin^2 \lambda} / \cos^6 \lambda, \tag{8.55}$$

where R_E is the Earth's radius, L is the McIlwain parameter, and index $_{\text{eq}}$ hereafter refers to values of the parameters at the magnetospheric equator. The generalization of our model for a dayside compressed magnetosphere could be made in a straightforward way following Sazhin (1987c).

As to the dependence of Π on λ we assume, similarly to Sazhin (1984), that it has the form:

$$\Pi = \Pi_{\text{eq}} \cos^{-n} \lambda, \tag{8.56}$$

where $n = 1$ corresponds roughly to plasmaspheric conditions and $n = 3\text{--}4$ corresponds to extraplasmaspheric conditions.

Substituting (8.34)–(8.37) and (8.54)–(8.56) into (8.51) and replacing the integration in (8.51) with respect to s by the integration with respect to λ (and the limits of integration $\pm s_{\text{ion}}$ by $\pm \arccos \sqrt{L^{-1}}$) we obtain the final expression for B_f^2 in the form:

$$B_f^2 = \frac{8\pi^{3/2} m_e c^4}{\ln(1/R_w) v_0^3} \frac{\mathrm{d} n_{ie}}{\mathrm{d}t} \frac{R_E L \Omega_{\text{eq}}^3}{\Pi_{\text{eq}}^4} \sum_{i=0}^{3} G_i, \tag{8.57}$$

where

$$G_i = Y_{\text{eq}}^3 \int_0^{\arccos \sqrt{L^{-1}}} \mathrm{d}\lambda \frac{(1 + 3\sin^2 \lambda)^{7/2}}{\cos^{35-4n} \lambda} I_i, \tag{8.58}$$

$$I_0 = \int_{-1}^0 \frac{\mathrm{d}\xi}{\xi^4} \exp\left(-\frac{qY}{\xi^2}\right) \nu_m, \tag{8.59}$$

$$I_1 = \int_{-1}^0 \frac{\mathrm{d}\xi}{\xi^4} \exp\left(-\frac{qY}{\xi^2}\right) \left\{ Y^{-1} \left[\left(\frac{\xi^2}{2} - qY\right) \nu_m + \left(\frac{qY}{\xi^2} - \frac{3}{2}\right) \tilde{\nu}_m \right] \right\}, \tag{8.60}$$

$$I_2 = \eta_b I_0, \tag{8.61}$$

$$I_3 = -qY\kappa \int_{-1}^0 \frac{\mathrm{d}\xi}{\xi^6} \exp\left(-\frac{qY}{\xi^2}\right) \nu_m, \tag{8.62}$$

$$\eta_b = -3Y^{-1} + 2(A_e - 1)\beta_e + 3Yr, \tag{8.63}$$

$$Y = \frac{Y_{\text{eq}} \sqrt{1 + 3\sin^2 \lambda}}{\cos^6 \lambda}, \tag{8.64}$$

$$q = q_{\text{eq}} \frac{1 + 3\sin^2 \lambda}{\cos^{12-2n} \lambda}, \tag{8.65}$$

$$\beta_e = \beta_{\text{eq}} \frac{\cos^{12-2n} \lambda}{1 + 3\sin^2 \lambda}, \tag{8.66}$$

ν_m and $\tilde{\nu}_m$ are defined by equations (8.52) and (8.53), κ is the same as in equation (8.42), and $q_{eq} = c^2\Omega^2_{eq}/(\Pi^2_{eq}w^2)$. Note that our definition of β_e is different from that of Sazhin (1984).

The expressions for ν_m and $\tilde{\nu}_m$ have particularly simple forms for $m = 2$:

$$\nu_2 = \frac{\xi^3}{2} - \frac{3\xi}{2}, \tag{8.67}$$

$$\tilde{\nu}_2 = \frac{3\xi^5}{10} - \frac{\xi^3}{2}. \tag{8.68}$$

The integrand in (8.58) is a rapidly decreasing function of λ. Hence, the values of G_i depend only slightly on the values of the upper limit of integration provided that $\arccos\sqrt{L^{-1}}$ is close to $\pi/2$, and we can replace it by $\pi/2$ as was done by Sazhin (1984, 1987c).

In the limit $Y^{-1}_{eq} \to 0$, the contribution of the terms G_1, G_2 and G_3 can be neglected when compared with that of G_0. The relative contribution of the sum $G_1 + G_2 + G_3$ for finite Y^{-1}_{eq} , $m = 2$, $L = 4$, $A_e = 1$, $n = 0$ and 3, and $q_{eq} = 0.05$ can be seen from Fig. 8.2 where we have shown the plots of G_0 and $G_\Sigma = \sum^3_{i=0} G_i$ versus Y^{-1}_{eq}. As follows from this figure, the plots of G_0 and G_Σ versus Y^{-1}_{eq} are roughly as close to each other as the curves ϕ_0 and ϕ_Σ versus Y^{-1} shown in Fig. 8.1, the values of G_0 being slightly lower than G_Σ. Similar curves but for $q_{eq} = 0.1$ and 0.02 show roughly the same behaviour as the curves shown in Fig. 8.2, with $G_{0(\Sigma)}$ decreasing with increasing q_{eq}. For all q_{eq} the maximum of wave intensity is attained at $Y^{-1} = Y^{-1}_{max} \approx q_{eq}/1.3$. Y^{-1}_{max} is slightly larger for G_Σ than for G_0.

The dependence of G_0 and G_Σ on n is not strong. In particular, the ratios of the maximal values of G_0 and G_Σ for $n = 0$ to the corresponding values of G_0 and G_Σ for $n = 3$ are approximately equal to 0.87 for all q_{eq} in the range between 0.02 and 0.1.

As in the case of a homogeneous plasma our assumption that most of the wave energy is concentrated at $Y^{-1}_{eq} \ll 1$ is marginally satisfied for $q_{eq} = 0.05$ (the case shown in Fig. 8.2). For smaller q_{eq} the reliability of this assumption increases.

Equations (8.38) and (8.57) derived in this section can be generalized to the case of oblique whistler-mode propagation by replacing ω by $\omega\cos\theta$, provided $|\theta| \ll 1$ (see Sazhin (1987e) for details). However, the assumption that waves propagate in one direction only should be retained, for otherwise the steady state could not be achieved at all (Vedenov & Ryutov, 1975).

Note that besides a stationary solution, the quasi-linear equations considered in this section can also have quasi-periodic solutions. One type of these

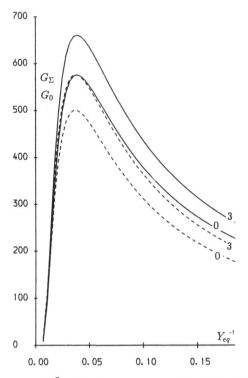

Fig. 8.2 Plots of $G_\Sigma = \sum_{i=0}^{3} G_i$ (see equations (8.58)–(8.62)) versus Y_{eq}^{-1} (solid) and G_0 (see equations (8.58) and (8.59)) versus Y_{eq}^{-1} (dashed) for $q = 0.05$, $m = 2$, $L = 4$, $n = 0$ and $n = 3$ (values of n indicated).

solutions was considered by Sazhin (1987f). However, the analysis of this solution is beyond the scope of this book. In the next section we consider other non-linear effects related to whistler-mode instabilities assuming the waves to be monochromatic.

8.2 Monochromatic waves

The non-linear evolution of the electron distribution function in the field of a monochromatic whistler-mode wave results in the formation of the so called ergodic distribution function. This function appears to be stable with respect to the wave by which it is formed ('mother wave') and thus can be considered as analogous to the quasi-linear plateau. However, contrary to the case considered in Section 8.1, the evolution of the distribution function towards an ergodic state is not monotonic (as described by the diffusion equation, e.g. equation (8.4)) but oscillating. In addition, the ergodic distribution function, being stable with respect to the 'mother wave', appears to be unstable with respect to whistler-mode waves at certain frequencies close, but not equal, to the frequency of the 'mother wave' ('daughter waves'). A theoretical study of all these processes was greatly stimulated by experiments with ground-based transmitters (see e.g. Roux & Solomon, 1970; Sudan & Ott, 1971; Dysthe, 1971; Bud'ko et al., 1972; Karpman, Istomin & Shklyar, 1974; Nunn, 1974; Karpman, 1974; Roux & Pellat, 1978). In what follows we will give in fact only a brief introduction to the problem, mainly following the paper by Bud'ko et al. (1972), restricting ourselves (as for the previous section) to considering whistler-mode waves propagating along the magnetic field in a homogeneous plasma.

Let us first consider the trajectories of individual electrons in the field of a monochromatic whistler-mode wave. In the system moving with the phase velocity of the wave, the electric field of the whistler-mode wave propagating along the external magnetic field vanishes (cf. Section 7.5). Thus the electron's motion in this system is controlled by the equation:

$$\frac{d\mathbf{v}}{dt} = -\frac{e}{m_e c}\left(\mathbf{v}\times\mathbf{B}_0 + \mathbf{v}\times\mathbf{B}_w\right), \tag{8.69}$$

where $\mathbf{B}_0$ and $\mathbf{B}_w$ are external and wave magnetic fields respectively. The components of $\mathbf{B}_w$ can be presented as:

$$\left.\begin{array}{l} B_x = |\mathbf{B}_w|\sin kz \\ B_y = |\mathbf{B}_w|\cos kz \end{array}\right\}. \tag{8.70}$$

The components of the electron's velocity are (cf. equation (7.107)):

$$\left.\begin{array}{l} v_x = v_\perp\cos\aleph \\ v_y = v_\perp\sin\aleph \\ v_z = v_\| \end{array}\right\}, \tag{8.71}$$

where $\aleph \approx \Omega t + \varsigma_e$. After substituting (8.70) and (8.71) into (8.69) we can present the latter equation in the form:

$$\left.\begin{array}{l} dv_\perp/dt = \Omega v_\| h\cos(kz+\aleph) \\ v_\perp\,(d\aleph/dt) = \Omega v_\perp - \Omega v_\| h\sin(kz+\aleph) \\ dv_\|/dt = -\Omega v_\perp h\cos(kz+\aleph) \end{array}\right\}, \tag{8.72}$$

where $h = |\mathbf{B}_w|/|\mathbf{B}_0|$. The last equation in the system (8.72) is equivalent to equation (7.114).

As follows from (8.72), the electron kinetic energy does not change during its motion in the system moving with the phase velocity of the wave. Hence, system (8.72) has an integral of motion:

$$W = \frac{m_e}{2}\left(v_\parallel^2 + v_\perp^2\right). \tag{8.73}$$

(Note that conservation of the electron energy in the system moving with the wave does not imply its conservation in the laboratory system.)

As can also be seen from (8.72), the wave influences the electron trajectory most strongly when $v_\parallel = dz/dt$ is close to $-\Omega/k$ (resonant electrons). Hence, it is convenient to introduce new variables:

$$\left.\begin{array}{l} u = v_\parallel + \Omega/k \\ \iota = kz + \aleph \end{array}\right\}, \tag{8.74}$$

in which the system (8.72) can be presented in a Hamiltonian form:

$$\left.\begin{array}{l} d\iota/dt = \partial\mathcal{H}/\partial u \\ du/dt = -\partial\mathcal{H}/\partial\iota \end{array}\right\}, \tag{8.75}$$

where

$$\mathcal{H} = \frac{ku^2}{2} + \Omega h v_\perp \sin\iota = \frac{ku^2}{2} + \Omega h \sin\iota \sqrt{\frac{2W}{m_e} - \left(u - \frac{\Omega}{k}\right)^2}. \tag{8.76}$$

As u is small for the resonant electrons, we can consider $u \ll \Omega/k$, and correspondingly:

$$\mathcal{H} = \frac{ku^2}{2} + \Omega h v_{\perp R}\sin\iota, \tag{8.77}$$

where

$$v_{\perp R} = \sqrt{\frac{2W}{m_e} - \left(\frac{\Omega}{k}\right)^2}. \tag{8.78}$$

As the Hamiltonian $\mathcal{H}$ is conserved during an electron's motion, it follows from (8.77) that:

$$\frac{d\Upsilon}{dt} = \frac{ku}{2} = \frac{1}{\tau\kappa_0}\sqrt{1 - \kappa_0^2 \sin^2\Upsilon}, \tag{8.79}$$

where

$$\Upsilon = \frac{\iota}{2} - \frac{3\pi}{4}, \tag{8.80}$$

$$\kappa_0^2 = \frac{2h v_{\perp R}\Omega}{\mathcal{H} + \Omega h v_{\perp R}}, \tag{8.81}$$

$$\tau = 1/\sqrt{hk\Omega v_{\perp R}}. \tag{8.82}$$

Equation (8.79) reduces to the corresponding equation which describes an electron's motion in the field of a Langmuir wave, if we replace the Lorentz force in (8.81) by the electrostatic force $e\mathbf{E}$ (O'Neil, 1965; Davidson, 1972).

Let us first assume that $|\kappa_0| < 1$. For this case, equation (8.79) has two solutions for $\dot{\Upsilon} \equiv d\Upsilon/dt$ as functions of Υ, symmetric with respect to the line $\dot{\Upsilon} = 0$ and oscillating with period π. For these solutions the amplitude of $\dot{\Upsilon}$ lies in the interval between $\pm\sqrt{1 - \kappa_0^2}/(\tau|\kappa_0|)$ and $\pm 1/(\tau|\kappa_0|)$. For $|\kappa_0| > 1$, equation (8.79) has a solution only for $|\Upsilon| \leq \arcsin(1/|\kappa_0|)$: this solution corresponds to ellipses with centres at the point $\Upsilon = n\pi$ and passing through the points $\Upsilon = n\pi \pm \arcsin(1/|\kappa_0|)$ and $\dot{\Upsilon} = \pm 1/(\tau|\kappa_0|)$ ($n = 0, \pm 1, \pm 2, ...$). In the case of the Langmuir wave, this solution refers to the particles trapped in the wave field. In the case of whistler-mode waves, it describes their phase trapping which is accompanied by the oscillation of $v_\parallel$ around v_R. By analogy with the Langmuir waves, these electrons with closed trajectories in the $(\dot{\Upsilon}, \Upsilon)$ plane are also termed trapped in order to distinguish them from electrons with non-closed trajectories, corresponding to $|\kappa_0| < 1$, which are termed untrapped. For $|\kappa_0| = 1$, equation (8.79) has the form $\dot{\Upsilon} = \pm \cos \Upsilon/(\tau|\kappa_0|)$, which corresponds to the superposition of two cosines shifted in phase by π. When an electron with such a trajectory approaches the point where $\dot{\Upsilon} = 0$, there is an equal probability for it to continue its movement to smaller Υ or to larger Υ, that is to be trapped (as in the case $|\kappa_0| > 1$), or to be untrapped (as in the case $|\kappa_0| < 1$). This makes it necessary to use statistical methods to describe the electron trajectories with $|\kappa_0|$ equal or close to 1 (Zaslavsky, 1970; Zaslavsky & Chirikov, 1972). The contribution of these electrons to the process of wave–particle energy exchange is, in most cases, not important (e.g. Karpman, 1974) and will not be considered.

Schematic shapes of these three types of electron trajectories in the field of a whistler-mode wave are shown in Fig. 8.3.

An implicit solution $\Upsilon(t)$ of equation (8.79) has the form:

$$F(\Upsilon(t), \kappa_0) - F(\Upsilon(0), \kappa_0) = \frac{t}{\kappa_0 \tau}, \tag{8.83}$$

where $F(\Upsilon, \kappa_0)$ is an elliptic integral of the first kind with modulus κ_0:

$$F(\Upsilon, \kappa_0) = \int_0^\Upsilon \frac{d\Upsilon'}{\sqrt{1 - \kappa_0^2 \sin^2 \Upsilon'}}. \tag{8.84}$$

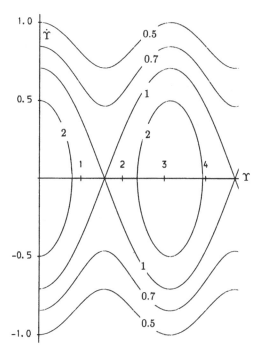

Fig. 8.3 An electron's trajectories in the field of a whistler-mode wave in the plane $(\dot{\Upsilon}, \Upsilon)$ for $\tau = \sqrt{2}$ s and $\kappa_0 = 0.5, 0.7, 1$ and 2 (curves indicated) (see equation (8.79)).

An alternative presentation of this integral is (Abramowitz & Stegun, 1964):

$$F(\Upsilon, \kappa_0) = \int_0^{\sin \Upsilon} \frac{\mathrm{d}\Upsilon'}{\sqrt{(1 - \Upsilon'^2)(1 - \kappa_0^2 \Upsilon'^2)}}. \qquad (8.85)$$

In most cases we are interested not in the details of electron motion in the wave field, but in some average characteristics of this motion, in particular, in the period of electron phase oscillations in the wave field. For the untrapped electrons, this period is given by:

$$T_{ut} = 2|\kappa_0|\tau \int_0^{\pi/2} \frac{\mathrm{d}\Upsilon}{\sqrt{1 - \kappa_0^2 \sin^2 \Upsilon}} = 2|\kappa_0|\tau K(\kappa_0), \qquad (8.86)$$

where $K(\kappa_0)$ is the complete elliptic integral of the first kind ($|\kappa_0| < 1$):

$$K(\kappa_0) = F(\pi/2, \kappa_0).$$

For the trapped electrons:

$$T_{tr} = 4|\kappa_0|\tau \int_0^{\arcsin(1/|\kappa_0|)} \frac{d\Upsilon}{\sqrt{1 - \kappa_0^2 \sin^2 \Upsilon}} = 4\tau K(1/\kappa_0). \qquad (8.87)$$

Equations (8.86) and (8.87) follow from (8.79) in a straightforward way. When $|\kappa_0| \to 1$, T_{tr} appears to be twice the value of T_{ut}.

Remembering the Taylor expansion for $K(\kappa_0)$ for small $|\kappa_0|$:

$$K(\kappa_0) = \frac{\pi}{2}\left(1 + \frac{\kappa_0^2}{4} + \dots\right), \qquad (8.88)$$

it follows from (8.87) that $T_{tr} = 2\pi\tau$ when $|\kappa_0| \ll 1$. The same result could also be obtained if we assumed from the start that the amplitude of the electron phase oscillations in the wave field is small.

The change of Υ during one period of oscillation for untrapped electrons is π while for trapped electrons it is zero. Hence, the average value of $\dot{\Upsilon}$ ($\overline{\dot{\Upsilon}}$) for one period of oscillation can be obtained from the equation:

$$\overline{\dot{\Upsilon}} = \begin{cases} \pi/\left(2|\kappa_0|\tau K(|\kappa_0|)\right) & (|\kappa_0| < 1) \\ 0 & (|\kappa_0| > 1). \end{cases} \qquad (8.89)$$

As follows from (8.89), the average velocity of the trapped electrons is not changed during their phase oscillation in the wave field, while $\overline{\dot{\Upsilon}}$ for the untrapped electrons depends on the value of κ_0.

Next we consider some properties of the electron distribution function. The latter is assumed to depend only on the longitudinal ($v_{\parallel}$) and transverse ($v_{\perp}$) velocities:

$$v_{\parallel} = u_0 + \frac{\omega - \Omega}{k}; \quad v_{\perp} = \sqrt{\frac{2W}{m_e} - \left(u_0 - \frac{\Omega}{k}\right)^2}$$

($v_{\parallel}$ refers to the longitudinal velocity in the reference system where the plasma is at rest, and u_0 is the initial value of the velocity u).

If we restrict ourselves to considering the evolution of the distribution function for resonant electrons, we can expand it in a series in powers of u_0, keeping only the terms of the order of u_0:

$$f(v_{\perp}, v_{\parallel}) = f(v_{\perp R}, v_R) + u_0 f'(v_{\perp R}, v_R), \qquad (8.90)$$

where

$$f'(v_{\perp R}, v_R) = \left(\frac{\partial f(v_{\perp R}, v_{\parallel})}{\partial v_{\parallel}} + \frac{\Omega}{k v_{\perp R}} \frac{\partial f(v_{\perp R}, v_{\parallel})}{\partial v_{\perp R}}\right)\Bigg|_{v_{\parallel}=v_R}. \qquad (8.91)$$

In view of (8.89) and (8.90), and changing the arguments of the distribution function from $(v_{\perp R}, v_R)$ to $(v_{\perp R}, \kappa_0)$ we obtain the expression for the time-averaged distribution function, $f_E(v_{\perp R}, \kappa_0)$, known as an ergodic distribution function:

$$
\begin{aligned}
&f_E(v_\perp, \kappa_0) \\
&= f(v_{\perp R}, v_R) + \begin{cases} \pi f'(v_{\perp R}, v_R)/[\tau k_0 |\kappa_0| K(|\kappa_0|)] & (|\kappa_0| < 1) \\ 0 & (|\kappa_0| > 1). \end{cases}
\end{aligned} \tag{8.92}
$$

When the state in which the electron distribution has the form (8.92) is reached, the interaction of whistler-mode waves with these electrons leads to no energy exchange between them.

Let us now consider the stability of whistler-mode waves at frequencies close, but not equal, to the frequency of the wave by which the electron distribution function was initially disturbed. Without presenting the details of the calculations we show the final expression for the instability increment (Bud'ko *et al.*, 1972; Gokhberg *et al.*, 1972; Karpman, 1974):

$$
\gamma = \gamma_L \Phi(\varrho), \tag{8.93}
$$

where γ_L is the linear growth rate of the whistler-mode waves (see equations (7.53) and (7.56)–(7.58)),

$$
\varrho = \frac{\delta\omega\bar{\tau}}{2}\left(1 + \frac{2\Omega}{\omega}\right), \tag{8.94}
$$

$$
\bar{\tau} = 1/\sqrt{kw_\perp \Omega h}, \tag{8.95}
$$

$\delta\omega$ is the difference between the frequency under consideration and the wave frequency ω ($|\delta\omega|$ is assumed to be much less than ω), and the function $\Phi(\varrho)$ has the following main properties: $\Phi(\varrho) \to 0$ when $|\varrho| \to 0$, $\Phi(\varrho) \to 1$ when $|\varrho| \to \infty$, $\Phi(\varrho) = \Phi(-\varrho)$, $\Phi(\varrho)$ has two symmetric maxima at $\varrho = \pm 1.5$ where $\Phi(\pm 1.5) \approx 1.4$ (see Fig. 3 in Karpman, 1974).

Expression (8.95) for the 'average' bounce period of the electrons was derived under the assumption that the unperturbed electron distribution function is bi-Maxwellian: $j = 0$ in expression (1.90). For arbitrary j in the same expression, $w_\perp$ should be replaced by $\sqrt{j+1}w_\perp$. When deriving (8.93) it was assumed that temperature, ion and finite electron density effects on whistler-mode propagation could be neglected.

Remembering the above-mentioned properties of $\Phi(\varrho)$ we can see from (8.93) and (8.94) that in the centre of the resonant region ($\delta\omega = 0$), the growth rate is zero ($\gamma = 0$) and outside the resonant region ($|\delta\omega| \gg \bar{\tau}^{-1}$),

$\gamma \to \gamma_L$ (in the latter region the electron distribution function is not perturbed by the monochromatic wave). The region of maximum $\Phi(\varrho)$ corresponds to the preferable excitation of the waves for these $\delta\omega$. Hence, we can expect that the ergodic distribution function is mostly unstable for:

$$\delta\omega = \frac{\pm 3}{\mp\left(1 + \frac{2\Omega}{\omega}\right)}. \tag{8.96}$$

Therefore we can conclude that the deformation of the electron distribution function by a monochromatic whistler-mode wave results in the generation of two satellite waves, the frequencies of which differ from the frequency of the initial wave by $\delta\omega$, determined by (8.96). This seems to be confirmed by the observed modulation of the monochromatic signals of ground-based transmitters in the magnetosphere (Likhter, Molchanov & Chmyrev, 1971).

It should be emphasized that the results presented in this section are only the basic ones and they do not reflect later developments in this field (see e.g. the review by Karpman (1974), as well as the numerous theoretical studies of the quasi-monochromatic whistler-mode emissions triggered by ground based transmitters, referred to e.g. in the review by Omura *et al.* (1991)).

Problems

Problem 8.1 Expression (8.57) predicts $B_f^2 > 0$ for $m = 0$ (which corresponds to the injection of isotropic electrons; see equation (8.45)), when we would expect wave damping, and for ω close to Ω, when we would expect that the instability could develop at anisotropies much larger than those predicted by equation (8.45) for any realistic value of m (see equation (7.53)). Explain the paradox.

Problem 8.2 When deriving (8.87) we took into account that

$$\kappa_0 \int_0^{\arcsin(1/|\kappa_0|)} \frac{d\Upsilon}{\sqrt{1 - \kappa_0^2 \sin^2 \Upsilon}} = K(1/\kappa_0)$$

for $\kappa_0 > 1$. Prove this identity using the definition of the complete elliptic integral of the first kind ($K(\kappa_0) = F(\pi/2, \kappa_0)$, $|\kappa_0| < 1$).

Problem 8.3 Remembering that (Abramowitz & Stegun, 1964):

$$K(\kappa_0) = \frac{\pi}{2}\left[1 + \left(\frac{1}{2}\right)^2 \kappa_0^2 + \left(\frac{1 \cdot 3}{2 \cdot 4}\right)^2 \kappa_0^4 + \ldots\right] \quad (|\kappa_0| < 1)$$

we have that

$$\frac{\pi}{2|\kappa_0|\tau K(|\kappa_0|)}$$

increases with increasing $|\kappa_0|$. Hence, we can expect that $\overline{\Upsilon}$ predicted by (8.89) for the untrapped electrons increases with decreasing $|\kappa_0|$. Illustrate this property of $\overline{\Upsilon}$ using the curves in Fig. 8.3.

9

Applications to the Earth's magnetosphere

As mentioned in the Introduction, interest in the theory of whistler-mode waves was stimulated mainly by the observations of these waves at ground-based stations and in the Earth's magnetosphere, their applications to the diagnostics of magnetospheric parameters and their role in the balance of the electron radiation belts. An overview (even a brief one) of theoretical models of all manifestations of whistler-mode waves in the magnetosphere is obviously beyond the scope of this book (it would require writing a separate monograph) and we will restrict ourselves to illustrating the application of the theoretical analysis developed in the previous chapters to interpreting only three particular phenomena. In Section 9.1 we consider the problem of the diagnostics of magnetospheric parameters with the help of whistlers generated by lightning discharges. Theoretical models of natural whistler-mode radio emissions observed in the vicinity of the magnetopause are discussed in Section 9.2. In Section 9.3 we apply one of the quasi-linear models described in Chapter 8 to the interpretation of mid-latitude hiss-type emissions observed in the inner magnetosphere. The approaches developed for these three illustrative examples can be extended with some modifications to several other related whistler-mode phenomena. These will be discussed in the appropriate sections.

9.1 Whistler diagnostics of magnetospheric parameters

Dynamic spectra of whistlers generated by lightning discharges are shown at the beginning of the book in Fig. I. As follows from this figure a typical feature of the whistler dynamic spectra is the whistler 'nose' which was interpreted in Section 2.2 in terms of the frequency dependence of the group delay time, t_g, of whistlers propagating from one hemisphere to the other provided that initially the waves were excited in a wide frequency range but during a very short period of time (lightning discharge). The value of t_g determined by expression (2.30) is minimal at a frequency ω corresponding to about 0.4 electron gyrofrequency at the magnetospheric equator Ω_{eq}. This frequency was called the *nose frequency*, ω_n, and it was used in a straightforward way for determining the field line along which whistlers propagate, using the condition $\omega_n \approx 0.4\Omega_{\text{eq}}$. Then if we assumed that the electron density was distributed along this field line according to a certain law (say, the diffusive equilibrium distribution), a comparison between direct measurements of $t_g(\omega = \omega_n)$ and the theoretical calculation of this parameter would give us the value of electron density at the magnetospheric equator, n_{eq} ($t_g \sim \sqrt{n_{\text{eq}}}$). Making similar comparisons for whistlers propagating along different field lines, we could get the distribution of n_{eq} over a wide area of the magnetospheric equatorial plane. Also, measurements of ω_n for different whistlers at successive moments of time would allow us to estimate the velocity of displacement of the magnetic field tube (duct) along which whistlers propagate, and, correspondingly, the values of the large-scale electric field in the magnetosphere E_M which is the main cause of this displacement. These methods of diagnostics of n_{eq} and E_M not only appeared to be theoretically possible, but have in fact been widely used for practical diagnostics, being complementary to the *in situ* measurements of n_{eq} as well as the plasma flow in the equatorial plane (e.g. Rycroft, 1975).

In subsection (*a*) we shall describe in more detail the basic principles of the above-mentioned 'traditional' diagnostics of n_{eq} and E_0 based on measurements or estimates of $t_g(\omega_n)$ within the simplest model of parallel whistler-mode propagation in a cold dense electron plasma in a dipole magnetic field (t_{g0}). In subsection (*b*) we consider different perturbations of t_{g0} and discuss the contribution of these perturbations to the improvement of the accuracy of traditional diagnostics. In subsection (*c*) we discuss the possibility of using whistlers for diagnostics of the magnetospheric electron temperature.

(*a*) *Basic principles of 'traditional' diagnostics*

As has already been mentioned, the 'traditional' diagnostics of n_{eq} are based on the measurements of the group delay time of whistlers at different frequencies and a comparison of the results of these measurements with theoretically calculated values of $t_g(\omega_n)$ for different n_{eq} provided we know the distribution of electron density along magnetospheric magnetic field lines (assumed to be dipolar). The identification of the magnetic field line along which the whistler propagates and the value of a large-scale electric field E_M are based on the measurement of ω_n and its change with time respectively. Although the approximate condition $\omega_n \approx 0.4\Omega_{eq}$ holds true for most whistlers, the actual accuracy of this relation is only about $\pm 0.05\Omega_{eq}$, and a more precise value of the coefficient multiplying Ω_{eq} depends on the choice of the electron distribution along the magnetic field lines and the value of L.

The measurement of $t_g(\omega_n)$ should, strictly speaking, have been based on a very accurate measurement of the time of the lightning discharge. The only practical way of measuring the latter parameter is to identify the so called sferic related to the corresponding discharge. Sferics are the waves generated during the same process and in the same frequency range as whistlers. However, in contrast to whistlers they propagate not through the magnetosphere, but inside the Earth–ionosphere waveguide (subionospheric propagation). During the process of this propagation they can reach virtually every point of the Earth's surface including the conjugate point, in the vicinity of which whistlers are mainly recorded. As a result we can see in the same spectrogram both whistlers and sferics. In contrast to whistlers, sferics show practically no dispersion as their group delay time depends only slightly on the frequency, being close to the distance of propagation divided by the velocity of light. If we neglect the group delay time of sferics altogether, then $t_g(\omega_n)$ can be determined directly from spectrograms of the type shown in Fig. I as a delay time of the whistler with respect to the corresponding causative sferic.

Practical implementation of this simple idea is strongly hindered by the fact that along with the causative sferic related to the whistler under consideration a number of other sferics generated by different lightning discharges are usually observed. This can be seen in Fig. I. One of the ways to identify the right sferic lies in the superposition of repeated occurrences of the same pattern of whistlers and measuring echo periods (see Smith, Carpenter & Lester, 1981). Just this method was used for identifying the causative sferic for whistlers indicated by the two arrows on the right in Fig. I. This sferic is indicated by the arrow on the left. Note that one lightning discharge can (and usually does) produce a number of whistlers propagating along different field lines.

Measurements of ω_n and $t_g(\omega_n)$ cannot give us much information about magnetospheric parameters unless we have a reasonably accurate model of electron density, or, equivalently, the electron plasma frequency distribution along the magnetic field line and the model of the magnetic field itself. As to the magnetic field model, it can be reasonably accurately approximated by a dipole model (see the discussion later in subsection (b)). However, the choice of the right model of electron density distribution appears to be one of the most sensitive elements of the method. In one of the first reviews on whistler diagnostics of magnetospheric parameters by Carpenter & Smith (1964) the simplest model of latitude dependence of the electron plasma frequency in the form (8.56) was suggested. It was claimed that by varying the parameter n in this equation from 1 to 4, one could adjust this expression to virtually every electron distribution both inside and outside the plasmasphere. Later the diffusive equilibrium (DE) model of electron distribution (Angerami & Thomas, 1964), rather than its approximation by (8.56) with $n = 1$, was mainly used for the diagnostics of the parameters inside the plasmasphere (e.g. Park, 1972, 1982; Sagredo & Bullough, 1972; Sazhin, Smith & Sazhina, 1990). However, even if we assume that the DE model itself is correct we still have uncertainties about the input parameters of this model. These parameters can vary significantly with time and in most cases they are unavailable for the time when whistlers were observed. The practical way out of this uncertainty is either to use 'average' parameters, such as in the DE–1 or DE–2 models (briefly described later) considered by Park (1972), or 'extreme' parameters, such as in the summer day or winter night models considered by Sagredo & Bullough (1972), or to try to work out more sophisticated numerical models applicable to any particular moment of time. The latter type of model was suggested e.g. by Moffett *et al.* (1989) and this seems to be one of the latest achievements in modelling the electron density and temperature distribution in the ionosphere and the magnetosphere. However, owing to the extreme complexity of this model most researchers still base their diagnostics on diffusive equilibrium models in spite of the restrictions of these models (cf. Sazhin, Balmforth, Moffett & Rippeth, 1992).

Although the group delay time of sferics is generally small, and most of the group delay time of whistlers appears during the process of whistler-mode propagation in the near equatorial magnetosphere, the corrections due to the finite group delay time of sferics and the group delay time of whistlers during their propagation through the ionosphere can considerably improve the accuracy of the diagnostics. For $L \approx 4$ the correction for subionospheric propagation of sferics may vary from 0.00 to 0.04 s (Park, 1972) depending on where the lightning discharge happens (equator or middle latitudes). That is, the whistler travel time through the ionosphere and the magnetosphere, t_g, is related to the travel time of the whistlers measured on the spectrogram with respect to sferic, τ, by the following condition:

$$0.00 \text{ s} \lesssim t_g - \tau \lesssim 0.04 \text{ s}. \tag{9.1}$$

The additional group delay time due to whistler propagation through both ionospheres (Δt_{ion}) is given by the following expression (Park, 1982):

$$\Delta t_{\text{ion}} \approx \frac{1.4\overline{f_oF2}}{\sqrt{\omega/2\pi}}, \tag{9.2}$$

where Δt_{ion} is given in s, $\overline{f_oF2}$ is the critical frequency of the $F2$ layer of the ionosphere in MHz, averaged for both hemispheres, and $\omega/2\pi$ is the wave cyclic frequency in Hz. For $\overline{f_oF2} = 4$ MHz and $\omega/2\pi \approx 5$ kHz (see Fig. I) we have $\Delta t_{\text{ion}} \approx 79$ ms which is quite noticeable in the measurements.

Finally, we should note that in most cases ω_n and $t_g(\omega_n)$ cannot be measured directly from whistler dynamic spectra and are to be estimated using different extrapolation techniques (Smith & Carpenter, 1961; Dowden & Allcock, 1971; Bernard, 1973; Rycroft & Mathur, 1973; Ho & Bernard, 1973; Corcuff and Corcuff, 1973; Sagredo, Smith & Bullough, 1973; Smith, Smith & Bullough, 1975; Tarcsai, 1975; Corcuff, 1977; Stuart, 1977a,b; Sazhin, Hayakawa & Bullough, 1992). Detailed analysis of these techniques is beyond the scope of this book.

In the next subsection we shall consider some other perturbations (apart from ionospheric) of whistler group delay times, which have normally been neglected although this was not (and sometimes could not be) justified.

(b) Perturbations of whistler group delay time

As was mentioned in subsection (a), the traditional method of diagnostics of n_{eq} and E_M with the help of whistlers is based on the comparison between experimentally observed and computed values of the group delay time for these whistlers, with the effects of the finite group delay time of sferics and the group delay time of whistlers in the process of their propagation through the ionosphere being taken into account. However, the magnetospheric plasma itself has almost always been assumed to be dense (displacement currents were neglected) and cold (although the finite electron temperature influenced the electron density distribution along the field lines), and the contribution of protons and other ions was neglected (these ions were considered only as a neutralizing background). At the same time it is not at first obvious whether we in fact can neglect all these effects when estimating the integral (2.30) and assume that the magnetospheric magnetic field can be approximated by a dipole.

In this subsection we concentrate our attention mainly on the corrections to t_{g0} due to the effects of finite electron density, the contribution of ions and the non-zero electron temperature based on equation (4.29). The value of t_g with all these corrections taken into account can be obtained after substituting (4.29) into (2.30). Remembering our assumption that the perturbations of v_{g0} due to these effects are small, we can rewrite the expression for t_g in the form:

$$t_g = t_{g0} + \Delta t_{gh} + \Delta t_{gc} + \Delta t_{gr}, \tag{9.3}$$

where:

$$t_{g0} = \int_{-s_{\text{ion}}}^{s_{\text{ion}}} \frac{\mathrm{d}s}{v_{g0}}, \tag{9.4}$$

$$\Delta t_{gh} = -\int_{-s_{\text{ion}}}^{s_{\text{ion}}} \frac{\tilde{b}_h \beta_e \mathrm{d}s}{v_{g0}}, \tag{9.5}$$

$$\Delta t_{gc} = -\int_{-s_{\text{ion}}}^{s_{\text{ion}}} \frac{\tilde{b}_c \nu^{-1} \mathrm{d}s}{v_{g0}}, \tag{9.6}$$

$$\Delta t_{gr} = -\int_{-s_{\text{ion}}}^{s_{\text{ion}}} \frac{\tilde{b}_r r \mathrm{d}s}{v_{g0}}. \tag{9.7}$$

Here $\Delta t_{gh}, \Delta t_{gc}, \Delta t_{gr}$ are corrections to t_{g0} due to non-zero electron temperature, finite density, and ion effects, the other notations being the same as in (2.30) and (4.29).

Assuming a dipole magnetic field model, changing variables from s to λ (magnetic latitude) and assuming the following dependence of Π on λ:

$$\Pi = \Pi_{\text{eq}} \kappa(\lambda), \tag{9.8}$$

where $\kappa(\lambda)$ is a function of λ, we can rewrite expressions (9.4)–(9.7) as:

$$t_{g0} = \frac{R_E L Y_{\text{eq}}^2 \sqrt{\nu_{\text{eq}}}}{c} \int_0^{\lambda_{\text{ref}}} \frac{\cos^4 \lambda\, \kappa(\lambda)(1 + 3\sin^2 \lambda)\mathrm{d}\lambda}{\left[Y_{\text{eq}}\sqrt{1+3\sin^2 \lambda} - \cos^6 \lambda\right]^{3/2}}, \tag{9.9}$$

$$\Delta t_{gh} = \frac{R_E L Y_{\text{eq}}^3 \sqrt{\nu_{\text{eq}}}\beta_{\text{eq}}}{c}$$
$$\times \int_0^{\lambda_{\text{ref}}} \frac{\cos^{28} \lambda\, \kappa^3(\lambda)\sqrt{1+3\sin^2 \lambda}\left[3Y + 0.5Y^2 + A_e(2 - 1.5Y - 0.5Y^2)\right]\mathrm{d}\lambda}{\left[Y_{\text{eq}}\sqrt{1+3\sin^2 \lambda} - \cos^6 \lambda\right]^{9/2}}, \tag{9.10}$$

$$\Delta t_{gc} = -\frac{R_E L}{2 Y_{\text{eq}}\sqrt{\nu_{\text{eq}}}c} \int_0^{\lambda_{\text{ref}}} \frac{\sqrt{1+3\sin^2 \lambda}\left[4\cos^6 \lambda - 3Y_{\text{eq}}\sqrt{1+3\sin^2 \lambda}\right]\mathrm{d}\lambda}{\left[Y_{\text{eq}}\sqrt{1+3\sin^2 \lambda} - \cos^6 \lambda\right]^{1/2}\kappa(\lambda)\cos^2 \lambda}, \tag{9.11}$$

$$\Delta t_{gr} = \frac{R_E L Y_{\text{eq}}^2 \sqrt{\nu_{\text{eq}}}}{2c} \int_0^{\lambda_{\text{ref}}} \frac{\kappa(\lambda)r(\lambda)(1+3\sin^2 \lambda)\mathrm{d}\lambda}{\cos^2 \lambda\left[Y_{\text{eq}}\sqrt{1+3\sin^2 \lambda} - \cos^6 \lambda\right]^{1/2}}, \tag{9.12}$$

where R_E is the Earth's radius, λ_{ref} is the dipole latitude corresponding to the ionospheric level, index $_{\text{eq}}$ refers to the values of the parameters in the equatorial plane of the magnetosphere, and L is the McIlwain parameter; when deriving (9.12) we took into account the dependence of the parameter r on λ.

In order to illustrate the significance of corrections Δt_{gh}, Δt_{gc} and Δt_{gr} and the reliability of our analysis we computed the values of t_{g0}, Δt_{gh}, Δt_{gc} and Δt_{gr} for the same values of parameters as were used by Tarcsai, Strangeways & Rycroft (1989) in their computation of the corrections Δt_{gc} and Δt_{gr}, based on a different approach to the problem. Namely, we considered two magnetic field lines at $L = 2.4$ and 3.5. For $L = 1.6$, also considered by Tarcsai *et al.*, the perturbations under consideration are smaller than for $L = 2.4$ and $L = 3.5$. Similarly to Tarcsai *et al.* we assume the DE–1 model of electron distribution (Park, 1982). In this model it is assumed that at the reference height of 1000 km the plasma temperature is equal to 1600 K, and the ion composition is assumed to be 90% O^+, 8% H^+, 2% He^+. The equatorial electron density is assumed to be $n_{\text{eq}} = 1212$ cm^{-3} for $L = 2.4$ and $n_{\text{eq}} = 914$ cm^{-3} for $L = 3.5$. These values of n_{eq} are greater by 20% than those in the DE–1 model considered by Park (1972), assuming the electron density at 1000 km altitude to be 10^4 cm^{-3} (see his Fig. 12). This choice of values of n_{eq} was made by Tarcsai *et al.* for the modelling of a magnetospheric duct with 20% density enhancement above the background (Tarcsai, private communication, 1988).

The main equations of the diffusive equilibrium model (which enable us to specify the function $\kappa(\lambda)$ in equations (9.9)–(9.12)) are given in Appendix A of Sazhin, Smith & Sazhina (1990). Without discussing the details of this model we only mention that it takes into account the rotation of the Earth and the increase of electron and ion temperatures along magnetospheric magnetic field lines. The latter increase is described by the following equation:

$$\frac{T_e}{T_{e\ \text{ref}}} = \left(\frac{\cos^2 \lambda}{\cos^2 \lambda_{\text{ref}}} \right)^n , \tag{9.13}$$

where index $_{\text{ref}}$ refers to the reference level taken at 1000 km, $n \neq -1$.

We have taken two values of the parameter n: $n = 0$ and $n = 2$. The case $n = 0$ corresponds to constant temperature along the field line. This case was considered by Tarcsai *et al.* and we shall directly compare our results with theirs. The case $n > 0$ seems to correspond to a more realistic temperature distribution (Seely, 1977; Strangeways, 1986, 1991).

In Fig. 9.1 we show the curves of t_{g0} versus frequency, ignoring the

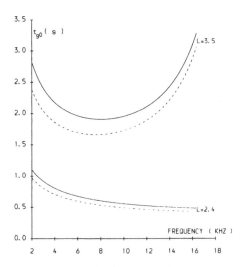

Fig. 9.1 Plots of t_{g0} (see equation (9.9)) versus frequency for $n = 0$ (dashed) and $n = 2$ (solid), and for different L (curves indicated) (Sazhin, Smith & Sazhina, 1990).

rotation of the Earth. For $L = 2.4$ the range of frequencies chosen is the same as in Fig. 1 of Tarcsai *et al.* In contrast to these authors, however, for $L = 3.5$ we have not restricted our curves to frequencies below the usually observed upper limit of $\Omega_{eq}/4\pi$ (≈ 10 kHz at $L = 3.5$). The traces otherwise are typical of observed whistler dispersion curves. Comparing the solid and dashed curves it is seen that the presence of the temperature gradient leads to an increase of 10–15% in t_{g0} (200–250 ms at $L = 3.5$). This is in agreement with Seely (1977) and Strangeways (1986), who also demonstrated the strong influence of temperature gradients on whistler analysis.

The plots of Δt_{gh} versus frequency in Fig. 9.2 for $L = 3.5$ (for $L = 2.4$ these corrections are smaller than 0.005 ms) show that this correction is much larger for the model with a temperature gradient than without it and rapidly increases with increasing frequency. It seems reasonable to ignore this term for small frequencies, but for large frequencies it can result in a significant increase of t_g. It is to be noted, however, that for $L = 3.5$ frequencies above 10 kHz are observed rather rarely in the case of ducted whistlers; therefore we can expect that the magnitude of the temperature effect (i.e. Δt_{gh}) is generally lower than about 3 ms. Hereafter we set the anisotropy of the electron distribution function A_e equal to 1. This assumption is justified in our analysis as v_g and t_g change only very slightly for values of A_e appropriate to magnetospheric conditions.

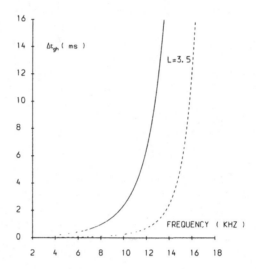

Fig. 9.2 The same as Fig. 9.1 but for Δt_{gh} (see equation (9.10)); the plots for $L = 2.4$ correspond to very small values of Δt_{gh} and are not shown in the figure (Sazhin, Smith & Sazhina, 1990).

Plots of Δt_{gc} versus frequency are shown in Fig. 9.3. In the same figure we have also shown the corresponding curves reproduced from Fig. 1 of Tarcsai *et al.* The accuracy of this reproduction was not high: about ± 3 ms. The curves corresponding to $n = 0$ and $L = 2.4$ agree within the accuracy of plotting with those given by Tarcsai *et al.* However, the curves corresponding to $n = 0$ and $L = 3.5$ lie slightly below theirs. The discrepancy between the curves is presumably due to the quadratic and higher-order terms that were not taken into account in our analysis. In contrast to the term Δt_{gh}, an increase of n results in a decrease of Δt_{gc}. Comparing Figs. 9.2 and 9.3, one can see that for $L = 3.5$ and $n = 2$ at frequencies above 10 kHz the contribution of the term Δt_{gh} can be of the same order of magnitude as, and even larger than, Δt_{gc}.

Plots of Δt_{gr} versus frequency as well as the corresponding curves reproduced from Fig. 1 of Tarcsai *et al.* are shown in Fig. 9.4. One can see an agreement between the plots for $n = 0$ and the corresponding plots of Tarcsai *et al.* within the accuracy of reproduction of the latter plots (± 3 ms) for small Δt_{gr}. A slight discrepancy between the curves which, however, does not influence the similarity of their behaviour, is again presumably due to quadratic and higher-order terms which were not taken into account in our analysis (cf. the curves for Δt_{gc} shown in Fig. 9.3). Similarly to Δt_{gh}, Δt_{gr}

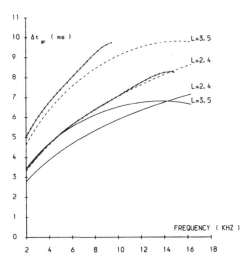

Fig. 9.3 The same as Fig. 9.1 but for Δt_{gc} (see equation (9.11)); the plots of this correction to t_{g0} for $n = 0$ derived by Tarcsai, Strangeways & Rycroft (1989) are also shown (dashed–dotted curves: the upper curve refers to $L = 3.5$, the lower one to $L = 2.4$) (Sazhin, Smith & Sazhina, 1990).

increases with increasing n, but in contrast to Δt_{gh} it decreases with increasing frequency. At low frequencies it can result in a considerable increase of t_g.

In a number of papers other corrections to t_g including those due to the rotation of the Earth (Sazhin, Smith & Sazhina, 1990), obliqueness of whistler-mode propagation (Tarcsai, Strangeways & Rycroft, 1989; Sazhin, Smith & Sazhina, 1990), day-time compression of the magnetosphere by the solar wind (Sazhin, Smith & Sazhina, 1990), deformation of magnetospheric magnetic field under the influence of the ring current (Sagredo & Bullough, 1972; Sazhin et al., 1991) have also been calculated. We cannot discuss all these corrections in this section (interested readers are referred to the above-mentioned original papers), and we only mention that the most significant among these corrections seems to be that due to the ring current: during strong storms the effects of the ring current could increase t_g by about 100 ms and even more. However, more effort is required before all these corrections can be introduced into the routine of practical diagnostics. For the time being we restrict ourselves to the corrections described by equation (9.3) and show how this equation might be used for whistler diagnostics of magnetospheric electron temperature.

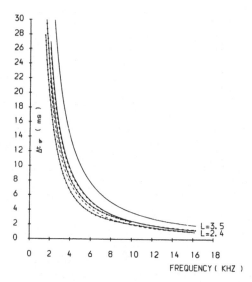

Fig. 9.4 The same as Fig 9.3 but for Δt_{gr} (see equation (9.12)) (Sazhin, Smith & Sazhina, 1990).

(c) A method for electron temperature diagnostics

The possibility of using whistlers for magnetospheric electron temperature diagnostics was probably first discussed by Scarf (1962) who estimated this temperature from the thermal attenuation of nose whistlers at the upper cut-off frequency. This method was developed by Liemohn & Scarf (1962a,b; 1964) but, as far as we know, was never used in practice, perhaps for two reasons. Firstly, it is difficult to decide whether the whistler upper cut-off frequency is determined by wave attenuation or by propagation effects. Secondly, the interpretation of the whistler cut-off frequency in Scarf's method is very sensitive to the anisotropy of the electron distribution function, which can, in general, be determined only by *in situ* measurements.

In an alternative approach to this problem McChesney & Hughes (1983) measured the electron density at the magnetospheric equator (n_{eq}) by whistler dispersion analysis, and in the topside ionosphere from *in situ* observations of LHR (low hybrid resonance) noise. The ratio of these densities was fitted to a diffusive equilibrium model of electron density distribution with temperature as a parameter. The main assumption was that the electron temperature did not change along the magnetospheric magnetic field line. However, this assumption seems to be incompatible with satellite measurements of electron temperature: equatorial temperatures can be up to a factor of 10 larger than those at ionospheric altitudes (Serbu & Maier, 1966; Seely, 1977).

A different approach was taken by Guthart (1965) who attempted to estimate the magnetospheric electron temperature from its effect on the whistler group velocity assuming a gyrotropic model of the electron distribution. He predicted that the thermal effect on whistler spectra should be largest at frequencies near the upper cut-off frequency of nose whistlers. However, the size of the effect was less than the experimental error associated with the whistler spectral analysis. This conclusion enabled Guthart to estimate an upper bound on the magnetospheric electron temperature of 2×10^4 K ≈ 1.7 eV. By contrast, Kobelev & Sazhin (1983) have argued that thermal effects in the vicinity of the plasmapause can be estimated by comparison of observed and theoretical whistler dispersion curves. They obtained values of electron temperature in the range 7–19 eV, depending on the choice of electron density distribution model along geomagnetic field lines. This temperature corresponded to an average temperature of all electrons: 'cold' ones with energies $\lesssim 1$ eV plus a small 'hot' component with energies of the order of 1 keV.

In this subsection we consider in more detail thermal effects on whistler spectra based on equations (9.3) and (9.9)–(9.12), following Sazhin, Smith & Sazhina (1990). The main idea of our method of diagnosing electron temperature is essentially the same as that of Guthart (1965) and Kobelev & Sazhin (1983), although particular details are different.

As follows from Fig. 9.2, the contribution of temperature effects to t_g is largest at higher frequencies. Hence, we expect that the whistlers most suitable for determining the magnetospheric electron temperature should be of the nose type with a well-defined upper branch. Selecting these whistlers, we first use them for the determination of L and n_{eq} following the basically 'traditional' approach to whistler diagnostics described in subsection (a), but with different corrections to t_{g0} described by equation (9.3) and the effect of the Earth's rotation taken into account. Following Park (1972) we consider two diffusive equilibrium models of electron distribution: DE–1 (which has already been described) and DE–2 (which differs from DE–1 in the temperature of thermal electrons at the reference level: 3200 K). However, in contrast to Park (1972), we consider not only an isothermal distribution of electrons along the magnetospheric magnetic field lines ($n = 0$ in equation (9.13)), but also take into account the possibility of a temperature increase along these field lines ($n = 1$ and $n = 2$ in the same equation). Two values of subionospheric travel time of sferics, τ, were taken: 0.00 s and 0.03 s (see condition (9.1)) and the group delay time of whistlers during the process of their propagation through the ionosphere was taken into account.

Having found L and n_{eq} for which the computed $t_g(\omega_n)$ coincide with the observed values within the above-mentioned models, we compute the group delay times t_g at the observed upper cut-off frequency $f_u \equiv \omega_u/2\pi$ within the same models. These times do not necessarily coincide with the observed group delay times at this frequency. We assume that the discrepancy is due to the fact that the 'effective' temperature of magnetospheric electrons at the equatorial region is different from that of the thermal electrons, described by the diffusive equilibrium model, due to the penetration of hot plasma sheet electrons into the inner magnetosphere (cf. the electron observations of Bahnsen *et al.*, 1985). This increased 'effective' temperature of magnetospheric electrons does not apparently influence the values of n_{eq} and L determined from the nose frequency and nose group delay time, but is to be taken into account in the analysis of the group delay time at the upper cut-off frequency (cf. Fig. 9.2). We can set this 'effective' temperature to be a free parameter and write equation (9.3) as:

$$t'_{gu} = \tilde{t}_{g0} + \tilde{w}_{\parallel}^2 \tilde{t}_{gh}, \tag{9.14}$$

where

$$\tilde{t}_{g0} = t_{g0} + \Delta t_{gc} + \Delta t_{gr}, \tag{9.15}$$

$$\tilde{t}_{gh} = \frac{R_E L Y_{eq}^3 \nu_{eq}^{3/2}}{2c}$$

$$\times \int_0^{\lambda_{ref}} \frac{\cos^{28}\lambda\, \kappa^3(\lambda)\sqrt{1+3\sin^2\lambda}\,[3Y+0.5Y^2+A_e(2-1.5Y-0.5Y^2)]\,\mathrm{d}\lambda}{\left[Y_{eq}\sqrt{1+3\sin^2\lambda}-\cos^6\lambda\right]^{9/2}},$$

$$\tag{9.16}$$

$\tilde{w}_{\parallel} = w_{\parallel}/c$, and $W_e(\mathrm{eV}) = 255076\tilde{w}_{\parallel}^2$; measuring the value of t_g at this frequency with the corrections (9.1) and (9.2) (t'_{gu}) we obtain the equation for $\tilde{w}_{\parallel}^2$ in the form:

$$\tilde{w}_{\parallel}^2 = \frac{t'_{gu} - \tilde{t}_{g0}}{\tilde{t}_{gh}}. \tag{9.17}$$

Now we attempt to apply this general method to the analysis of two particular whistlers recorded at Halley at 20:00 UT on 24 June 1977 and indicated by two arrows on the right in Fig. I. As has already been mentioned, for these whistlers we identified the causative sferic by the superposition of repeated occurrences of the same pattern of whistlers, and by measuring

Table 9.1 *Estimated electron energy* W_e *and average error due to finite resolution of whistler spectrograms,* L *and* n_{eq} *for the second whistler indicated in Fig. I, for different models of plasma density and temperature distributions and different corrections taken into account. The values of* W_e *given in brackets were obtained under the assumption that the sferic was excited near the equator of the Earth, i.e. the residual subionospheric travel time* $t - \tau$ *was equal to zero. Otherwise, the value of* $t - \tau = 0.03$ *s was taken for the subionospheric correction. The main characteristics of the whistler under consideration are the following: observed nose frequency* $f_n \equiv \omega_n/2\pi = 3867$ *Hz, nose group delay time* $t_g(\omega_n) = 2.705$ *s, upper cut-off frequency* $f_u = 4570$ *Hz, and group delay time at this frequency* $t_{gu} = 2.746$ *s.*

	DE–1	DE–2
$n = 0$	$W_e = 1.4 \pm 1.4(1.6 \pm 1.4)$ eV $L = 4.40,\ n_{eq} = 378.7$ cm^{-3}	$W_e = 1.9 \pm 1.3(2.0 \pm 1.4)$ eV $L = 4.38,\ n_{eq} = 416.2$ cm^{-3}
$n = 1$	$W_e = 1.1 \pm 1.5(1.2 \pm 1.5)$ eV $L = 4.45,\ n_{eq} = 306.8$ cm^{-3}	$W_e = 1.9 \pm 1.5(2.0 \pm 1.6)$ eV $L = 4.43,\ n_{eq} = 321.9$ cm^{-3}
$n = 2$	$W_e = 2.0 \pm 1.7(2.2 \pm 1.8)$ eV $L = 4.50,\ n_{eq} = 225.6$ cm^{-3}	$W_e = 3.6 \pm 1.8(3.8 \pm 1.8)$ eV $L = 4.49,\ n_{eq} = 232.0$ cm^{-3}

echo periods (Smith *et al.*, 1981). The selected whistlers had well-defined upper branches, so they seemed to be potentially suitable for the study of magnetospheric electron temperature effects. The whistlers were scaled using AVDAS (Advanced VLF Data Analysis System, see Smith & Yearby (1987)). Then, following the procedure described earlier, we estimated the values of L and n_{eq} using the DE–1 and DE–2 models with $n = 0$, 1 and 2.

The results for the first whistler appeared to be essentially negative: the uncertainty of measurements did not encourage us to make an estimate of the temperature. This agrees with the result of Guthart (1965). The results of the calculations of electron energy W_e, L and n_{eq} for the second whistler and for different models of electron density and temperature distribution are shown in Table 9.1. In the same table we also show the average error in the determination of W_e due to the limited resolution of the spectrograms, with respect to both frequency and group delay time (electron temperature was calculated not only for nose and cut-off points, but also for the nose points shifted by $\pm\Delta f$, $\pm\Delta t$, where Δf and Δt are the frequency and time resolution of AVDAS: 39 Hz and 7 ms respectively).

As one can see from Table 9.1, the estimated temperature of magnetospheric electrons inferred from the second whistler is about a few electronvolts for both models DE–1 and DE–2 and all the values of n and $t_g - \tau$ under consideration. Predicted values of W_e are larger for the DE–2 model than for the DE–1 model and increase with increasing n and decreasing $t - \tau$. The values of W_e for the second whistler are compatible with, although somewhat lower than, those obtained by Kobelev & Sazhin (1983) from a less rigorous analysis of whistler spectrograms.

Also, we analysed our whistlers using the DE–3 and DE–4 models of the electron distribution. For both of these models the ion composition was assumed to be 50% O^+, 40% H^+, 10% He^+; the electron temperature at the reference height was assumed to be equal to 1600 K for the DE–3 model and to 800 K for the DE–4 model (Park, 1972). The results did not differ much from those obtained for the DE–1 and DE–2 models. In particular, for the DE–3 model we obtained values of W_e for the second whistler equal to 1.6 eV, 1.4 eV, and 2.2 eV for $n = 0, 1$ and 2, respectively. Similar values of W_e for the DE–4 model were all equal to 0.9 eV (L for both models was in the range 4.4–4.5). As in the case of the DE–1 and DE–2 models, the results for the first whistler appeared to be negative.

Although this preliminary test of our method of temperature diagnostics is somewhat encouraging, before this method can be recommended for practical applications we need to be able to specify more accurately the model of electron density and temperature distribution in the magnetosphere, have a better estimate for the effect of ducted ray paths and increase the precision of determining whistler parameters. The first steps in this direction have already been made by Hamar *et al.* (1990) who suggested a new method (the matched filter technique) for the analysis of whistler fine structure.

Equation (9.3) has also been recently used for the analysis of the group delay of whistler-mode signals generated by NAA (24.0 kHz) and NSS (21.4 kHz) Navy transmitters and recorded at Faraday, Antarctica ($L = 2.3$) (Sazhin *et al.*, 1992). The contribution of the sum $\Delta t_{gh} + \Delta t_{gc} + \Delta t_{gr}$ appeared to be of the order of 10 ms for the signals from both stations and could be taken into account when analysing these signals.

9.2 Whistler-mode emissions in the vicinity of the magnetopause

Besides whistler-mode waves propagating in the magnetosphere and generated by lightning discharges or ground-based transmitters, the same type of wave can be generated directly in the magnetospheric plasma owing to the development of whistler-mode instabilities or non-linear wave–wave or wave–particle–wave interactions. Waves generated directly in the magnetospheric plasma in the frequency range from tens of Hz to hundreds of kHz are known as ELF/VLF (extremely low frequency/ very low frequency) emissions. These waves can propagate in different wave modes, in general, but in what follows we shall concentrate our attention on those emissions which propagate in the whistler-mode and call them whistler-mode emissions. Experimental and theoretical studies of different types of whistler-mode emissions have been reported in thousands of papers and have been summarized in numerous review papers and monographs (e.g. Helliwell, 1965; Rycroft, 1972; Bullough, Hughes & Kaiser, 1974; Kaiser & Bullough, 1975; Likhter, 1979; Al'pert, 1980; Sazhin, 1976, 1982a, 1983b). In this chapter we concentrate our attention on only a few types of whistler-mode emission, for which we can best illustrate the applicability of the theoretical methods of analysis worked out in the previous chapters. Namely, in this section we consider a specific type of whistler-mode emission observed in the immediate vicinity of the magnetopause, while in the next section we refer to some types of whistler-mode emission observed in the inner magnetosphere.

Plasma waves in the vicinity of the magnetopause at ELF/VLF frequencies have been studied by many researchers (e.g. LaBelle *et al.*, 1987; LaBelle & Treumann, 1988; Farrugia *et al.*, 1988; Tsurutani *et al.*, 1981, 1989) and it is now known that the intensification of these waves is a general feature of magnetopause crossings. As follows from the wave spectrogram of Farrugia *et al.* (1988), enhanced wave activity near the magnetopause during the passage of an FTE (flux transfer event) was particularly pronounced at frequencies below the local electron gyrofrequency, i.e. in the whistler-mode frequency range. This result is essentially consistent with our results also based on AMPTE–UKS plasma wave data. Although a detailed statistical study of the UKS data for magnetopause crossings has not yet been completed, the crossings of the dayside magnetopause by the AMPTE–UKS satellite on orbit 26, day 276 and orbit 29, day 283 (i.e. 2 and 9 October 1984) were generally accompanied by enhanced whistler-mode activity.

The whistler-mode activity during these crossings was studied in detail in our previous papers (Sazhin, Walker & Woolliscroft, 1990a,b,c) in which we tried to identify the type of observed waves (i.e. to confirm that these are whistler-mode waves and to estimate their wave normal angle), to identify the energy source of these waves, and to explain the concentration of their energy in the vicinity of the magnetopause. The main results of these papers will be briefly discussed below (cf. Sazhin, Walker & Woolliscroft, 1991).

(a) Spin modulation effect

The most reliable methods of determining the wave mode and finding the wave normal angle have been carried out using either the analysis of wave polarization (e.g. Kimura & Matsuo, 1982) or complete direction finding techniques (e.g. Muto *et al.*, 1987). In particular, Kimura & Matsuo (1982) used spin modulation measurements from a rocket experiment to determine the wave normal angle of auroral hiss emissions in the upper ionosphere. To determine the wave normal angle of whistler-mode waves recorded on board the AMPTE–UKS satellite in the vicinity of the magnetopause we use an approach in some respects similar to that of Kimura & Matsuo (1982). An **E** field antenna was operated on board this satellite, measuring wave power in the plane perpendicular to the spacecraft axis of rotation (see Darbyshire *et al.*, 1985). In contrast to Kimura & Matsuo (1982) no additional measurements of wave normal direction (such as its direction with respect to the axis of rotation) were presumed and we took into account non-zero temperature effects on whistler-mode polarization at wave normal angles θ close to the resonance cone in a cold plasma, θ_R, where they lead to qualitative changes of the character of this polarization (see equations (6.52)–(6.56)).

For this analysis, data obtained between 14:10 and 14:20 UT on orbit 29, day 283, 1984 of the mission were considered. At this time, the spacecraft was situated Earthward of the magnetopause at a distance of 10.3 Earth radii, in the noon–dusk sector of the magnetosphere ($\sim$ 13:00 LT) and at a magnetic latitude of $-10.4°$. The axis of satellite rotation was at an angle $\phi_r = \pi/3$ (within an accuracy of about $\pm 1°$) with respect to the ambient magnetic field. The wave frequency ω was approximately 0.4Ω and the ratio $\bar{R} = W_{\max}/W_{\min}$, where $W_{\max(\min)}$ is the maximum (minimum) wave power spectral density measured during a satellite rotation, was found to lie mainly between 3 and 4.

Theoretical estimates of $\bar{R}$ were calculated starting from the polarization equations and using both the quasi-electrostatic and quasi-longitudinal approximations (see equations (5.44), (5.45), (6.52) and (6.53)). For propagation at angles $\theta = \theta_R$ (quasi-electrostatic approximation) the calculated value of $\bar{R}$ was of the order of 20–2000, too high for our experimental measurements. Treating the problem within the quasi-longitudinal approximation and neglecting the contribution of finite electron density, non-zero electron temperature and the contribution of ions (the contribution of these effects would only slightly change the value of $\bar{R}$ at the expense of considerable complication of the equations) gives the following expression for $\bar{R}$ (Sazhin, Walker & Woolliscroft, 1990a):

$$\bar{R} = \max(\bar{R}_0, \bar{R}_0^{-1}), \qquad (9.18)$$

where:

$$\bar{R}_0 = \frac{W_0(\bar{\lambda} = 0)}{W_0(\bar{\lambda} = \pi/2)} = \cos^2 \phi_r + \frac{2 \sin^2 \phi_r \sin^2 \theta}{Y^2(1 + \cos^2 \theta) - 4Y \cos \theta + 1 + \cos^2 \theta},$$

$$\bar{\lambda} = \frac{1}{2}\arctan\left\{\left[\frac{(Y\cos\theta-1)^2}{(Y-\cos\theta)^2}\sin 2\delta\cos\phi_r + \sin 2\delta\cos\phi_r\right.\right.$$

$$+ \left.\frac{2\sin\theta\sin\delta\sin\phi_r}{Y-\cos\theta}\right]\times\left[\frac{(Y\cos\theta-1)^2}{(Y-\cos\theta)^2}(\sin^2\delta\cos^2\phi_r - \cos^2\delta)\right.$$

$$+ \left.\left.\sin^2\delta - \cos^2\delta\cos^2\phi_r - \frac{\sin\theta\cos\delta\sin 2\phi_r}{Y-\cos\theta} - \frac{\sin^2\theta\sin^2\phi_r}{(Y-\cos\theta)^2}\right]^{-1}\right\},$$

and δ indicates the orientation of the projection of the wave refractive index on the plane perpendicular to the external magnetic field $\mathbf{B}_0$ (δ varies from 0 to 2π).

As already mentioned, magnetometer and plasma wave data show that for the time period under consideration $\phi_r = \pi/3$ and $Y = 2.5$, and so expression (9.18) gives us $\bar{R}$ as a function of θ and δ. For θ not close to 0 (roughly $|\theta| > 0.2$ rad) $\bar{R}$ can vary by more than an order of magnitude (from about 1 to more than 10) when δ changes from 0 to 2π, which seems to contradict the experimental findings provided it is assumed that δ changes randomly. However, for small θ ($|\theta| < 0.2$ rad), the values of $\bar{R}$ lie close to 4 (± 2) for all δ, which agrees with the experimental results. For $\theta = 0$, $\bar{R} = 4$ and does not depend on δ. Hence, on the basis of the spin modulation measurements, we can conclude that the waves under consideration are whistler-mode waves propagating almost parallel to the magnetic field.

(b) Generation of waves

There are several possible mechanisms for observed whistler-mode wave generation in the vicinity of the magnetopause. The most obvious one involves the transfer of energy from energetic electrons to waves, in other words whistler-mode instabilities. These instabilities were analysed using data from the magnetometer and electron experiments on board AMPTE–UKS (Ward et al., 1985). The electron experiment measured electron count rates in the energy range 12 eV to 25 keV for the data collected on day 283 and 12 eV to 640 eV on day 276, a complete distribution being measured every 5 s (Shah, Hall & Chaloner, 1985). From these data, the density and distribution function of electrons were computed. The electron density during these wave observations was about 1 cm^{-3}, which corresponded to $\Pi_\Sigma^2/\Omega^2 \approx 15$, where Π_Σ is the total electron plasma frequency. It was shown that a reasonable approximation to the observed distribution function could be a superposition of two bi-Maxwellian distributions with equal densities and anisotropies ($T_\perp/T_\parallel$) close to 2, and drifting in opposite directions with respect to the magnetic field with an average drift energy of about 10 eV. This allows us to use equations (7.65) (with equations (7.74) and (7.75) taken into account), (7.101), (5.48) and (6.62) for the increment of whistler-mode instability, $\Im D$, and whistler-mode refractive index, derived for an anisotropic plasma consisting of different electron populations with the contribution of beams taken into account. The average parallel temperature $(T_\parallel)$ of the electron population corresponding to the first bi-Maxwellian distribution was shown to be roughly equal to 2 eV (cold electrons), while $T_\parallel$ of the second population was shown to be roughly equal to 200 eV. The stability of plasma with this electron distribution function with respect to the generation of whistler-mode waves at different frequencies and wave normal angles was analysed and it was shown that the instability most likely developed for the waves propagating parallel or almost parallel to the magnetic field at frequencies below half electron gyrofrequency. This result is consistent with the results of the analysis of spin modulation effects on wave power measurements considered in the previous subsection. Wave observations at $\omega > 0.5\Omega$ could be explained by assuming larger values for anisotropy of the electron distribution function, or that the waves could penetrate from a region of stronger magnetic field.

(c) Wave trapping

As already mentioned, a typical feature of whistler-mode observations in the vicinity of the magnetopause was the well-defined maximum of their intensity in the immediate vicinity of the Earthward side of the magnetopause. It seems implausible to relate this maximum to the local excitation of these waves, as the electrons responsible for their generation were observed in a much larger region. An alternative explanation was related to the peculiarities of their propagation in the vicinity of the magnetopause. This interpretation appears to be more straightforward since the magnetopause is a natural discontinuity at which waves can be reflected. Reflection of the waves is accompanied by concentration of their energy in the reflection layer. Additional enhancement of wave energy could be related to possible wave trapping in the vicinity of the magnetopause. This process is briefly considered below (following Sazhin, Walker & Woolliscroft, 1990c).

As follows from the analysis of spin modulation of whistler-mode waves in the vicinity of the magnetopause, their wave normal angle θ is in many cases close to zero and so we can use the quasi-longitudinal approximation for those waves with $\theta^2 \ll 1$. In view of the application of our results to the magnetopause, where $\Pi_\Sigma \gg \Omega$ and wave frequencies are well above the proton gyrofrequency, we can neglect the relativistic effects on whistler-mode propagation as well as the effects of finite electron density and the contribution of ions (cf. equation (5.48)). The influence of electron anisotropy on whistler-mode propagation is not significant, so it can be set equal to 1 when estimating thermal effects on whistler-mode propagation. Also as a first approximation we can neglect the contribution of the non-zero electron beam velocity. Finally, we restrict ourselves to considering waves at $\theta^2 \ll 1$ and write the whistler-mode dispersion in the form (5.2), taken in the non-relativistic approximation and for $A_e = 1$:

$$N^2 = N_{0\parallel}^2 [1 + a_\parallel \beta_e + (a_0 + a_\theta \beta_e)\theta^2], \tag{9.19}$$

where: $N_{0\parallel}^2 = \nu Y^2/(Y-1)$, $a_0 = Y/[2(Y-1)]$, $a_\parallel = Y^2/(Y-1)^3$, $a_\theta = Y^3(-4 + 13Y - 6Y^2)/[2(Y-1)^4(2Y-1)]$, $\nu = \Pi_\Sigma^2/\Omega^2$, $\beta_e = 0.5(w/c)^2\nu$, and c is the velocity of light. When deriving (9.19) we assumed that the unperturbed electron distribution function is defined by equation (1.90) with $j = 0$ and $w \equiv w_\perp = w_\parallel$. Also we assumed that $|a_\parallel \beta_e| \ll 1$, $|a_\theta \beta_e| \ll 1$ and $\nu \gg 1$. These conditions are satisfied in the magnetopause region where $\beta_e \approx 10^{-3}$, provided ω is not too close to Ω.

For the conditions near the magnetopause, we can assume that the gradients of the parameters in the direction perpendicular to the magnetic field are well above the gradients in the direction parallel to this field. Hence, according to Snell's law the parameter $N^2 \cos^2 \theta$ should be conserved during the process of wave propagation. For $\theta^2 \ll 1$ this could be written as:

$$N_{0\parallel}^2 [1 + a_\parallel \beta_e + (a_0 + a_\theta \beta_e - 1)\theta^2] = \text{const.} \tag{9.20}$$

Let us define ω_T as the frequency at which:

$$a_0 + a_\theta \beta_e - 1 = 0. \tag{9.21}$$

As $|a_\theta \beta_e| \ll 1$ at ω close to $\Omega/2$ and $a_0 = 1$ at $\omega = \Omega/2$, we can put $\omega = \Omega/2$ when estimating a_θ in (9.21). As a result we obtain: $a_\theta = -8/3 \approx -2.67$. Substituting this value of a_θ into (9.21) it follows that this equation is satisfied at

$$\omega = \omega_T \equiv 0.5\Omega(1 + 2.67\beta_e). \tag{9.22}$$

In view of the fact that $a_0 + a_\theta \beta_e > 1$ for $\omega > \omega_T$ and $a_0 + a_\theta \beta_e < 1$ for $\omega < \omega_T$, we can see from (9.20) that a decrease of $|\theta|$ should be accompanied by a decrease of $N_\parallel^2 \equiv N_{0\parallel}^2(1 + a_\parallel \beta_e)$ for $\omega > \omega_T$ and an increase for $\omega < \omega_T$. Another interpretation of equation (9.20) is the following: if during the process of wave propagation $N_\parallel^2$ increases (decreases) then θ^2 should decrease (increase) for $\omega > \omega_T$ and should increase (decrease) for $\omega < \omega_T$. If during the process of wave propagation the wave reaches the point where $\theta = 0$ then it will be reflected (θ changes its sign) provided that the derivative of $N_\parallel^2$ does not change its sign at this point. If we have successive reflections of the wave, so that its path appears to be snake-like, then the wave is trapped. This trapping is obviously accompanied by the concentration of wave energy in the areas of trapping.

From considerations similar to those given above, we can see that whistler-mode waves at $\theta^2 \ll 1$ can be trapped between the positions x_1 and x_2 with refractive indices $N(x_1)$ and $N(x_2)$, where the x axis is assumed to be perpendicular to the external magnetic field direction, if:

$$N_\parallel^2(x) < \min\left[N_\parallel^2(x_1), N_\parallel^2(x_2)\right], \tag{9.23}$$

where $x_1 < x < x_2$ for $\omega > \omega_T$, or

$$N_\parallel^2(x) > \max\left[N_\parallel^2(x_1), N_\parallel^2(x_2)\right], \tag{9.24}$$

where $x_1 < x < x_2$ for $\omega < \omega_T$.

Using electron, magnetometer and plasma wave observations on board the AMPTE–UKS satellite we have shown that either condition (9.23) or (9.24) is satisfied for the crossings of the magnetopause at orbit 26, day 276 at about 14:26 UT and at orbit 29, day 283 at about 14:19, 14:25, 14:34 and 14:43 UT. However, some peaks of wave spectral density recorded on day 283 for the crossings at about 14:04 and 14:39 UT seem not to be directly related to wave trapping. They could be explained either by the contribution of the effects of local generation of the waves, or by assuming that the propagation pattern was more sophisticated than the simple model described above. We note in particular that the topology of the actual magnetopause is more complicated than in the model presented here. This may well account for intensifications of the waves which are not exactly at the inner edge of the magnetopause on some occasions. Equally it could account for those magnetopause crossings which are not accompanied by wave intensifications. Thus, although whistler-mode trapping in the vicinity of the magnetopause seems to be relevant to many cases of wave intensification in this region, it cannot be considered as the only mechanism responsible for this intensification and a full description must take into account the wave generation as well.

The results of this section lead us to the following main conclusions.

(1) From an analysis of the spin modulation effect in the wave power measured in the ELF/VLF frequency range on board the AMPTE–UKS satellite in the vicinity of the magnetopause it is concluded that the observed waves are whistler-mode waves propagating mainly parallel or almost parallel to the ambient magnetic field.

(2) The energy source of these waves is shown to be most likely due to the anisotropy of the electron distribution function ($T_\perp > T_\parallel$).

(3) The localized intensification of these waves near the Earthward boundary of the Earth's magnetopause can often be explained in terms of wave trapping in this region.

9.3 Mid-latitude hiss-type emissions

At an early stage of the investigation of natural radio waves at mid-latitude stations, electromagnetic emissions at frequencies of a few kHz with essentially structureless (hiss-type) spectra were discovered (e.g. Watts, 1957; Dowden, 1961). Since that time these emissions have been extensively studied both from ground-based stations (e.g. Jørgensen, 1966; Harang, 1968; Ondoh *et al.*, 1981; Hayakawa *et al.*, 1986, 1988; Hayakawa, 1989a) and satellites (Koons & McPherron, 1972; Bullough, Hughes & Kaiser, 1974; Likhter, 1979; Likhter & Sazhin, 1980; Ondoh *et al.*, 1981; Hayakawa, 1989b). It is not our aim to give a detailed analysis of the different morphological properties of mid-latitude hiss-type emissions and their possible interpretations. Instead we concentrate our attention on those properties which can be understood in terms of the theoretical background developed in previous chapters.

The OVI–14 satellite observed mid-latitude hiss in the energetic electron slot region ($L = 2.5$–3.6), in the frequency range 3.9–10.4 kHz. The most intense emissions were observed at the frequency 5.6 kHz (Koons & McPherron, 1972). The observation of emissions at frequencies close to 5 kHz was confirmed later by Ondoh *et al.* (1981) using data from the satellites ISIS–1 and ISIS–2. These emissions have a bandwidth of up to several kHz (Hayakawa *et al.*, 1986, 1988).

Observations of mid-latitude hiss-type emissions at ground-based stations usually followed the development of substorms: emissions were observed 3 to 11 hours after the substorm onset as measured by AE index (Hayakawa *et al.*, 1986, 1988; Hayakawa, 1989a). Substorm-related dawnside emissions often showed a steady rise in frequency at a rate between 0.2 and 2.5 kHz/hour (Hayakawa *et al.*, 1986; Hayakawa, 1989a), while the frequency of duskside emissions first increased and then gradually decreased. The rate of the latter increase and decrease varied from one event to the other from a fraction of a kHz/hour to several kHz/hour (Hayakawa *et al.*, 1988).

The modelling of these emissions is based on the theoretical background given in Chapters 7 and 8. We begin by discussing the generation mechanisms based on the linear theory given in Chapter 7.

From the frequency range of mid-latitude hiss-type emissions and their localization we can infer that their frequencies are below the electron gyrofrequency at the equatorial magnetosphere. Remembering also the electromagnetic nature of these waves we expect these emissions to propagate in the whistler mode. As was the case for the whistler-mode emissions discussed in Section 9.2, the most probable energy source of these waves is the kinetic energy of the electrons, which is converted into wave energy via the development of the whistler-mode instability. The most likely source of such an instability is the pitch-angle anisotropy of energetic electrons trapped in the magnetosphere, whereby their effective temperature in the direction perpendicular to the magnetic field becomes greater than the corresponding temperature in the direction parallel to this field; i.e. the parameter A_e, defined in Sections 1.5 and 1.6, is greater than 1. Such an anisotropy of the electron distribution function could be achieved either by electron precipitation from the loss cone, or by the acceleration of electrons drifting from the plasma sheet to the inner magnetosphere (Hess, 1968). Anisotropic electron distributions have actually been measured *in situ* in the magnetosphere (e.g. Bahnsen *et al.*, 1985). The conditions for the development of the instability appear to be different for different wave normal angles, θ. *In situ* measurement of these angles is not possible in most cases, and so we must rely on indirect evidence.

As follows from Sections 7.3 and 7.4, the conditions for the development of the quasi-electrostatic whistler-mode instability are more stringent than those for the quasi-longitudinal instability. The most favourable conditions for the development of the latter instability are obtained for $\theta = 0$ (the instability at a given frequency develops at the lowest A_e) (see equation (7.65)). Hence, we conclude that the mid-latitude hiss-type emissions are likely to be generated at wave normal angles close to zero.

Assuming wave normal angles equal to zero, and restricting ourselves to considering the low-temperature limit (zero-order approximation), we can use expression (7.42) for the increment of the instability. This expression predicts that $\tilde{\gamma} > 0$ (waves grow) when $A_e > \bar{A}_{e0}$ defined by (7.38). The inequality $A_e > \bar{A}_{e0}$ can be rewritten in a different form:

$$\omega < \omega_{\mathrm{up}} \equiv \frac{\Omega(A_e - 1)}{A_e}. \tag{9.25}$$

As follows from condition (9.25), the upper cut-off of the instability increases from zero to Ω as A_e increases from 1 to infinity. Observed values of A_e in the magnetosphere are generally below 2, which implies that $\omega_{\mathrm{up}} \lesssim \Omega/2$.

Condition (9.25) determines the sign of $\tilde{\gamma}$. However, the absolute value of $\tilde{\gamma}$ depends on a number of other parameters besides A_e, and especially on the parameter ξ_{00}. As follows from (7.43) and (7.44),

$$\xi_{00} = -\frac{(Y-1)^{3/2}}{Y\sqrt{2\beta_e}}. \tag{9.26}$$

Neglecting the variations of β_e along the magnetic field line, we have from (9.26) that, for a given ω, $|\xi_{00}|$ is minimal where Ω is minimal, i.e. on the magnetospheric equator. At the same time, as follows from (7.42), the minimal value of $|\xi_{00}|$ roughly corresponds to the maximal value of $\tilde{\gamma}$. Hence, mid-latitude hiss-type emissions are most likely to be generated in the immediate vicinity of the magnetospheric equator.

On the other hand, $|\xi_{00}|$ determined by (9.26) increases with decreasing ω, if all other parameters on the right hand side of (9.26) are held constant. Hence, on the basis of (7.42) we can expect that at $\omega < \omega_{\rm up}$ (see equation (9.25)) $\tilde{\gamma}$ first increases with decreasing ω owing to the increase of the term $A_e + (1 - A_e)Y$, but when ω decreases further, $\tilde{\gamma}$ begins to decrease rapidly owing to increasing $|\xi_{00}|$. Thus we can expect that the frequency of the generated waves is limited not only from above by the frequency $\omega_{\rm up}$ determined by (9.25), but also from below by the frequency at which $\tilde{\gamma}$ becomes too low to provide sufficient energy to the generated waves.

Thus we see that linear theory alone can explain the frequency range of mid-latitude hiss-type emissions, and also clarify the energy source of these emissions. However, linear theory can only establish the tendency for the waves to be amplified or to be damped, and it cannot in general be used to estimate the wave amplitude. Such an estimate can be made with the quasi-linear models discussed in Section 8.1.

The steady-state model discussed in subsection 8.1(c) seems to be directly applicable to mid-latitude hiss-type emissions, as their amplitude remains almost unchanged during the bounce period of the waves in a field tube. The steady state could be achieved by a balance between the numbers of electrons penetrating to the inner magnetosphere from an external source and the numbers precipitating into a loss cone.

Let us assume that the energy of electrons $W \equiv m_e w^2/2$ drifting from the plasma sheet towards the inner magnetosphere is proportional to L^{-3} (Southwood & Kivelson, 1975) which can be qualitatively predicted from the conservation of the first adiabatic invariant. In this case, neglecting the terms G_1, G_2, and G_3 when compared with G_0, we obtain from equation (8.57) (Sazhin, 1984):

$$B_f^2 \sim n_{\rm eq} L^{5.5}. \tag{9.27}$$

As follows from (9.27), we can expect that B_f^2 is maximal near the inner boundary of the plasmapause where both $n_{\rm eq}$ and L are large (see Fig. 2

of Sazhin, 1984). This agrees with the direction finding results reported by Hayakawa *et al.* (1986).

The frequency at which B_f^2 is maximal can be determined by the following expression (see the discussion in subsection 8.1(c)):

$$\omega = \omega_{\max} \equiv \frac{\Omega_{\text{eq}}^3 c^2}{1.3\Pi_{\text{eq}}^2 w^2}, \tag{9.28}$$

where Π_{eq} is the angular electron plasma frequency at the magnetospheric equator, Ω_{eq} is the angular electron gyrofrequency there, and w is the thermal velocity of the energetic electrons penetrating into the inner magnetosphere.

As follows from equation (9.28), the value of $\omega_{\max}$ depends on three parameters Π_{eq}, Ω_{eq} and v_0, none of which is available from *in situ* measurements in general. In practice either we use indirect evidence about the values of these parameters, or, remembering our previous conclusion that the emissions are generated near the inner boundary of the plasmapause, we can assume that Ω_{eq} and Π_{eq} in (9.28) are equal to the average values of Ω_{eq} and Π_{eq} determined from averaged measurements of electron density near the plasmapause. For example, in the case of emissions recorded on 19 December 1978 and reported by Hayakawa *et al.* (1986), the L value of the plasmapause, as determined from direction finding measurements, was approximately equal to 2.9. Then using the statistical results obtained by Chappell, Harris & Sharp (1970) we can assume that the electron density in the equatorial magnetosphere was approximately equal to 2×10^3 cm^{-3}. For these particular conditions we can simplify (9.28) to:

$$f_{\max}(\text{kHz}) \equiv \omega_{\max}/2\pi \approx 60/W(\text{keV}). \tag{9.29}$$

Remembering that substorm activity is connected with the injection of energetic plasma from the plasma sheet to the inner magnetosphere, we can relate the time delay of the emissions with respect to the onset of the substorm to the time required for the electrons to drift from the midnight sector to the morning sector ($\text{LT} = 6^{\text{h}}$) where emissions were observed. This time can be estimated by the equation (Fälthammar, 1973):

$$T(\text{sec}) = \frac{0.675 \times 10^9}{(0.7 + 0.3 \sin\alpha_{\text{eq}}) \, L \, W(\text{eV})}, \tag{9.30}$$

where α_{eq} is the equatorial pitch-angle of the drifting electrons.

For qualitative analysis we can neglect a weak dependence of T on α_{eq} and assume $\alpha_{\text{eq}} = \pi/4$. Remembering that, for the event of 19 December 1978, $L \approx 2.9$, we can simplify (9.30) to:

$$T(\text{hours}) \approx \frac{70}{W(\text{eV})}. \tag{9.31}$$

From (9.29) and (9.31) we obtain:

$$T(\text{hours}) \approx 1.2\, f_{\max}(\text{kHz}). \tag{9.32}$$

In view of (9.32) we expect that emissions at a frequency of 4 kHz should be generated about 5 hours after the substorm onset, which is roughly compatible with the observations (see Fig. 1 in Hayakawa *et al.*, 1986).

It follows from equation (9.32) that the electrons responsible for higher frequency emissions (less energetic) drift more slowly than the electrons responsible for lower frequency emissions (more energetic). Thus we expect that electrons with different energies penetrating into the midnight magnetosphere will generate emissions with increasing frequencies in the dawn sector. The rate of increase in frequency given by equation (9.32) is approximately 0.8 kHz/hour. Both this rate and its increase with time, which also follows from equation (9.32), are compatible with the observations of emissions on 19 December 1978 (see Fig. 2 of Hayakawa *et al.*, 1986). However, for other events considered by Hayakawa *et al.* (1986) the agreement between the experimental results and those which follow from equations similar to equation (9.32) (but with different coefficients relevant to the events under consideration) appears to be less good. Hence, more effort is required before we can provide a quantitative theory of these events. The frequency drift of duskside emissions can also be explained in terms of the quasi-linear model discussed in subsection 8.1(c), although in this case the theory is more complicated (Hayakawa *et al.*, 1988).

Essentially the same model as described above could be applied to the interpretation of hiss-type emissions at frequencies below 1 kHz which are observed inside the plasmapause and are known as plasmaspheric hiss (e.g. Thorne *et al.*, 1973; Solomon *et al.*, 1988; Storey *et al.*, 1991). However, in this case linear theory predicts a preference for the generation of plasmaspheric hiss at $\theta = 0$, which was not observed by the ISEE–1 satellite (Storey *et al.*, 1991). The obliquely propagating waves could be explained by the quasi-linear theory of oblique whistler-mode emissions, which predicts equal amplitudes for emissions propagating at different θ, provided $|\theta| \ll 1$ (see Sazhin (1987e) for details).

The spectrum of dawnside mid-latitude hiss-type emissions can sometimes acquire a fine structure, forming the so called chorus emissions (see the review by Sazhin & Hayakawa, 1992). The theoretical background of this process has recently been considered by Nunn & Sazhin (1991).

References

Abramowitz, M. & Stegun, I. A. (eds.) (1964). *Handbook of Mathematical Functions.* Washington: National Bureau of Standards (US Government Printing Office).

Airoldi, A. C. & Orefice, A. (1982). Relativistic dielectric tensor of a Maxwellian plasma for electron cyclotron waves at arbitrary propagation angles. *Journal of Plasma Physics*, **27**, 515–24.

Akhiezer, A. I., Akhiezer, I. A., Polovin, R. V., Sitenko, A. G. & Stepanov, K. N. (1975). *Plasma Electrodynamics.* Oxford: Pergamon.

Alexandrov, A. F., Bogdankevich, L. S. & Rukhadze, A. A. (1978). *Introduction to Plasma Electrodynamics.* Moscow: Vishaia Shkola (in Russian).

Al'pert, Ya. L. (1980). 40 years of whistlers. *Journal of Atmospheric and Terrestrial Physics*, **42**, 1–20.

Al'pert, Ya. L. (1990). *Space Plasma* (2 vols). Cambridge University Press.

Anderson, D., Askne, J. & Lisak, M. (1975). Wave propagation in an absorptive and strongly dispersive medium. *Physical Review A*, **12**, 1546–52.

Anderson, D., Askne, J. & Lisak, M. (1976). The temporal velocity of whistler wave packets. *Plasma Physics*, **18**, 163–4.

André, M. (1985). Dispersion surfaces. *Journal of Plasma Physics*, **33**, 1–19.

Andronov, A. A. & Trakhtengertz, V. Yu. (1964). Instability of one-dimensional packets and absorption of electromagnetic waves in a plasma. *Soviet Physics JETP*, **18**, 698–702.

Angerami, J. J. & Thomas, J. O. (1964). Studies of planetary atmospheres, 1. The distribution of electrons and ions in the Earth's exosphere. *Journal of Geophysical Research*, **69**, 4537–60.

Aubry, M. P., Bitoun, J. & Graff, Ph. (1970). Propagation and group velocity in a warm magnetoplasma. *Radio Science*, **5**, 635–45.

Backus, G. (1960). Linearised plasma oscillations in arbitrary electron velocity distributions. *Journal of Mathematical Physics*, **1**, 178–91.

Bahnsen, A., Jespersen M., Neubert T., Canu P., Borg H. & Frandsen, P. E. (1985). Morphology of keV-electrons in the Earth's magnetosphere as observed by GEOS-1. *Annales Geophysicae*, **3**, 19–25.

Baynham, A. C. & Boardman, L. S. (1971). *Plasma Effects in Semiconductors: Helicon and Alfvén Waves.* London: Taylor & Francis Ltd.

Bernard, L. P. (1973). A new nose extension method for whistlers. *Journal of Atmospheric and Terrestrial Physics*, **35**, 871–80.

Bespalov, P. A. & Trakhtengertz, V. Yu. (1986). Cyclotron instability of the Earth radiation belts. *Reviews of Plasma Physics*, ed. M. A. Leontovich (English translation), **10**, 155–292.

Bornatici, M., Ghiozzi, G. & de Chiara, P. (1990). Weakly relativistic dielectric tensor in the presence of temperature anisotropy. *Journal of Plasma Physics*, **44**, 319–35.

Breizman, B. N. (1990). Collective interaction of relativistic electron beams with plasmas. *Reviews of Plasma Physics*, ed. B. B. Kadomtsev (English translation), **15**, 61–162.

Briggs, R. J. (1964). *Electron-stream Interaction with Plasmas*. Cambridge, Massachusetts: M.I.T. Press.

Brillouin, L. (1960). *Wave Propagation and Group Velocity*. San Diego, California: Academic Press.

Brinca, A. L. (1973). Approximations to the plasma dispersion function. *Journal of Plasma Physics*, **10**, 123–33.

Budden, K. G. (1983). Approximations in magnetoionic theory. *Journal of Atmospheric and Terrestrial Physics*, **45**, 213–8.

Budden, K. G. (1985). *The Propagation of Radio Waves*. Cambridge University Press.

Bud'ko, N. I., Karpman, V. I. & Pokhotelov, O. A. (1972). Nonlinear theory of the monochromatic circular polarized VLF and ULF waves in the magnetosphere. *Cosmic Electrodynamics*, **3**, 147–64.

Bullough, K., Hughes, A. R. W. & Kaiser, T. R. (1974). Spacecraft studies of VLF emissions. In *Magnetospheric Physics*, ed. B. M. McCormac, pp. 231–40. Dordrecht, Boston & London: D. Reidel.

Burtis, W. J. & Helliwell, R. A. (1976). Magnetospheric chorus: occurrence patterns and normalized frequency. *Planetary and Space Science*, **24**, 1007–24.

Burton, R. K. (1976). Critical electron pitch angle anisotropy necessary for chorus generation. *Journal of Geophysical Research*, **81**, 4779–81.

Carpenter, D. L. (1988). Remote sensing of the magnetospheric plasma by means of whistler mode signals. *Reviews of Geophysics*, **26**, 535–49.

Carpenter, D. L. & Smith, R. L. (1964). Whistler measurements of electron density in the magnetosphere. *Reviews of Geophysics*, **2**, 415–41.

Chappell, C. R., Harris, K. K. & Sharp, G. W. (1970). A study of the influence of magnetic activity on the location of the plasmapause as measured by OGO–5. *Journal of Geophysical Research*, **75**, 50–6.

Corcuff, P. (1977). Méthodes d'analyse des sifflements électroniques: 1. Application à des sifflements théoriques. *Annales de Géophysique*, **33**, 443–54.

Corcuff, P. & Corcuff, Y. (1973). Détermination des paramètres $f_n - t_n$ caractéristiques des sifflements radioélectriques reçus au sol. *Annales de Géophysique*, **29**, 273–8.

Coroniti, F. V., Kurth, W. S., Scarf, F. L., Krimigis, S. M., Kennel, C. F. & Gurnett, D. A. (1987). Whistler mode emissions in the Uranian radiation belts. *Journal of Geophysical Research*, **92**, 15,234–48.

Cuperman, S. (1981). Electromagnetic kinetic instabilities in multicomponent space plasmas: theoretical predictions and computer simulation experiments. *Reviews of Geophysics and Space Physics*, **19**, 307–43.

Curtis, S. A. (1978). A theory for chorus generation by energetic electrons during substorms. *Journal of Geophysical Research*, **83**, 3841–8.

Darbyshire, A. G., Gershuny, E. J., Jones, S. R., Norris, A. J., Thompson, J. A., Whitehurst, G. A., Wilson G. A. & Woolliscroft, L. J. C. (1985). The UKS wave experiment. *IEEE Transactions on Geoscience and Remote Sensing*, GE-**23**, 311–14.

Das, I. M. L. & Singh, R. P. (1982). Effect of dc electric field on whistler-mode propagation. *Journal of Geophysical Research*, **87**, 2369–76.

Davidson, R. C. (1972). *Methods in Nonlinear Plasma Theory*. New York: Academic Press.

De Jagher, P. C. & Sluijter, F. W. (1987). Three-pole approximations for the plasma dispersion function. *Plasma Physics*, **29**, 677–8.

Derfler, H. (1967). Growing wave and instability criteria for hot plasmas. *Physics Letters*, **24A**, 763–4.

Derfler, H. (1969). The frequency cusp, a new means for discriminating between growing waves and instabilities in hot plasmas. In *Ninth International Conference on Phenomena in Ionized Gases*, Bucharest, Romania, ed. G. Musa *et al.*, p. 431.

Dowden, R. L. (1961). Simultaneous observations of VLF noise (hiss) at Hobart and Macquarie Island. *Journal of Geophysical Research*, **66**, 1587–8.

Dowden, R. L. & Allcock, G. McK. (1971). Determination of nose frequency of non-nose whistlers. *Journal of Atmospheric and Terrestrial Physics*, **33**, 1125–9.

Dysthe, K. B. (1971). Some studies of triggered whistler emissions. *Journal of Geophysical Research*, **76**, 6915–31.

Etcheto, J. & Gendrin, R. (1970). About the possibility of VLF Čerenkov emission in the ionosphere by artificial electron beams. *Planetary and Space Science*, **18**, 777–84.

Etcheto, J., Gendrin, R., Solomon, J. & Roux A. (1973). A self-consistent theory of magnetospheric ELF hiss. *Journal of Geophysical Research*, **78**, 8150–66.

Fälthammar, C.-G. (1973). Motion of charged particles in the magnetosphere. In *Cosmical Geophysics*, ed. A. Egeland, Ø. Holter & A. Omholt, pp. 121–42. Oslo: Universitetsforlaget.

Farrugia, C. J., Rijnbeek, R. P., Saunders, M. A., Southwood, D. J., Rodgers, D. J., Smith, M. F., Chaloner, D. S., Hall, D. S., Christiansen, P. J. & Woolliscroft, L. J. C. (1988). A multi-instrument study of flux transfer event structure. *Journal of Geophysical Research*, **93**, 14,465–77.

Fried, B. D. & Conte, S. D. (1961). *The Plasma Dispersion Function*. New York: Academic Press.

Fried, B. D., Hedrick, C. L. & McCune, J. (1968). Two-pole approximation for the plasma dispersion function. *Physics of Fluids*, **11**, 249–52.

Gail, W. B., Inan, U. S., Helliwell, R. A. & Carpenter, D. L. (1990). Gyroresonant wave–particle interactions in a dynamic magnetosphere. *Journal of Geophysical Research*, **95**, 15,103–12.

Galeev, A. A. & Sagdeev, R. Z. (1979). Nonlinear plasma theory. *Reviews of Plasma Physics*, ed. M. A. Leontovich (English translation), **7**, 1–180. New York: Consultants Bureau.

Galeev, A. A., Sagdeev, R. Z., Shapiro, V. D. & Shevchenko, V. I. (1980). Is renormalization necessary in the quasilinear theory of Langmuir oscillations? *Soviet Physics JETP*, **52**, 1095–9.

Gendrin, R. (1960). Le guidage des sifflements radioélectriques par le champ magnétique terrestre. *Comptes rendus hebdomadaires des séances de l'Académie des sciences*, **251**, 1085–7.

Gendrin, R. (1981). General relationships between wave amplification and particle diffusion in a magnetoplasma. *Reviews of Geophysics*, **19**, 171–84.

Ginzburg, V. L. (1970). *The Propagation of Electromagnetic Waves in Plasma*. Oxford: Pergamon.

Gokhberg, M. B., Karpman, V. I. & Pokhotelov, O. A. (1972). On the nonlinear theory of Pc–1 evolution. *Reports of the Academy of Science of the USSR*, **204**, 848–50 (in Russian).

Gokhberg, M. B., Morgunov, B. A. & Pokhotelov, O. A. (1988). *Seismo-electromagnetic Phenomena*. Moscow: Nauka (in Russian).

Gurnett, D. A. & Frank, L. A. (1972). VLF hiss and related plasma observations in the polar magnetosphere. *Journal of Geophysical Research*, **77**, 172–90.

Gurnett, D. A., Kurth, W. S., Cairns, I. H. & Granroth, L. J. (1990). Whistlers in Neptune's magnetosphere: evidence of atmospheric lightning. *Journal of Geophysical Research*, **95**, 15,967–76.

Guthart, H. (1965). Nose whistler dispersion as a measure of magnetosphere electron temperature. *Radio Science*, **69D**, 1417–24.

Hamar, D., Tarcsai, Gy., Lichtenberger, J., Smith, A. J. & Yearby, K. H. (1990). Fine structure of whistlers recorded digitally at Halley, Antarctica. *Journal of Atmospheric and Terrestrial Physics*, **52**, 801–10.

Harang, L. (1968). VLF emissions observed at stations close to the auroral zone and at stations on lower latitudes. *Journal of Atmospheric and Terrestrial Physics*, **30**, 1143–60.

Hasegawa, A. (1975). *Plasma Instabilities and Nonlinear Effects*. Berlin: Springer-Verlag.

Hashimoto, K., Kimura, I. & Kumagai, H. (1977). Estimation of electron temperature by VLF waves propagation in the direction of the resonance cone. *Planetary and Space Science*, **25**, 871–8.

Hayakawa, M. (1989a). Further study of the frequency drift of dawnside mid-latitude VLF emissions associated with magnetic disturbances. *Planetary and Space Science*, **37**, 269–81.

Hayakawa, M. (1989b). Satellite observation of low-latitude VLF radio noises and their association with thunderstorms. *Journal of Geomagnetism and Geoelectricity*, **41**, 573–95.

Hayakawa, M., Parrot, M. & Lefeuvre, F. (1972). The wave normal of ELF hiss emissions observed on board GEOS–1 at the equatorial and off-equatorial regions of the plasmasphere. *Journal of Geophysical Research*, **77**, 172–90.

Hayakawa, M., Tanaka, Y., Sazhin, S. S., Okada, T. & Kurita, K. (1986). Characteristics of dawnside mid-latitude VLF emissions associated with substorms as deduced from the two-stationed direction finding measurements. *Planetary and Space Science*, **34**, 225–43.

Hayakawa, M., Tanaka, Y., Sazhin, S. S., Tixier, M. & Okada, T. (1988). Substorm-associated VLF emissions with frequency drift observed in the premidnight sector. *Journal of Geophysical Research*, **93**, 5685–700.

Hayes, J. (1961). Damping of plasma oscillations in the linear theory. *Physics of Fluids*, **4**, 1387–92.

Hayes, J. (1963). On non-Landau damped solutions to the linearised Vlasov equation. *Nuovo Cimento*, **30**, 1048–63.

Helliwell, R. A. (1965). *Whistlers and Related Ionospheric Phenomena.* Stanford, California: Stanford University Press.

Hess, W. N. (1968). *The Radiation Belt and Magnetosphere.* Waltham, Massachusetts: Blaisdell.

Ho, D. & Bernard, L. C. (1973). A fast method to determine the nose frequency and minimum group delay of whistlers when the causative spheric is unknown. *Journal of Atmospheric and Terrestrial Physics,* **35**, 881–7.

Holter, Ø. & Kildal, A. (1973). Waves in Plasma. In *Cosmical Geophysics,* ed. A. Egeland, Ø. Holter & A. Omholt, pp. 247–65. Oslo: Universitetsforlaget.

Horne, R. B. (1989). Path integrated growth of electrostatic waves: the generation of terrestrial myriametric radiation. *Journal of Geophysical Research,* **94**, 8895–909.

Horne, R. B. & Sazhin, S. S. (1990). Quasielectrostatic and electrostatic approximations for whistler-mode waves in the magnetospheric plasma. *Planetary and Space Science,* **38**, 311–18.

Jacquinot, J. & Leloup, C. (1971). Electron cyclotron electromagnetic instabilities in weakly relativistic plasma. *Physics of Fluids,* **14**, 2440–6.

Jancel, R. & Wilhelmsson, H. (1991). Quasi-linear parabolic equations: some properties of physical significance. *Physica Scripta,* **43**, 393–415.

Jørgensen, T. S. (1966). Morphology of VLF hiss zones and their correlation with particle precipitation events. *Journal of Geophysical Research,* **71**, 1367–75.

Kadomtsev, B. B. (1965) *Plasma Turbulence.* New York: Academic Press.

Kaiser, T. R. & Bullough, K. (1975). VLF and ELF emissions. *Annales de Géophysique,* **31**, 137–41.

Kaner, E. A. & Skobov, V. G. (1971). *Plasma Effects in Metals: Helicon and Alfvén Waves.* London: Taylor & Francis Ltd.

Karpman, V. I. (1974). Nonlinear effects in the ELF waves propagating along the magnetic field in the magnetosphere. *Space Science Reviews,* **16**, 361–88.

Karpman, V. I., Alekhin, Ju. K., Borisov, N. D. & Rjabova, N. A. (1973). Electrostatic waves with frequencies exceeding the gyrofrequency in the magnetosphere. *Astrophysics and Space Science,* **22**, 267–78.

Karpman, V. I., Istomin, Ja. I. & Shklyar, D. R. (1974). Nonlinear frequency shift and self-modulation of the quasi-monochromatic whistlers in the inhomogeneous plasma (magnetosphere). *Planetary and Space Science,* **22**, 859–71.

Kennel, C. F. (1966). Low frequency whistler mode. *Physics of Fluids,* **9**, 2190–202.

Kennel, C. F. & Petschek, H. E. (1966). Limit on stably trapped particle fluxes. *Journal of Geophysical Research,* **71**, 1–28.

Kimura, I. & Matsuo, T. (1982). Wave normal direction of auroral hiss observed by the S-310JA-5 rocket. *Memoirs of the National Institute of Polar Research, Tokyo, Special Issue* No. 22, pp. 185–95.

Kobelev, V. V. & Sazhin, S. S. (1983). Estimate of the temperature of magnetospheric electrons from the shape of whistler spectrograms. *Soviet Technical Physics Letters,* **9**, 369–70.

Koons, H. C. & McPherron, D. A. (1972). Observation of very low frequency whistler-mode waves in the region of the radiation belt slot. *Journal of Geophysical Research,* **77**, 3475–85.

Korn, G. A. & Korn, T. M. (1968). *Mathematical Handbook for Scientists and Engineers.* New York: McGraw-Hill Book Company.

Krivenski, V. & Orefice, A. (1983). Weakly relativistic dielectric tensor and dispersion functions of a Maxwellian plasma. *Journal of Plasma Physics*, **30**, 125–31.

LaBelle, J. & Treumann, R. A. (1988). Plasma waves at the dayside magnetopause. *Space Science Reviews*, **47**, 175–202.

LaBelle, J., Treumann, R. A., Haerendel, G., Bauer, O. H., Paschmann, G., Baumjohann, W., Lühr, H., Anderson, R. R., Koons, H. C. & Holzworth, R. H. (1987). AMPTE IRM observations of waves associated with flux transfer events in the magnetosphere. *Journal of Geophysical Research*. **92**, 5827–43.

Lacina, J. (1972) Longitudinal waves in plasma and physical mechanism of Landau damping. *Plasma Physics*, **14**, 605–16.

Landau, L. D. (1946). On the vibrations of the electronic plasma. *Journal of Physics*, **10**, 25–34.

Laval, G. & Pesme, D. (1983). Inconsistency of quasilinear theory. *Physics of Fluids*, **26**, 66–8.

Lee, J. C. & Crawford, F. W. (1970). Stability analysis of whistler amplification. *Journal of Geophysical Research*, **75**, 85–96.

Liemohn, H. B. & Scarf, F. L. (1962a). Exospheric electron temperatures from nose whistler attenuation. *Journal of Geophysical Research*, **67**, 1785–9.

Liemohn, H. B. & Scarf, F. L., (1962b). Whistler attenuation by electrons with an $E^{-2.5}$ distribution. *Journal of Geophysical Research*, **67**, 4163–7.

Liemohn, H. B. & Scarf, F. L. (1964). Whistler determination of electron energy and density distribution in the magnetosphere. *Journal of Geophysical Research*, **69**, 883–904.

Lifshitz, E. M. & Pitaevsky, L. P. (1979). *Physical Kinetics*. Moscow: Nauka Publishing House (in Russian).

Likhter, Ja. I. (1979). ELF and VLF noise intensity and spectra in the magnetosphere. In *Wave Instabilities in Space Plasma*, ed. P. J. Palmadesso & K. Papadopoulos, pp. 3–19. Dordrecht, Boston & London: D. Reidel.

Likhter, Ja. I., Molchanov, O. A. & Chmyrev, V. M. (1971). Modulation of spectrum and amplitudes of low-frequency signal in the magnetospheric plasma. *Soviet Physics JETP, Letters to the editor*, **14**, 325–7.

Likhter, Ja. I. & Sazhin S. S. (1980). On the frequency shift in modulated VLF emissions. *Journal of Atmospheric and Terrestrial Physics*, **42**, 381–4.

Lutomirski, R. F. (1970). Physical model of cyclotron damping. *Physics of Fluids*, **13**, 149–53.

McCabe, J. H. (1984). Continued fraction expansions for the plasma dispersion function. *Journal of Plasma Physics*, **32**, 479–85.

McChesney, J. & Hughes, A. R. W. (1983). Temperatures in the plasmasphere determined from VLF observations. *Journal of Atmospheric and Terrestrial Physics*, **45**, 33–9.

Mann, G. & Baumgärtel, K. (1988). Coronal magnetic field strengths determined from fiber bursts. In *Proceedings of an International Workshop on Reconnection in Space Plasma*, Potsdam, vol. 2, pp. 153–5.

Martín, P., Donoso, G. & Zamudio-Cristi, J. (1980). A modified asymptotic Padé method. Application to multiple approximation for the plasma dispersion function Z. *Journal of Mathematical Physics*, **21**, 280–5.

Martín, P. & González, M. A. (1979). New two-pole approximation for the plasma dispersion function Z. *Physics of Fluids*, **22**, 1413–4.

Melrose, D. B. (1986). *Instabilities in Space and Laboratory Plasmas.* Cambridge University Press.

Menietti, J. D., Tsintikidis, D., Gurnett, D. A. & Curran, D. B. (1991). Modeling of whistler ray paths in the magnetosphere of Neptune. *Journal of Geophysical Research,* **96,** Supplement, 19,117–22.

Mikhailovskii, A. B. (1975). Electromagnetic instabilities in a non-Maxwellian plasma. *Reviews of Plasma Physics,* ed. M. A. Leontovich (English translation), **6,** 77–159. New York: Consultants Bureau.

Moffett, R. J., Bailey, G. J., Quegan, S., Rippeth, Y., Samson, A. M. & Sellek, R. (1989). Modelling the ionospheric and plasmaspheric plasma. *Philosophical Transactions of the Royal Society of London,* **A328,** 255–70.

Muto, H., Hayakawa, M., Parrot, M. & Lefeuvre, F. (1987). Direction finding of half-gyrofrequency VLF emissions in the off-equatorial region of the magnetosphere and their generation and propagation. *Journal of Geophysical Research,* **92,** 7538–50.

Namikawa, T., Hamabata, H. & Tanabe, K. (1981). The effect of electron temperature anisotropy on the propagation of whistler waves. *Journal of Plasma Physics,* **26,** 83–93.

Németh, G., Ág, A. & Paris, Gy. (1981). Two sided Padé approximations for the plasma dispersion function. *Journal of Mathematical Physics,* **22,** 1192–5.

Nunn, D. (1974). A self-consistent theory of triggered VLF emissions. *Planetary and Space Science,* **22,** 349–78.

Nunn, D. (1990). The numerical simulation of VLF nonlinear wave–particle interactions in collision free plasmas using the Vlasov hybrid simulation technique. *Computer Physics Communications,* **60,** 1–25.

Nunn, D. & Sazhin, S. S. (1991). On the generation mechanism of hiss-triggered chorus. *Annales Geophysicae,* **9,** 603–13.

Ohmi, N. & Hayakawa, M. (1986). On the generation of quasi-electrostatic half-electron-gyrofrequency VLF emissions in the magnetosphere. *Journal of Plasma Physics,* **35,** 351–73.

Olver, F. W. J. (1974). *Asymptotics and Special Functions,* Chapter 4. San Diego: Academic Press.

Omura, Y., Nunn, D., Matsumoto, H. & Rycroft, M. J. (1991). A review of observational, theoretical and numerical studies of VLF triggered emissions. *Journal of Atmospheric and Terrestrial Physics,* **53,** 351–68.

Ondoh, T., Nakamura, Y., Watanabe, S. & Murakami, T. (1981). Narrow band 5 kHz hiss observed in the vicinity of the plasmapause. *Planetary and Space Science,* **29,** 65–72.

O'Neil, T. (1965). Collisionless damping of nonlinear plasma oscillations. *Physics of Fluids,* **8,** 2255–62.

Park, C. G. (1972). Methods of determining electron concentrations in the magnetosphere from nose whistlers. *Technical Report 3454-1.* Stanford, California: Radioscience Laboratory, Stanford Electronics Laboratories, Stanford University.

Park, C. G. (1982). Whistlers. In *CRC Handbook of Atmospherics,* vol. 2, ed. H. Volland, pp. 21–77. Boca Raton, Florida: CRC Press.

Petviashvili, V. I. & Pokhotelov, O. A. (1991). *Solitary Waves in Plasma and Atmosphere.* London: Gordon & Breach.

Poppe, G. P. M. & Wijers, C. M. J. (1990). More efficient computation of the complex error function. *ACM (Association of Computing Machinery) Transactions on Mathematical Software*, **16**, 38–46.

Pritchett, P. L. (1984). Relativistic dispersion, the cyclotron maser instability, and auroral kilometric radiation. *Journal of Geophysical Research*, **89**, 8957–70.

Ratcliffe, J. A. (1959). *The Magneto-ionic Theory and its Applications to the Ionosphere*. Cambridge University Press.

Rawer, K. & Suchy, K. (1976). Remarks concerning the dispersion equation of electromagnetic waves in a magnetized cold plasma. *Journal of Atmospheric and Terrestrial Physics*, **38**, 395–8.

Robinson, P. A. (1986). Relativistic plasma dispersion functions. *Journal of Mathematical Physics*, **27**, 1206–14.

Robinson, P. A. (1987a). Relativistic plasma dispersion functions: series, integrals, and approximations. *Journal of Mathematical Physics*, **28**, 1203–5.

Robinson, P. A. (1987b). Thermal effects on parallel-propagating electron cyclotron waves. *Journal of Plasma Physics*, **37**, 149–62.

Robinson, P. A. & Newman, D. L. (1988). Approximation of the dielectric properties of Maxwellian plasmas: dispersion functions and physical constants. *Journal of Plasma Physics*, **40**, 553–66.

Romanov, Yu. A. & Filippov, G. F. (1961). The interaction of fast electron beams with longitudinal plasma waves. *Soviet Physics JETP*, **13**, 87–92.

Rönnmark, K. (1990). Quantitative methods for waves in space plasma. *Space Science Reviews*, **54**, 1–73.

Roux, A. & Pellat, R. (1978). A theory of triggered emissions. *Journal of Geophysical Research*, **83**, 1433–41.

Roux, A. & Solomon, J. (1970). Mécanismes non linéaires associés aux interactions ondes–particules dans la magnétosphère. *Annales de Géophysique*, **26**, 279–97.

Roux, A. & Solomon, J. (1971). Self-consistent solution of the quasilinear theory: application to the spectral shape and intensity of VLF waves in the magnetosphere. *Journal of Atmospheric and Terrestrial Physics*, **33**, 1457–71.

Rowlands, J., Shapiro, V. D. & Shevchenko, V. I. (1966). Quasilinear theory of plasma cyclotron instability. *Soviet Physics JETP*, **23**, 651–60.

Rycroft, M. J. (1972). VLF emissions in the magnetosphere. *Radio Science*, **7**, 811–30.

Rycroft, M. J. (1975). A review of *in situ* observations of the plasmapause. *Annales de Géophysique*, **31**, 1–16.

Rycroft, M. J. (1991). Interactions between whistler-mode waves and energetic electrons in the coupled system formed by the magnetosphere, ionosphere and atmosphere. *Journal of Atmospheric and Terrestrial Physics*, **53**, 849–58.

Rycroft, M. J. & Mathur, A. (1973). The determination of the minimum group delay of a non-nose whistler. *Journal of Atmospheric and Terrestrial Physics*, **35**, 2177–82.

Saenz, A. W. (1965). Long-term behavior of the electric potential and stability in the linearized Vlasov theory. *Journal of Mathematical Physics*, **6**, 859–75.

Sagdeev, R. Z. & Galeev A. A. (1969). *Nonlinear Plasma Theory*. New York: W. A. Benjamin, Inc.

Sagredo, J. L. & Bullough, K. (1972). The effect of the ring current on whistler propagation in the magnetosphere. *Planetary and Space Science*, **20**, 731–46.

Sagredo, J. L., Smith, I. D. & Bullough, K. (1973). The determination of whistler nose-frequency and minimum delay and its implication for the measurement of the east–west electric field and tube content in the magnetosphere. *Journal of Atmospheric and Terrestrial Physics*, **35**, 2035–46.

Sato, M. (1984). Transformation approximation for the plasma dispersion function and application to electrostatic waves. *Journal of Plasma Physics*, **31**, 325–31.

Sazhin, S. S. (1976). VLF emissions in the Earth's magnetosphere. In *Geomagnetic Research*, No. 18, ed. O. A. Troshichev, pp. 24–53. Moscow: Nauka (in Russian).

Sazhin, S. S. (1982a). *Natural Radio Emissions in the Earth's Magnetosphere*. Moscow: Nauka (in Russian).

Sazhin, S. S. (1982b). A physical model for oblique whistler-mode instabilities. *Annales de Géophysique*, **38**, 111–18.

Sazhin, S. S. (1983a). Whistler-mode propagation at frequencies near the electron gyrofrequency. *Journal of Plasma Physics*, **29**, 217–22.

Sazhin, S. S. (1983b). Diagnostic possibilities of natural radio emissions. In *Magnetospheric Research*, No. 2, ed. M. I. Pudovkin, pp. 5–27. Moscow, Geophysical Committee (in Russian).

Sazhin, S. S. (1984). A model for hiss-type mid-latitude VLF emissions. *Planetary and Space Science*, **31**, 487–93.

Sazhin, S. S. (1985). Whistler-mode polarization in a hot anisotropic plasma. *Journal of Plasma Physics*, **34**, 213–26.

Sazhin, S. S. (1986a). On whistler-mode group velocity. *Annales Geophysicae*, **4**, 155–60.

Sazhin, S. S. (1986b). Quasielectrostatic wave propagation in a hot anisotropic plasma. *Planetary and Space Science*, **34**, 497–509.

Sazhin, S. S. (1987a). An approximate theory of electromagnetic wave propagation in a weakly relativistic plasma. *Journal of Plasma Physics*, **37**, 209–30.

Sazhin, S. S. (1987b). A physical model of parallel whistler-mode propagation in a weakly relativistic plasma. *Journal of Plasma Physics*, **38**, 301–7.

Sazhin, S. S. (1987c). A model for day-time ELF emissions. *Planetary and Space Science*, **35**, 139–43.

Sazhin, S. S. (1987d). A kinetic model of parallel electric field in the magnetosphere. *Annales Geophysicae*, **5A**, 273–80 (Erratum: **6A**(1), 139; 1988).

Sazhin, S. S. (1987e). Quasilinear models of oblique whistler-mode instabilities. *Planetary and Space Science*, **35**, 753–8.

Sazhin, S. S. (1987f). An analytical model of quasiperiodic ELF–VLF emissions. *Planetary and Space Science*, **35**, 1267–74.

Sazhin, S. S. (1988a). Oblique whistler-mode growth and damping in a hot anisotropic plasma. *Planetary and Space Science*, **36**, 663–7.

Sazhin, S. S. (1988b). An improved quasilongitudinal approximation for whistler-mode waves. *Planetary and Space Science*, **36**, 1111–19.

Sazhin, S. S. (1988c). An extrapolation of the solution of the parallel whistler-mode dispersion equation. *Astrophysics and Space Science*, **145**, 163–6.

Sazhin, S. S. (1988d). Almost-parallel electromagnetic wave propagation at frequencies near the electron plasma frequency. *Astrophysics and Space Science*, **145**, 377–80.

Sazhin, S. S. (1988e). An improved quasielectrostatic approximation. *Planetary and Space Science*, **36**, 123–4.

Sazhin, S. S. (1988f). On the polarization of quasielectrostatic whistler-mode waves in the magnetospheric plasma. *Annales Geophysicae*, **6**, 177–9.

Sazhin, S. S. (1989a). Effects of ions and finite electron density on quasi-electrostatic whistler-mode propagation. *Astrophysics and Space Science*, **158**, 107–15.

Sazhin, S. S. (1989b). A physical model of quasi-electrostatic whistler-mode propagation. *Astrophysics and Space Science*, **161**, 171–4.

Sazhin, S. S. (1989c). Parallel whistler-mode propagation in a weakly relativistic plasma. *Physica Scripta*, **40**, 114–16.

Sazhin, S. S. (1989d). Approximate methods of the solution of the parallel whistler-mode dispersion equation. *Planetary and Space Science*, **37**, 311–14.

Sazhin, S. S. (1989e). Improved quasilinear models of parallel whistler-mode instability. *Planetary and Space Science*, **37**, 633–47.

Sazhin, S. S. (1990a). Storey angle for whistler-mode waves. *Planetary and Space Science*, **38**, 327–31.

Sazhin, S. S. (1990b). A new approximate solution of the parallel whistler-mode dispersion equation. *Astrophysics and Space Science*, **172**, 235–47.

Sazhin, S. S. (1991a). Whistler-mode polarization in a rarefied plasma. *Planetary and Space Science*, **39**, 725–8.

Sazhin, S. S. (1991b). Landau damping of low frequency whistler-mode waves. *Annales Geophysicae*, **9**, 690–5.

Sazhin, S. S. (1992). The propagation of damped or growing whistler-mode waves. *Planetary and Space Science*, **40**, 985–8.

Sazhin, S. S., Balmforth, H. F., Moffett, R. J. & Rippeth, Y. (1992). Modified models of electron distribution in the magnetosphere at $L = 2.3$. *Planetary and Space Science*, **40**, 671–9.

Sazhin, S. S., Bullough, K., Smith, A. J. & Saxton, J. M. (1991). On the influence of the ring current on whistler group delay time in the magnetosphere. *Annales Geophysicae*, **9**, 21–9.

Sazhin, S. S. & Hayakawa, M (1992). Magnetospheric chorus emissions: a review. *Planetary and Space Science*, **40**, 681–97.

Sazhin, S. S., Hayakawa, M. & Bullough, K. (1992). Whistler diagnostics of magnetospheric parameters: a review. *Annales Geophysicae*, **10**, 293–308.

Sazhin, S. S. & Horne, R. B. (1990). Quasi-longitudinal approximation for whistler-mode waves in the magnetospheric plasma. *Planetary and Space Science*, **38**, 1551–3.

Sazhin, S. S., Ponyavin, D. I. & Varshavski, S. P. (1979). Some features of whistler propagation in magnetospheric plasma. *Radiophysics and Quantum Electronics*, **22**, 547–50.

Sazhin, S. S. & Sazhina, E. M. (1988). Some particular cases of oblique whistler-mode propagation in a hot anisotropic plasma. *Journal of Plasma Physics*, **40**, 69–85.

Sazhin, S. S., Smith, A. J., Bullough, K., Clilverd, M. A., Saxton, J. M., Strangeways, H. J. & Tarcsai, Gy. (1992). Group delay times of whistler-mode signals from VLF transmitters observed at Faraday, Antarctica. *Journal of Atmospheric and Terrestrial Physics*, **54**, 99–107.

Sazhin, S. S., Smith, A. J. & Sazhina, E. M. (1990). Can magnetospheric electron temperature be inferred from whistler dispersion measurements? *Annales Geophysicae*, **8**, 273–85.

Sazhin, S. S. & Strangeways, H. J. (1989). Ray tracing in inhomogeneous plasma. *Planetary and Space Science*, **37**, 339–47.

Sazhin, S. S., Sumner, A. E. & Temme, N. M. (1992). Relativistic and non-relativistic analysis of whistler-mode waves in a hot anisotropic plasma. *Journal of Plasma Physics*, **47**, 163–74.

Sazhin, S. S. & Temme, N. M. (1990). Relativistic effects on parallel whistler-mode propagation and instability. *Astrophysics and Space Science*, **166**, 301–13.

Sazhin, S. S. & Temme, N. M. (1991a). The threshold of parallel whistler-mode instability. *Annales Geophysicae*, **9**, 30–3.

Sazhin, S. S. & Temme, N. M. (1991b). Marginal stability of parallel whistler-mode waves (asymptotic analysis). *Annales Geophysicae*, **9**, 304–8 (Erratum: p. 500).

Sazhin, S. S. & Walker, S. N. (1989). Marginal stability of oblique whistler-mode waves. *Planetary and Space Science*, **37**, 223–7.

Sazhin, S. S., Walker, S. N. & Woolliscroft, L. J. C. (1990a). On spin-modulation diagnostics of whistler-mode wave normal angles in the vicinity of the Earth's magnetopause. *Planetary and Space Science*, **38**, 333–9.

Sazhin, S. S., Walker, S. N. & Woolliscroft, L. J. C. (1990b). Oblique whistler-mode waves in the presence of electron beams. *Planetary and Space Science*, **38**, 791–805.

Sazhin, S. S., Walker, S. N. & Woolliscroft, L. J. C. (1990c). On whistler-mode wave trapping in the vicinity of the Earth's magnetopause. *Annales Geophysicae*, **8**, 583–9.

Sazhin, S. S., Walker, S. N. & Woolliscroft, L. J. C. (1991). Observations and theory of whistler-mode waves in the vicinity of the Earth's magnetopause. *Advances in Space Research*, **11**(9), 33–6.

Scarf, F. L. (1962). Landau damping and the attenuation of whistlers. *Physics of Fluids*, **5**, 6–13.

Seely, N. T. (1977). Whistler propagation in a distorted quiet-time model magnetosphere. *Technical Report 3472-1*. Stanford, California: Radioscience Laboratory, Stanford Electronics Laboratories, Stanford University.

Serbu, G. P. & Maier, E. J. R. (1966). Low energy electrons measured on IMP2. *Journal of Geophysical Research*, **71**, 3755–66.

Shah, H. M., Hall, D. S. & Chaloner, C. P. (1985). The electron experiment on the AMPTE UKS. *IEEE Transactions on Geoscience and Remote Sensing*, GE-**23**, 293–300.

Shkarofsky, I. P. (1966). Dielectric tensor in Vlasov plasmas near cyclotron harmonics. *Physics of Fluids*, **9**, 561–70.

Shkarofsky, I. P. (1986). New representation of dielectric tensor elements in magnetized plasma. *Journal of Plasma Physics*, **35**, 319–31.

Smith, R. L. & Carpenter, D. L. (1961). Extension of nose whistler analysis. *Journal of Geophysical Research*, **66**, 2582–6.

Smith, A. J., Carpenter, D. L. & Lester, M. (1981). Longitudinal variations of plasmapause radius and the propagation of VLF noise within small ($\Delta L = 0.5$) extensions of the plasmasphere. *Geophysical Research Letters*, **8**, 980–3.

Smith, A. J., Smith, I. D. & Bullough, K. (1975). Methods of determining whistler nose-frequency and minimum group delay. *Journal of Atmospheric and Terrestrial Physics*, **37**, 1179–92.

Smith, A. J., Yearby, K. H. (1987). AVDAS – A microprocessor-based VLF signal acquisition, processing and spectral analysis facility for Antarctica. *British Antarctic Survey Bulletin*, **75**, 1–15.

Solomon, J., Cornilleau-Werlin N., Korth A. & Kremser G. (1988). An experimental study of ELF/VLF hiss generation in the Earth's magnetosphere. *Journal of Geophysical Research*, **93**, 1839–47.

Southwood, D. J. & Kivelson, M. G. (1975). An approximate analytic description of plasma bulk parameters and pitch angle anisotropy under adiabatic flow in a dipolar magnetospheric field. *Journal of Geophysical Research*, **80**, 2069–73.

Stix, T. (1962). *The Theory of Plasma Waves*. New York: McGraw-Hill Book Company.

Stix, T. (1990). Waves in plasma: highlights from the past and present. *Physics of Fluids*, **B2**, 1729–1743.

Storey, L. R. O. (1953). An investigation of whistling atmospherics. *Philosophical Transactions of the Royal Society of London*, **A246**, 113–41.

Storey, L. R. O., Lefeuvre, F., Parrot, M., Cairó, L. & Anderson, R. R. (1991). Initial survey of the wave distribution function for plasmaspheric hiss observed by ISEE–1. *Journal of Geophysical Research*, **96**, 19,469–89.

Strangeways, H. J. (1986). A model for the electron temperature variation along geomagnetic field lines and its effect on electron density profiles and VLF paths. *Journal of Atmospheric and Terrestrial Physics*, **48**, 671–83.

Strangeways, H. J. (1991). The upper cut-off frequency of nose whistlers and implications for duct structure. *Journal of Atmospheric and Terrestrial Physics*, **53**, 151–69.

Stuart, G. F. (1977a). Systematic errors in whistler extrapolation – 1. Linear Q analysis and the extrapolation factor. *Journal of Atmospheric and Terrestrial Physics*, **39**, 415–25.

Stuart, G. F. (1977b). Systematic errors in whistler extrapolation – 2. Comparison of methods. *Journal of Atmospheric and Terrestrial Physics*, **39**, 427–31.

Sturrock, P. A. (1958). Kinematics of growing waves. *Physical Review*, **112**, 1488–503.

Subramaniam, V. V. & Hughes, W. F. (1986). A macroscopic interpretation of Landau damping. *Journal of Plasma Physics*, **36**, 127–33.

Suchy, K. (1972). The velocity of a wave packet in an anisotropic absorbing medium. *Journal of Plasma Physics*, **8**, 33–51.

Sudan, R. N. & Ott, E. (1971). A theory of triggered VLF emissions. *Journal of Geophysical Research*, **76**, 4463–76.

Tanaka, M. (1989). Description of a wave packet propagating in anomalous dispersion media – a new expression of propagation velocity. *Plasma Physics and Controlled Fusion*, **31**, 1049–67.

Tanaka, M., Fujiwara, M. & Ikegami, H. (1986). Propagation of a Gaussian wave packet in an absorbing medium. *Physical Review A*, **34**, 4851–8.

Tarcsai, Gy. (1975). Routine whistler analysis by means of accurate curve fitting. *Journal of Atmospheric and Terrestrial Physics*, **37**, 1447–57.

Tarcsai, Gy., Strangeways, H. J. & Rycroft, M. J. (1989). Error sources and travel time residuals in plasmaspheric whistler interpretation. *Journal of Atmospheric and Terrestrial Physics*, **51**, 249–58.

Temme, N. M., Sumner, A. E. & Sazhin, S. S. (1992). Analytical and numerical analysis of the generalized Shkarofsky function. *Astrophysics and Space Science*, **194**, 173–196.

Thorne, R. M., Smith, E. J., Burton, R. K. & Holzer, R. E. (1973). Plasmaspheric hiss. *Journal of Geophysical Research*, **78**, 1581–96.

Timofeev, A. V. (1989). Cyclotron oscillations of an equilibrium plasma. *Reviews of Plasma Physics*, ed. B. B. Kadomtsev (English translation), **14**, 63–252. New York: Consultants Bureau.

Tokar, W. W. L. & Gary, S. P. (1985). The whistler mode in a Vlasov plasma. *Physics of Fluids*, **28**, 1063–8.

Tonks, L. & Langmuir, I. (1929). Oscillations in ionized gases. *Physical Review*, **33**, 195–210 (Erratum: **33**, 990).

Trocheris, M. (1965). Sur les modes normaux des oscillations de plasma. *Nuclear Fusion*, **5**, 299–314.

Tsai, S. T., Wu, C. S., Wang, Y. D. & Kang, S. W. (1981). Dielectric tensor of a weakly relativistic, nonequilibrium and magnetized plasma. *Physics of Fluids*, **24**, 2186–90.

Tsang, K. T. (1984). Electron-cyclotron maser and whistler instabilities in a relativistic electron plasma with loss cone distribution. *Physics of Fluids*, **27**, 1659–64.

Tsurutani, B. T., Brinca, A. L., Smith, E. J., Okida, R. T., Anderson R. R. & Eastman, T. E. (1989). A statistical study of ELF–VLF plasma waves at the magnetopause. *Journal of Geophysical Research*, **94**, 1270–80.

Tsurutani, B. T. & Smith, E. T. (1974). Postmidnight chorus: a substorm phenomenon. *Journal of Geophysical Research*, **79**, 118–27.

Tsurutani, B. T., Smith, E. J., Anderson, R. R., Ogilive K. W., Scudder, J. D., Baker, D. N. & Bame, S. J. (1982). Lion roars and nonoscillatory drift mirror waves in the magnetosheath. *Journal of Geophysical Research*, **87**, 6060–72.

Tsurutani, B. T., Smith, E. J., Thorne, R. M., Anderson, R. R., Gurnett, D. A., Parks, G. K., Lin, C. S. & Russel, C. T. (1981). Wave particle interaction at the magnetopause: contribution to the dayside aurora. *Geophysical Research Letters*, **8**, 183–6.

Tsytovich, V. N. (1972). *An Introduction to the Theory of Plasma Turbulence*. Oxford: Pergamon.

Vedenov, A. A. (1968). *Theory of Turbulent Plasma*. London: Iliffe Books Ltd.

Vedenov, A. A. & Ryutov, D. D. (1975). Quasilinear effects in two-stream instabilities. *Reviews of Plasma Physics*, ed. M. A. Leontovich (English translation), **6**, 1–76. New York: Consultants Bureau.

Villalón, E., Burke, W. J., Rothwell, P. L. & Silevich, M. B. (1989). Quasi-linear wave-particle interactions in the Earth's radiation belts. *Journal of Geophysical Research*, **94**, 15,243–56.

Vlasov, A. A. (1938). Vibrational properties of an electron gas. *Zhurnal Experimental'noi i Teoreticheskoi Fiziki*, **8**, 291–318 (in Russian).

Walker, A. D. M. (1992). *Plasma Waves in the Magnetosphere*. Berlin: Springer-Verlag.

Ward, A. K., Bryant, D. A., Edwards, T., Parker, D. J., O'Hea, A., Patrick, T. J., Sheather, P. H., Barnsdale, K. P. & Cruise, A. M. (1985) The AMPTE–UKS spacecraft. *IEEE Transactions on Geoscience and Remote Sensing*, GE-**23**, 202–11.

Watts, J. M. (1957). An observation of audio-frequency electromagnetic noise during a period of solar disturbances. *Journal of Geophysical Research*, **62**, 199–206.

Wharton, C. B. & Trivelpiece, A. W. (1966). Waves in laboratory plasma. In *Plasma Physics in Theory and Application*, ed. W. B. Kunkel, pp. 233–75. New York: McGraw-Hill Book Company.

Winglee, R. M. (1983). Interrelation between azimuthal bunching and semirelativistic maser cyclotron instabilities. *Plasma Physics*, **25**, 217–35.

Xu, J. S. & Yeh, K. C. (1990). Propagation of a VLF electromagnetic wave packet in a magnetoplasma. *Journal of Geophysical Research*, **95**, 10,481–93.

Yoon, P. H. & Davidson, R. C. (1990). Alternative representation of the dielectric tensor for a relativistic magnetized plasma in thermal equilibrium. *Journal of Plasma Physics*, **43**, 269–81.

Zaslavsky, G. M. (1970). *Statistical Irreversibility in Nonlinear Systems*. Moscow: Nauka (in Russian).

Zaslavsky, G. M. & Chirikov B. V. (1972). Stochastical instability of nonlinear oscillations *Soviet Physics. Uspekhi*, **14**, 549–68.

Zaslavsky, G. M. & Sagdeev R. Z. (1988). *An Introduction to Nonlinear Physics*. Moscow: Nauka (in Russian).

Solutions to the problems

Problem 1.1   The condition for $|\Re Z|$ to be maximal is $d\Re Z/d\xi = 0$. Remembering the definition of $\Re Z$ (see (1.21)) we have:

$$\frac{d\Re Z}{d\xi} = -2(1 + \xi \Re Z(\xi)).$$

As follows from this equation, $d\Re Z/d\xi = 0$ when $\Re Z(\xi) = -1/\xi$. Remembering that the maximum of $\Re Z(\xi)$ is achieved when $|\xi_0| = 0.924$ we have $|\Re Z|_{\max} = 1/0.924 \approx 1.082$.

Problem 1.2   Introducing the variable $\tau = t/\xi_0$ we obtain:

$$\Re Z(\xi_0) = \frac{-1}{\sqrt{\pi}} \int_{-\infty}^{+\infty} \frac{\exp(-\xi_0^2 \tau^2) d\tau}{1 - \tau}. \tag{1}$$

In the limit $\xi_0^2 \gg 1$ the main contribution to this integral comes from $\tau < 1$. This allows us to write:

$$(1 - \tau)^{-1} = 1 + \tau + \tau^2 + \tau^3 + \tau^4 + \tau^5 + \tau^6 + \dots . \tag{2}$$

Having substituted (2) into (1) we obtain:

$$\Re Z(\xi_0) = \frac{-1}{\sqrt{\pi}} \int_{-\infty}^{+\infty} (1 + \tau + \tau^2 + \tau^3 + \tau^4 + \tau^5 + \tau^6 + \dots) \exp(-\xi_0^2 \tau^2) d\tau$$

$$= -\frac{1}{\xi_0} - \frac{1}{2\xi_0^3} - \frac{3}{4\xi_0^5} - \frac{15}{8\xi_0^7} - \dots - \frac{1 \cdot 3 \cdot 5 \cdot \dots \cdot (2n-1)}{2^n \xi_0^{2n+1}} - \dots .$$

The first three terms in this expansion coincide with those given by (1.22). The fourth one is equal to $-15/8\xi_0^7$.

Problem 1.3

$$\begin{Vmatrix} a_{xx} & a_{xy} & a_{xz} \\ a_{yx} & a_{yy} & a_{yz} \\ a_{zx} & a_{zy} & a_{zz} \end{Vmatrix} \times \begin{Vmatrix} E_x \\ E_y \\ E_z \end{Vmatrix} = 0,$$

where:

$$a_{xx} = N^2(\sin^2\theta\cos^2\phi - 1) + \epsilon_{xx},$$
$$a_{xy} = N^2\sin^2\theta\sin\phi\cos\phi + \epsilon_{xy},$$
$$a_{xz} = N^2\sin\theta\cos\theta\cos\phi + \epsilon_{xz},$$
$$a_{yx} = N^2\sin^2\theta\sin\phi\cos\phi + \epsilon_{yx},$$
$$a_{yy} = N^2(\sin^2\theta\sin^2\phi - 1) + \epsilon_{yy},$$
$$a_{yz} = N^2\sin\theta\cos\theta\sin\phi + \epsilon_{yz},$$
$$a_{zx} = N^2\sin\theta\cos\theta\cos\phi + \epsilon_{zx},$$
$$a_{zy} = N^2\sin\theta\cos\theta\sin\phi + \epsilon_{zy},$$
$$a_{zz} = -N^2\sin^2\theta + \epsilon_{zz}.$$

Problem 1.4 Formally these expressions will remain the same except that $\Pi_{xy}^{n,\alpha}$, $\Pi_{yx}^{n,\alpha}$, $\Pi_{yz}^{n,\alpha}$, $\Pi_{zy}^{n,\alpha}$ will change their signs.

Problem 1.5 The proportionality of ϵ_{ij}^t to c^{-2} comes from the relativistic nature of the Maxwell equations, while $\epsilon_{ij}^r \sim c^{-2}$ was derived from relativistic corrections in the Vlasov equation. Hence, both corrections ϵ_{ij}^t and ϵ_{ij}^r are in fact relativistic by their nature. We call ϵ_{ij}^t non-relativistic thermal corrections merely to follow historical tradition, keeping in mind that these terms do not take into account the relativistic corrections in the Vlasov equation.

Problem 2.1

$$v_g = \frac{2c\sqrt{\nu Y^2 + Y - 1}(Y-1)^{3/2}}{\nu Y^3 + 2(Y-1)^2}.$$

In the limit $\nu \gg 1$ this expression reduces to (2.26) if we neglect the contribution of ions.

Problem 2.2 Substituting $\cos\theta = \cos\theta_{G0} = 2/Y$ into (2.41) and (2.42) we obtain $v_{g\parallel} = c/2\sqrt{\nu}$, $v_{g\perp} = 0$. Neither of these components of $\mathbf{v}_g$ depends on the wave frequency ω.

Problem 2.3 Let us rotate the coordinate system (x, y, z) around the y axis by an angle θ so that the new z axis (z') is along $\mathbf{k}$, the new x axis (x') is perpendicular to $\mathbf{k}$, and the new y axis (y') coincides with the old one. The components of $\mathbf{E}$ and $\mathbf{B}$ along the y' axis coincide with those along the y axis (see equations (2.69) and (2.73)). At the same time $\mathbf{E}$ and $\mathbf{B}$ along the x' axis are determined by the equations:

$$E_{x'} = E_x\cos\theta - E_z\sin\theta,$$

$$B_{x'} = B_x\cos\theta - B_z\sin\theta.$$

Remembering (2.70) and (2.74), these expressions for $E_{x'}$ and $B_{x'}$ can be written in a more explicit form:

$$E_{x'} = E_x \left(\cos\theta - \frac{\sin^2\theta}{Y - \cos\theta} \right) = E_x \frac{Y\cos\theta - 1}{Y - \cos\theta},$$

$$B_{x'} = B_x \left(\cos\theta + \frac{\sin^2\theta}{\cos\theta} \right) = \frac{B_x}{\cos\theta}.$$

Comparing these expressions with (2.69) and (2.73) we obtain:

$$E_{y'}/E_{x'} = B_{y'}/B_{x'} = -\mathrm{i}.$$

which proves our statement that both the electric and magnetic fields of the whistler-mode wave are circularly polarized in the plane perpendicular to **k**.

Problem 3.1

$$N^2 = N_{0l}^2 (1 + \tilde{a}_{\beta l\parallel}\beta_e),$$

where

$$N_{0l}^2 = 1 - \frac{\nu Y^2}{Y + 1},$$

$$\tilde{a}_{\beta l\parallel} = \tilde{a}_{l\parallel t} + \tilde{a}_{l\parallel R},$$

$$\tilde{a}_{l\parallel t} = \frac{Y^2[Y - A_e(1 + Y)]}{(Y + 1)^3},$$

$$\tilde{a}_{l\parallel R} = \frac{Y^2(1 + 4A_e)}{2(Y + 1 - \nu Y^2)(Y + 1)}.$$

Problem 3.2   The first relation follows from the identity:

$$\frac{b - 1}{(1 + s)^q (1 + bs)^p} = \frac{b}{(1 + s)^{q-1}(1 + bs)^p} - \frac{1}{(1 + s)^q (1 + bs)^{p-1}}.$$

The second relation follows from the following presentation of $\mathcal{F}_{q,p}$:

$$\zeta\mathcal{F}_{q,p} = -\int_0^\infty e^{-as/(1+s)} f(s)\mathrm{d}e^{-\zeta s}$$

after integrating the latter integral by parts. The third follows from the second (with p replaced by $p - 1$) and the first.

Problem 3.3   In the limit $|b| \ll 1$ we have:

$$\frac{1}{(1 - \mathrm{i}bt)^p} = \frac{1}{(1 - \mathrm{i}t)^p} + \frac{\mathrm{i}(b - 1)pt}{(1 - \mathrm{i}t)^{p+1}}.$$

In view of this equation the expression for $\mathcal{F}_{q,p}$ can be written in terms of the Shkarofsky function $\mathcal{F}_q$:

$$\mathcal{F}_{q,p} = \mathcal{F}_{q+p} + p(b-1)\frac{\mathrm{d}}{\mathrm{d}z}\mathcal{F}_{q+p+1}.$$

Substituting the latter equation into (3.10) and neglecting terms of the order of $(b-1)^2$, we obtain equation (3.13).

Problem 4.1 Let us rewrite equation (4.17) in the form:

$$\frac{(Y-1)^3\tau/\tilde{w}_\parallel^2 + [-(Y-1)+(1-A_e)(Y-1)\nu Y^2]}{\nu Y^2[A_e + (1-A_e)Y]} = Q(\tau).$$

Remembering that $Q(\tau)$ is finite (see Fig. 4.6) we can see that in the limit $A_e + (1-A_e)Y \to 0$ (metastable propagation) this equation is satisfied when the numerator in its right-hand side is equal to zero, i.e. when

$$(Y-1)^3\tau/\tilde{w}_\parallel = Y - 1 - (1-A_e)(Y-1)\nu Y^2.$$

Remembering the expression for A_e for metastable propagation (equation (4.6)) and the definition of τ (equation (4.18)), we reduce this equation to equation (2.8).

Problem 4.2 The contribution of the term $-1/2\xi_1^3$ in expansion (4.3) has the same order of magnitude with respect to $|\xi_1| \gg 1$ as the contribution of the term $-1/\xi_1$ in the same expansion if we substitute this expansion into expression (4.25).

Problem 5.1 Remembering the definitions of $\beta_\parallel$, $\beta_\perp$ and β_e we have $\beta_\parallel = 2\beta_e$ and $\beta_\perp = 2A_e\beta_e$. Neglecting the contribution of quadratic terms with respect to β_e we can simplify the dispersion equation of Namikawa, Hamabata & Tanabe (1981) to:

$$\left(1 + \frac{N^2}{X}\right)^2$$
$$= \frac{Y^2 N^4}{X^2}\cos^2\theta\left[1 + A_e\beta_e + 2\beta_e\left(A_e\beta_e - 4\beta_e\right)\cos^2\theta + \frac{N^2}{X}3A_e\beta_e\cos^2\theta\right].$$
$$(1)$$

Assuming condition (5.1) is valid we can write:

$$N^2 = N_0^2(1+\varpi),\qquad\qquad (2)$$

where N_0 is the whistler-mode refractive index in a cold plasma, and $|\varpi| \ll 1$. Substituting (2) into (1), keeping only linear terms with respect to β_e and ϖ, and restricting our analysis to low-frequency waves in a dense plasma

$(Y \gg 1, X \gg 1)$, when $N_0^2 = N_{0d}^2 = X/(Y \cos \theta)$ (θ is not close to $\pi/2$), we obtain:

$$\varpi = -\frac{\beta_e}{2} \left[A_e + 2 + (A_e - 4) \cos^2 \theta \right]. \qquad (3)$$

This expression for ϖ is equal to $\tilde{a}_\beta \beta_e$ in equation (5.16) taken in the limit $Y \gg 1$.

Problem 5.2 The focusing of the waves at $Y^{-1} \approx 0.5$ and $\nu = 0.25$ in a plasma with low β_e results from $\tilde{a}_\theta \to \infty$ at these Y^{-1} and ν and thus could not be predicted in a cold plasma theory. Note that when $Y^{-1} = 0.5$ and $\nu = 0.25$ then $\tilde{a}_\theta \beta_e \theta^2 = \infty$ for any non-zero β_e and θ. Hence, for these values our theory is no longer valid.

Problem 6.1 When $\beta_e \ll 1$ and $a_1^t \neq 0$ then the electrostatic equation $A = 0$ reduces to $a_1^t \beta_e N^2 + A_{00} \theta' = 0$ (see equation (6.1)). At $\theta' = \theta - \theta_R < 0$ this equation is not satisfied for any $N^2 > 0$, because $a_1^t < 0$. In the electrostatic equation for θ in the immediate vicinity of θ_R the term A_0 is of the same order of magnitude as or less than $a_1^t \beta_e N^2$. Hence in this equation we in fact neglect the term $B_0 N^2$ when compared with $a_1^t \beta_e N^6$, which cannot be justified, in general, when β_e is small. However, at θ' greater than 0, but not in the immediate vicinity of 0, N^2 strongly increases with increasing θ'. As A_0 increases with increasing θ' as well ($A_0 = A_{00} \theta'$) we can expect that the contribution of the term $B_0 N^2$ becomes less important when compared with the term $A_0 N^4$. This is just what follows from the numerical analysis of Horne & Sazhin (1990) (cf. their Figs. 1–3 and 11–12).

Problem 6.2 Having substituted expressions (1.91), (1.94) and (1.96) into (1.43) we write the electrostatic dispersion equation ($A = 0$) as:

$$1 + X \left[(A_e - 1) \sin^2 \theta - \frac{4 \sin^2 \theta}{\alpha^2 j! w_\perp^{2j+2}} \sum_{n=-\infty}^{n=+\infty} n^2 \mu_j^n Z(\xi_n) \right.$$

$$\left. + \frac{4 w_\| \sin \theta \cos \theta}{\alpha j! w_\perp^{2j+2}} \sum_{n=-\infty}^{n=+\infty} n \mu_j^n Z'(\xi_n) + \frac{2 w_\|^2 \cos^2 \theta}{j! w_\perp^{2j+2}} \sum_{n=-\infty}^{n=+\infty} \xi_n \mu_j^n Z'(\xi_n) \right] = 0.$$

$$(1)$$

Then using the expansion (Abramovitz & Stegun, 1964)

$$\exp(z \cos \theta) = \sum_{n=-\infty}^{n=+\infty} I_n(z) \cos(n\theta)$$

and its derivatives with respect to θ, we can simplify equation (1) to:

$$1 + \frac{2X}{N^2 \tilde{w}_\parallel^2} + \frac{2X}{j! w_\perp^{2j+2}} \sum_{n=-\infty}^{n=+\infty} \mu_j^n Z(\xi_n) \left(-\frac{2 \sin^2 \theta}{\alpha^2} n^2 \right.$$

$$\left. - \frac{4 \sin \theta \cos \theta}{\alpha} n w_\parallel \xi_n - 2w_\parallel^2 \cos^2 \theta \xi_n^2 \right) = 0. \tag{2}$$

Remembering the definitions of α and ξ_n (see equations (1.91)–(1.96)) we obtain:

$$\frac{n \sin \theta}{\alpha} + w_\parallel \cos \theta \xi_n = \frac{\omega}{k}. \tag{3}$$

In view of (3), equation (2) reduces to equation (6.70). Note that equation (6.70) is equivalent to the corresponding equation derived by Karpman *et al.* (1973) (see their equation (2.2)).

Problem 7.1   Remembering that in the limit $|A_e - 1| \ll 1$ the dispersion equation for parallel whistler-mode propagation in a weakly relativistic plasma can be written in the form (3.13), we can write the condition for marginal stability of the waves as:

$$\Im F_{5/2} + \Delta A_e (2 - N^2) \left(\Im F_{7/2} - \Im F_{5/2} \right) = 0, \tag{1}$$

where $\Delta A_e = A_e - 1$ and $|\Delta A_e| \ll 1$. Substituting (7.135) into condition (1) we reduce it to:

$$\Delta A_e = \frac{1}{\left(1 - \sqrt{1 - \mu} I_{5/2}/I_{3/2} \right) (2 - N^2)}. \tag{2}$$

In a weakly relativistic plasma when $\tilde{r} \ll 1$, $a \gg 1$, $|z| \gg 1$ we have the argument of Bessel functions $2\sqrt{a}\sqrt{a-2} \gg 1$ and equation (2) further simplifies to:

$$A_e = 1 + \frac{1}{(N^2 - 2)\left(\sqrt{1-\mu} - 1\right)}. \tag{3}$$

As follows from (3), the condition $|\Delta A_e| \ll 1$ is satisfied when $N^2 \gg 1$. Hence we can neglect 2 when compared with N^2 in equation (3) and reduce the latter to (7.26).

Problem 7.2   The change of energy of an individual electron under the influence of $\mathbf{E}_\parallel$ averaged over an electron gyroperiod is described by the following equation:

$$\frac{d}{dt} \left(\frac{m_e v_\parallel}{2} \right) = v_{\parallel 1} \frac{d}{dt} m_e v_{\parallel 1} + v_{\parallel 0} \frac{d}{dt} m_e v_{\parallel 2} + \dots, \tag{1}$$

where $v_{\|0}$, $v_{\|1}$ and $v_{\|2}$ are zero- , first- and second-order electron velocities $v_\|$. The expression for $v_{\|1}$ can be obtained from the equation:

$$\frac{dv_{\|1}}{dt} = \frac{e|\mathbf{E}_{\|0}|}{m_e} \cos\left(\omega t - k_\| v_{\|0} t + \varsigma_\| - \lambda_e \sin \Omega t\right), \tag{2}$$

where we used equation (7.109) for $\mathbf{r}$ and assumed that $k_\perp$ lies in the (x,z) plane, $\mathbf{r}_0 = 0$ and $\varsigma_e = 0$; $\lambda_e \equiv k_\perp v_{\perp 0}/\Omega$. Equation (2) can be rearranged to

$$\frac{dv_{\|1}}{dt} = \frac{e|\mathbf{E}_{\|0}|}{m_e} \Big[\cos\left(\omega t - k_\| v_{\|0} t + \varsigma_\|\right) \cos\left(\lambda_e \sin \Omega t\right)$$

$$- \sin\left(\omega t - k_\| v_{\|0} t + \varsigma_\|\right) \sin\left(\lambda_e \sin \Omega t\right)\Big]. \tag{3}$$

Remembering our assumption that $|\lambda_e| \ll 1$ and using the following expansions:

$$\cos\left(\lambda_e \sin \Omega t\right) = J_0 + 2J_2 \cos 2\Omega t + 2J_4 \cos 4\Omega t + \ldots \tag{4}$$

$$\sin\left(\lambda_e \sin \Omega t\right) = 2J_1 \cos \Omega t + 2J_3 \cos 3\Omega t + \ldots \tag{5}$$

we can simplify (3) to:

$$\frac{dv_{\|1}}{dt} = \frac{e|\mathbf{E}_{\|0}|}{m_e} \cos\left(\omega t - k_\| v_{\|0} t + \varsigma_\|\right). \tag{6}$$

For simplicity, we take $v_{\|1} = 0$ at $t = 0$ and write the solution of (6) as:

$$v_{\|1} = \frac{e|\mathbf{E}_{\|0}|}{m_e} \frac{\sin(\hat{\alpha} t + \varsigma_\|) - \sin \varsigma_\|}{\hat{\alpha}}, \tag{7}$$

where $\hat{\alpha} = \omega - k_\| v_{\|0}$. From (7) it follows that:

$$z_1 = \int_0^t v_{\|1} dt = \frac{e|\mathbf{E}_{\|0}|}{m_e} \left[\frac{-\cos(\hat{\alpha} t + \varsigma_\|) + \cos \varsigma_\|}{\hat{\alpha}^2} - \frac{t \sin \varsigma_\|}{\hat{\alpha}}\right]. \tag{8}$$

Assuming that $k_\| z_1 \ll 1$, we obtain from (6), after adding the term $k_\| z_1$ in the argument of cos:

$$\frac{dv_{\|2}}{dt} = -\frac{k_\| e^2 |\mathbf{E}_{\|0}|^2}{m_e^2} \left[\frac{-\cos(\hat{\alpha} t + \varsigma_\|) + \cos \varsigma_\|}{\hat{\alpha}^2} - \frac{t \sin \varsigma_\|}{\hat{\alpha}}\right] \sin(\hat{\alpha} t + \varsigma_\|). \tag{9}$$

Having substituted (6), (7) and (9) into (1) we obtain, after some trigonometric expansions and averaging over $\varsigma_\|$:

$$\left\langle \frac{d}{dt} \frac{m_e v_\|^2}{2}\right\rangle_{\varsigma_\|} = \frac{e^2 |\mathbf{E}_{\|0}|^2}{2m_e} \left(-\frac{\omega \sin \hat{\alpha} t}{\hat{\alpha}^2} + t \cos \hat{\alpha} t + \frac{\omega t \cos \hat{\alpha} t}{\hat{\alpha}}\right). \tag{10}$$

Multiplying both sides of (10) by $n_e f_0(v_\perp, v_\parallel)$ and integrating over the distribution of initial velocities we obtain:

$$\left\langle \frac{dW_\parallel}{dt} \right\rangle_{\varsigma_\parallel, \mathbf{v}} = \frac{\pi n_e e^2 |\mathbf{E}_{\parallel 0}|^2}{m_e} \int_0^\infty v_\perp dv_\perp \int_{-\infty}^{+\infty} dv_\parallel f_0(v_\perp, v_\parallel)$$

$$\times \left(-\frac{\omega \sin \hat{\alpha} t}{\hat{\alpha}^2} + t \cos \hat{\alpha} t + \frac{\omega t \cos \hat{\alpha} t}{\hat{\alpha}} \right), \tag{11}$$

where $W_\parallel$ is the parallel energy density of the electrons.

Replacing $f_0(v_\perp, v_\parallel)$ by $g(v_\perp, \hat{\alpha})$ as defined by (7.119), changing the variable of integration from $v_\parallel$ to $\hat{\alpha}$, and neglecting the terms vanishing at $t \to \infty$ (thus effectively averaging over $v_\parallel$), we obtain:

$$\left\langle \frac{dW_\parallel}{dt} \right\rangle_{\varsigma_\parallel, \mathbf{v}} = -\frac{\pi n_e e^2 |\mathbf{E}_{\parallel 0}|^2}{m_e |k_\parallel|} \int_0^\infty v_\perp dv_\perp \int_{-\infty}^{+\infty} \frac{g(v_\perp, \hat{\alpha}) \sin \hat{\alpha} t}{\hat{\alpha}^2} d\hat{\alpha}. \tag{12}$$

When deriving (12) we took into account a principal part of the integral over $v_\parallel$, which is possible as the integrand in (11) is well behaved at $\hat{\alpha} = 0$. Having substituted (7.121) into (12) we obtain expression (7.123) in which $W_2 \equiv W_\parallel$.

Problem 8.1   The main source of the free energy of the electrons responsible for wave generation is not in the anisotropy of injected electrons, but in the reduced number of electrons inside the loss cone when these electrons are already trapped in the inner magnetosphere (see equation (8.46)). All the equations in Section 8.1 have been derived on the assumption that $\omega \ll \Omega$ and so they cannot be applied for ω close to Ω.

Problem 8.2   Remembering the definition of $K(\kappa_0)$ we have:

$$K(1/\kappa_0) = \int_0^{\pi/2} \frac{d\Upsilon}{\sqrt{1 - (1/\kappa_0^2) \sin^2 \Upsilon}} \tag{1}$$

for $\kappa_0 > 1$. Changing the variable of integration in (1) from Υ to $\Upsilon' = \arcsin(\sin(\Upsilon)/\kappa_0)$ we obtain the identity required.

Problem 8.3   The value of $\overline{\Upsilon}$ can be estimated as $(\max(\dot{\Upsilon}) + \min(\dot{\Upsilon}))/2$. Hence, using Fig. 8.3 we have $\overline{\Upsilon} \approx 0.65$ for $\kappa_0^2 = 0.7$ and $\overline{\Upsilon} \approx 0.85$ for $\kappa_0^2 = 0.5$, which confirms the increase of $\overline{\Upsilon}$ with decreasing $|\kappa_0|$.

Index